Advances in

VIRUS RESEARCH

VOLUME 40

Advances in

VIRUS RESEARCH

Edited by

KARL MARAMOROSCH

Department of Entomology
Rutgers University
New Brunswick, New Jersey

FREDERICK A. MURPHY

Center for Infectious Diseases
Centers for Disease Control
Atlanta, Georgia

AARON J. SHATKIN

Center for Advanced Biotechnology and Medicine
Piscataway, New Jersey

VOLUME 40

ACADEMIC PRESS, INC.

Harcourt Brace Jovanovich, Publishers
San Diego   New York   Boston
London   Sydney   Tokyo   Toronto

This book is printed on acid-free paper. ∞

Academic Press, Inc.
San Diego, California 92101

United Kingdom Edition published by
ACADEMIC PRESS LIMITED
24-28 Oval Road, London NW1 7DX

Library of Congress Catalog Card Number: 53-11559

ISBN 0-12-039840-0 (alk. paper)

PRINTED IN THE UNITED STATES OF AMERICA
91 92 93 94   9 8 7 6 5 4 3 2 1

CONTENTS

The 5′-Untranslated Region of Picornaviral Genomes

Vadim I. Agol

Effects of Defective Interfering Viruses on Virus Replication and Pathogenesis *in Vitro* and *in Vivo*

Laurent Roux, Anne E. Simon, and John J. Holland

Structure and Function of the HEF Glycoprotein of Influenza C Virus

Georg Herrler and Hans-Dieter Klenk

Bunyaviridae: Genome Organization and Replication Strategies

Michèle Bouloy

CONTENTS

ADVANCES IN VIRUS RESEARCH, VOL. 40

REGULATION OF GENE EXPRESSION IN THE HUMAN IMMUNODEFICIENCY VIRUS TYPE 1

Bryan R. Cullen

Howard Hughes Medical Institute and Department of Microbiology and Immunology
Duke University Medical Center
Durham, North Carolina 27710

I. Introduction

The pathogenic retrovirus human immunodeficiency virus type 1 (HIV-1) is the prototype of a group of primate lentiviruses that also includes HIV-2 and a number of related but distinct simian immunodeficiency viruses. In addition, HIV-1 is distantly related to several nonprimate lentiviruses, including visna–maedi virus, equine infectious anemia virus, bovine immunodeficiency virus, caprine encephalitis virus, and feline immunodeficiency virus (Gojobori *et al.*, 1990). Lentiviruses exhibit a tropism for T lymphocytes and/or cells of the monocyte/macrophage lineage and share the ability to induce long-term chronic disease states in their hosts. These retroviruses also display a level of genetic complexity that distinguishes them from the simpler type C family of oncogenic retroviruses. This chapter outlines the current understanding of gene regulation in HIV-1. In particular, I focus on the various cellular and viral trans-activators that together serve to modulate both the quantity and quality of HIV-1-specific gene expression.

II. Genetic Organization of HIV-1

The avian leukemia virus (ALV), a representative type C retrovirus, displays a relatively simple genetic organization (reviewed by Var-

1

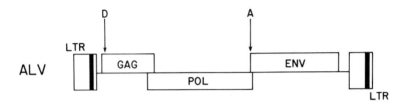

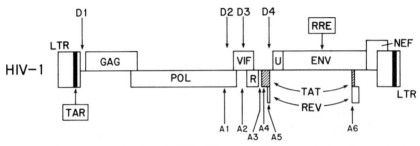

FIG. 1. Genetic organization of HIV-1. Shown are the genomic structures of avian leukosis virus (ALV), a representative "simple" retrovirus, and HIV-1, a member of the complex lentivirus subgroup of the retroviridae. Indicated are the locations of known viral genes as well as known splice donor (D) and acceptor (A) sites. Also shown are the locations of the RNA target sites for the HIV-1 Tat and Rev trans-activators. LTR, Long terminal repeat; TAR, trans-activation response element; RRE, Rev response element. Reproduced with permission from Cullen and Greene (1990).

mus, 1988) (Fig. 1). The viral *gag* gene encodes the virion core proteins, while *pol* encodes enzymes required for reverse transcription of the viral RNA genome and integration of the resultant DNA provirus into the host chromatin. Both the Gag and Pol proteins are translated from the full-length viral genomic RNA, the *pol* reading frame being accessed by a cis-acting sequence that induces ribosomal frame-shifting. The *env* gene encodes the viral envelope glycoprotein and is translated from a spliced mRNA species. The frequency of this splicing event is regulated by cis-acting sequences present within the viral RNA genome and does not appear to be subject to regulation by a virally encoded factor (Katz *et al.*, 1988). Of note, ALV contains only a single splice donor/acceptor combination.

Although the HIV-1 genome (Fig. 1) is considerably more complex than that of ALV, HIV-1 does encode functionally analogous structural genes, again, termed *gag*, *pol*, and *env*. The Env protein of HIV-1 is a primary determinant of the relatively narrow cell tropism of this virus, which is confined to lymphocytes and monocytic cells that bear the CD4 cell surface glycoprotein receptor for HIV-1. In addition,

HIV-1 encodes two proteins (Vif and Vpu) involved in the maturation and release of HIV-1 virions and two proteins (Vpr and Nef) that are of uncertain function and not required for efficient viral replication in culture. HIV-1 also encodes two nuclear regulatory proteins, Tat and Rev, that are essential for viral replication. Of note, HIV-1 displays at least four functional splice donors and six splice acceptors (Fig. 1). HIV-1 is therefore, in theory, capable of deriving a large number of distinct viral mRNA species by posttranscriptional processing of the initial genome-length transcript. As many as 30 HIV-1 mRNA species have, in fact, been identified (Muesing et al., 1985; Schwartz et al., 1990).

III. Role of Cellular Transcription Factors

The recognition sequences for several constitutively expressed or inducible host cell transcription factors have been identified within the HIV-1 long terminal repeat (LTR) promoter element (Fig. 2). Of particular importance are the binding sites for Sp1 and the TATA factor TFIID (Jones et al., 1986), as well as for the inducible transcription factor NF-κB (Nabel and Baltimore, 1987). In addition, the HIV-1 LTR contains binding sites for the nuclear transcription factors NFAT-1, USF, AP-1, and LBP (Garcia et al., 1987; Jones et al., 1988; Shaw et al., 1988; Lu et al., 1989).

The constitutively expressed cellular transcription factors Sp1 and TFIID play an important role in mediating promoter function in many cellular and viral genes (reviewed by Lewin, 1990). The HIV-1 promoter possesses three adjacent functional Sp1 binding sites as well as a typical TATA element. Both the Sp1 and TFIID binding interactions appear critical for HIV-1 LTR promoter function, and deletion of these binding domains therefore results in a defective HIV-1 provirus (Jakobovits et al., 1988; Leonard et al., 1989).

The inducible transcription factor NF-κB serves as a pleiotropic mediator of both tissue-specific and inducible gene expression in a wide range of human cell types (Lenardo and Baltimore, 1989). Activation of NF-κB, which occurs on treatment of resting human T cells with either lectin or specific antigen, is a key step in the activation of T-cell-specific genes such as interleukin 2 (IL-2) and the α chain of the IL-2 receptor (Bohnlein et al., 1988). Although the two NF-κB binding sites observed in the HIV-1 LTR are important for maximal expression from the HIV-1 LTR, they are not essential for viral replication. Indeed, HIV-1 is able to replicate effectively in cells that do not express significant levels of NF-κB (Maddon et al., 1986) and the deletion of

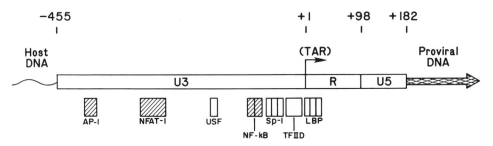

FIG. 2. cis-Acting elements within the 5' long terminal repeat (LTR) of an integrated HIV-1 provirus. LTRs are divided into three subsections, termed U3 (unique 3'), R (repeat), and U5 (unique 5'). The border between U3 and R marks the site of transcription initiation (arrow). The size of each HIV-1 LTR subsection is given in base pairs. Also indicated are the locations of the single binding sites for NFAT-1, USF, AP-1, and TFIID, the two adjacent binding sites for NF-κB and the three adjacent target sequences for the cellular factors Sp1 and LBP. The trans-activation response (TAR) element maps to positions 1–59, but is believed to be recognized as an RNA, rather than DNA, sequence (see Section IV). Target sequences for inducible transcription factors are indicated by hatching. Reproduced with permission from Cullen and Malim (1990).

both LTR NF-κB recognition sites has been reported to have relatively little effect on the ability of HIV-1 to replicate in activated primary blood lymphocytes (Leonard *et al.*, 1989).

Clearly, however, the functional interaction of NF-κB with its target sequence in the HIV-1 LTR does result in the enhanced expression of linked genes when measured in either T cells or macrophages (Nabel and Baltimore, 1987; Griffin *et al.*, 1989). It therefore seems probable that the NF-κB motifs present in the HIV-1 LTR primarily serve to boost the rate of viral gene transcription in response to the activation of NF-κB. Of particular interest is the observation that the induction of NF-κB activity noted on the *in vitro* stimulation of T cells by certain lymphokines occurs concomitantly with a significant induction of HIV-1-specific gene expression, suggesting a cause-and-effect relationship (Tong-Starksen *et al.*, 1987; Bohnlein *et al.*, 1988; Osborn *et al.*, 1989).

A second nuclear factor induced by activation of T cells, termed NFAT-1, also interacts with a specific binding site within the HIV-1 LTR (Shaw *et al.*, 1988) (Fig. 2). While it is tempting to speculate that NFAT-1 may also play a role in the activation of HIV-1 gene expression, no evidence exists regarding the functional significance of NFAT-1. Similarly, the importance of a binding site for the inducible transcription factor AP-1, located toward the 5' end of the LTR U3 (unique 3') region, remains to be established.

Two additional constitutive HIV-1 DNA–protein binding interactions have also been defined (Fig. 2). These involve binding of the cellular transcription factors USF and LBP. Both protein–DNA binding events are readily detectable *in vitro*, using gel retardation and DNase footprinting analysis, and *in vivo*, using DNase hypersensitivity analysis (Garcia *et al.*, 1987; Hauber and Cullen, 1988; Jones *et al.*, 1988). However, the functional significance of these interactions remains unclear, although some data do suggest that USF may act as a modest silencer of the HIV-1 LTR, giving rise to the earlier terminology, "negative regulatory factor" (Garcia *et al.*, 1987; Lu *et al.*, 1989). Mutational analyses have so far failed to demonstrate any direct role for the tripartite LBP interaction in regulating HIV-1-specific gene expression *in vivo* (Malim *et al.*, 1989c).

In addition to transcription factors encoded by the host cell itself, superinfecting viruses may also encode factors able to enhance HIV-1-specific gene expression in trans. Viral gene products encoded by the human T-cell leukemia virus type I (HTLV-I), several members of the herpesvirus family, the adenoviruses, and hepatitis B virus have all been shown to enhance HIV-1 LTR-dependent gene expression in transfected cells (Mosca *et al.*, 1987; Siekevitz *et al.*, 1987; Nabel *et al.*, 1988; Seto *et al.*, 1988; Barry *et al.*, 1990). Because of the somewhat narrow cell tropism of HIV-1, it is unclear whether any of these observed activation events have any relevance to the spread of HIV-1 *in vivo*. Two viruses that are known to infect human T-cell populations, and hence are strong candidates as cofactors in the pathogenesis of HIV-1-induced disease, are HTLV-I and human herpes virus 6 (HHV-6). The HTLV-I Tax trans-activator has been shown to induce NF-κB activity in HTLV-I-infected T cells (Siekevitz *et al.*, 1987), while herpes virus early gene products are believed to activate HIV-1 LTR-specific gene expression via both NF-κB-dependent and independent pathways (Gimble *et al.*, 1988; Nabel *et al.*, 1988). Of particular importance is the observation that dual infection of cultured CD4+ human T cells with HHV-6 and HIV-1 results in enhanced replication of HIV-1 and accelerated cell death (Lusso *et al.*, 1989).

IV. EARLY HIV-1 GENE EXPRESSION

The cellular transcription factors described in Section III are sufficient to induce a low or basal level of HIV-1 gene expression in infected cells. Evidence suggests that these initial transcripts reach the infected cell cytoplasm exclusively in the form of the fully spliced mRNA species that encode the viral regulatory proteins. Tat, Rev, and Nef (see Section V). The viral Tat trans-activator serves to dramatically increase HIV-1 LTR-directed gene expression (Arya *et al.*, 1985; Sodroski *et al.*, 1985) and thereby initiates a powerful positive feedback loop, leading to very high levels of HIV-1-specific RNA and protein synthesis. A functional copy of the viral *tat* gene is essential for HIV-1 replication *in vitro*, as is an intact copy of the cis-acting viral sequence responsive to Tat, the so-called trans-activation response (TAR) element (Fisher *et al.*, 1986; Leonard *et al.*, 1989).

The *tat* gene is divided into two coding exons, which together predict the synthesis of an 86-amino-acid protein. However, the 72-amino-acid first coding exon of Tat, which is flanked at its 3' end by a highly conserved translation termination signal, is sufficient to encode a fully active Tat protein (Cullen, 1986). Current evidence suggests that this truncated protein may also be physiologically relevant (Malim *et al.*, 1988; Schwartz *et al.*, 1990). At least two conserved functional protein domains have been identified in Tat. The first is a highly cysteine-rich domain believed to be important in metal ion conjugation and the dimerization of Tat (Frankel *et al.*, 1988). The second is a highly basic sequence element which has been shown to determine the nuclear localization of the Tat protein (Hauber *et al.*, 1989; Ruben *et al.*, 1989) and which has been proposed to function in the sequence-specific binding of Tat to the viral TAR element (see below). Targeted mutagenesis of either of these regions has been shown to ablate the trans-activation function of Tat.

The cis-acting TAR element coincides with a 59-nucleotide (nt) RNA stem–loop located at the 5' end of all viral mRNA species (Fig. 3). TAR function has been shown to be both orientation and position dependent (Rosen *et al.*, 1985; Hauber and Cullen, 1988; Selby *et al.*, 1989). The most important features of TAR, in addition to the structural integrity of the stem itself, are the terminal 6-nt loop and a 3-nt bulge located four nucleotides 5' of the loop (Feng and Holland, 1988; Berkhout and Jeang, 1989). Recent data suggest that Tat binds to TAR in a sequence-specific manner at the site of the RNA bulge–loop (Dingwall *et al.*, 1989; Roy *et al.*, 1990). Although mutations in both the bulge–loop and the terminal loop of TAR dramatically affect the level of trans-activation by Tat *in vivo*, only mutations of the bulge–loop appear to affect Tat binding *in vitro*. However, a 68-kDa cellular factor has been shown to bind to the terminal loop of TAR with the appropriate sequence specificity (Marciniak *et al.*, 1990), raising the possibility that this cellular factor may be directly involved in mediating Tat function or might instead facilitate the sequence-specific interaction of Tat with TAR *in vivo* (Fig. 3). The latter hypothesis appears to be supported by the observation that heterologous RNA–protein binding sites substituted for TAR can mediate significant trans-activation of HIV-1

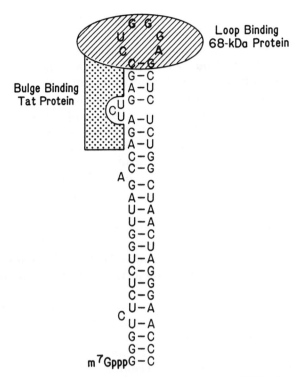

FIG. 3.   Structure of the HIV-1 trans-activation response (TAR) element. The TAR element is the cis-acting target sequence of the HIV-1 Tat protein. The TAR RNA stem–loop forms the 5′ end of all HIV-1 mRNA species. The Tat protein is believed to bind to a bulge–loop in TAR—possibly as a dimer—while a 68-kDa cellular protein is believed to bind to the terminal loop. The indicated interaction between Tat and the cellular factor is hypothetical. Reproduced with permission from Cullen (1990).

LTR-directed gene expression by chimeric proteins consisting of TAT fused to the homologous sequence-specific RNA-binding protein (Southgate *et al.*, 1990).

Although it is evident that the purpose of the Tat trans-activator is to raise the level of HIV-1-specific gene expression by two to three orders of magnitude, the mechanisms by which this activation occurs remain controversial. The major effect of Tat is clearly to increase the rate of transcription from the HIV-1 LTR (Hauber *et al.*, 1987). Current evidence suggests that the interaction of Tat with the nascent TAR RNA sequence results in both an increase in the rate of transcription initiation and a drop in the level of premature transcription termination (Kao *et al.*, 1987; Berkhout *et al.*, 1989; Laspia *et al.*, 1989). This has led to the hypothesis that TAR could be acting as an "RNA enhancer element," while Tat would then be at least functionally equivalent to an enhancer-binding protein (Sharp and Marciniak, 1989). The recent demonstration of significant Tat activity in a defined *in vitro* transcription system (Marciniak *et al.*, 1990) appears likely to represent a key step in the eventual unraveling of the mechanism of action of this novel regulatory protein.

While the transcriptional activation of viral RNA transcription is the major action of Tat in most experimental systems, it is clearly not the only effect of this small trans-activator. A number of reports have noted that the effect of Tat on the level of expression of genes linked to the viral LTR, when measured at the protein level, can be significantly, even dramatically, higher than the effect determined at the level of steady-state mRNA (Cullen, 1986; Feinberg *et al.*, 1986; Rosen *et al.*, 1986; Wright *et al.*, 1986). Although the molecular basis for this second, posttranscriptional, component of the bimodal action of Tat remains unclear, it also appears to be mediated by the sequence-specific interaction of Tat with the viral TAR RNA element. Results obtained using microinjection of preformed TAR containing RNA molecules into *Xenopus* oocytes show that this posttranscriptional effect also occurs in the cell nucleus, yet can be segregated from the transcriptional action of Tat (Braddock *et al.*, 1989). Thus far, Tat has not been shown to modulate the nuclear export of TAR-containing RNA species and thus a more complex mechanism of action appears likely. One hypothesis is that Tat could affect the cytoplasmic compartmentalization, and hence the translational utilization, of TAR-containing transcripts (Braddock *et al.*, 1989).

The second early gene product of HIV-1, termed Nef, is a 27-kDa myristilated phosphoprotein that is associated with cytoplasmic membrane structures in expressing cells. Nef has been reported to possess the GTPase, autophosphorylation, and GTP-binding properties typical of the G protein family of signal transduction proteins (Guy *et al.*, 1987); however, this observation has not been confirmed (Kaminchik *et al.*, 1990). Unlike Tat and Rev, the Nef gene product is not required for HIV-1 replication in culture. It has, in fact, been proposed that expression of Nef results in an inhibition of HIV-1 LTR-specific gene expression and viral replication (Ahmad and Venkatesan, 1988; Niederman *et al.*, 1989). However, these negative effects of Nef remain controversial, as others have observed no effect of the Nef protein on either viral replication or gene expression (Hammes *et al.*, 1989; Kim *et al.*, 1989b; Bachelerie *et al.*, 1990). The role of the Nef gene product in the HIV-1 replication cycle therefore remains unclear. However, the fact that the Nef open reading frame is reasonably conserved between all primate lentivirus species suggests that this protein is likely to play a significant role in the viral life cycle in the infected host. Recent data, in fact, suggest that a functional Nef gene product can markedly enhance viral replication and pathogenicity in rhesus macaques infected with a cloned isolate of simian immunodeficiency virus (Kestler *et al.*, 1991).

V. Transition to Viral Structural Gene Expression

The HIV-1 genome encodes two classes of viral mRNA species. The first of these consists of a multiply spliced ~2-kb mRNA class that encodes the viral regulatory proteins Tat, Nef, and Rev. The second class consists of the unspliced (~9-kb) and incompletely spliced (~4-kb) viral mRNAs that encode the virion structural proteins (Fig. 4). In the absence of Rev, only the fully spliced class of HIV-1 mRNAs is functionally expressed (Feinberg *et al.*, 1986). In fact, *rev*⁻ mutants of HIV-1 are incapable of inducing the synthesis of the viral structural proteins and are, therefore, replication defective. An analysis of the time course of HIV-1 infection of human T lymphocytes reveals a similar phenomenon (Kim *et al.*, 1989a). Initially, only the 2-kb class of viral mRNAs is detected in HIV-1-infected cells; however, as the level of viral gene expression increases (due to the action of the Tat

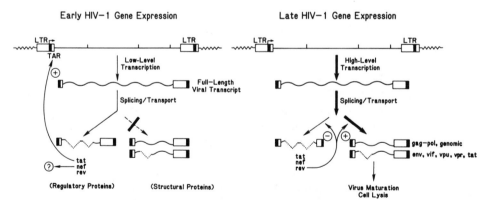

Fig. 4. Temporal regulation of HIV-1 gene expression. Cellular transcription factors mediate an initial low level of proviral transcription, which results in the expression of the fully spliced mRNAs that encode the viral regulatory proteins. The Tat protein acts to greatly amplify the expression of all the viral gene products. Subsequently, the Rev protein induces the transition from this early regulatory phase to the late phase of viral gene expression marked by the high-level synthesis of the incompletely spliced mRNAs that encode the viral structural proteins. LTR, Long terminal repeat. Reproduced with permission [Cullen and Greene (1989)].

protein), a switch to the synthesis of the viral structural gene mRNAs is observed. This effect, which reflects the action of the viral Rev trans-activator, occurs concomitantly with an essentially equivalent reduction in the synthesis of the fully spliced mRNA species that encode the viral regulatory proteins (Malim et al., 1988; Kim et al., 1989a). Therefore, Rev functions as a negative regulator of its own synthesis and also mediates the establishment of an equilibrium between viral structural and regulatory protein synthesis (Fig. 4).

The switch from the early regulatory phase of HIV-1 gene expression to the late structural phase appears to require the expression of a critical level of the Rev protein. Cultured cells that are latently or nonproductively infected by HIV-1 have been shown to constitutively express a low level of viral mRNA that is primarily of the 2-kb class (Pomerantz et al., 1990). Treatment of these cells with agents that result in the activation of NF-κB leads to enhanced transcription from the HIV-1 LTR and induces the expression of the viral structural proteins. It is therefore hypothesized that latency in this context is due to the expression of a subcritical level of the Rev trans-activator, a level which, in turn, reflects the lack of cellular transcription factors critical for efficient HIV-1 LTR-dependent gene expression. The primary role of the Rev regulatory pathway therefore may be to prevent the premature progression of the viral replication cycle to the late or lytic phase in cells which are incapable of supporting the required level of viral mRNA and protein synthesis. The Rev trans-activator, or lack thereof, may therefore be critical to the maintenance of a nonproductive or latent HIV-1 infection in resting cell populations in vivo (Pomerantz et al., 1990).

While Rev is absolutely required for the cytoplasmic expression of unspliced HIV-1 RNA species, it appears to have little effect on the pattern of HIV-1 RNA expression in the cell nucleus (Felber et al., 1989; Malim et al., 1989b). In particular, high levels of unspliced viral transcripts can be detected in the nucleus even in the absence of Rev. In common with other retroviruses, splice sites present in the HIV-1 genome appear to be inefficiently utilized by the cellular splicing machinery. It has therefore been hypothesized that splicing factors may be able to assemble on the HIV-1 primary transcript, but are then only poorly able to carry out the actual splicing step (Chang and Sharp, 1989; Malim et al., 1989b). Instead, this interaction appears to result in the retention of these incompletely spliced viral transcripts within the nucleus. The Rev protein is believed to function by activating the nuclear export of these sequestered viral RNA species either by antagonizing their interaction with these splicing factors (Chang and Sharp, 1989) or by directly facilitating their interaction with a component of a cellular RNA transport pathway (Felber et al., 1989; Malim et al.,

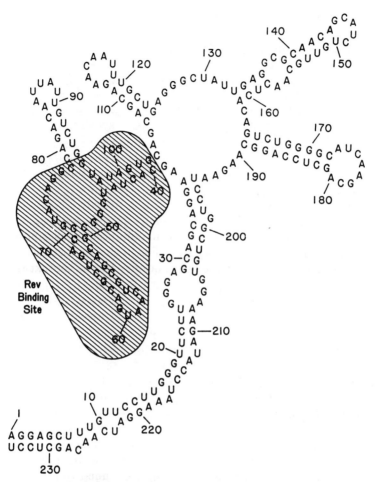

FIG. 5.   Structure of the HIV-1 Rev response element (RRE). The RRE, the cis-acting target sequence of the HIV-1 Rev protein, is located within the viral *env* coding region. Although the entire RRE is required for full biological activity *in vivo*, the indicated stem–loop subdomain is sufficient for full Rev binding *in vitro* and also functions as a minimal RRE *in vivo*. Modified with permission from Malim *et al.* (1990).

1989b). In contrast, in the absence of Rev, viral mRNAs are eventually fully spliced prior to their transport to the cytoplasm.

The action of the Rev trans-activator is specific for unspliced HIV-1 transcripts and thus far has not been shown to affect the splicing or transport of cellular RNAs. The specificity of this response is conferred by a highly structured 234-nt RNA target sequence, the Rev response element (RRE), located within the *env* gene of HIV-1 (Rosen *et al.*, 1988; Malim *et al.*, 1989b) (Figs. 1 and 5). The Rev protein has been shown to bind to the RRE with high affinity *in vitro* (Daly *et al.*, 1989;

Zapp and Green, 1989). Although the entire 234-nt RRE is required for full biological activity *in vivo* (Malim *et al.*, 1990), it is now clear that a 66-nt stem–loop subdomain of the RRE (Fig. 5) is both necessary and sufficient for high-affinity binding of Rev (Heaphy *et al.*, 1990; Malim *et al.*, 1990) and is also sufficient for significant biological activity *in vivo*. It is hypothesized that the remainder of the RRE functions to stabilize the RNA structure of the Rev binding site and/or to facilitate presentation of this RNA sequence *in vivo*.

The *rev* gene consists of two coding exons that together predict a protein of 116 amino acids. Rev is localized to the nuclei and, particularly, the nucleoli of expressing cells (Felber *et al.*, 1989; Malim *et al.*, 1989a). Rev is phosphorylated at two serine residues *in vivo*; however, this posttranslational modification does not appear to be essential for Rev function (Malim *et al.*, 1989a). Two distinct protein domains that are essential for Rev function have been defined. The first is an ~35-amino-acid sequence characterized by a high number of arginine residues (Malim *et al.*, 1989a). This sequence displays significant identity to the "Arg-rich" RNA recognition motif proposed by Lazinski *et al.* (1989) and functions as the sequence-specific RNA binding domain of Rev. Mutations in this domain result in Rev proteins displaying a recessive negative phenotype. The central core of this protein domain has also been shown to function as the Rev protein nuclear/nucleolar localization signal (Malim *et al.*, 1989a).

It seems probable that Rev also contains a protein sequence element that interacts directly with a component of the nuclear RNA transport machinery. Mutational analysis has suggested that a second Rev protein domain centered on amino acid 80 may serve this function. Interestingly, Rev proteins mutated in this latter domain are not only defective, but also inhibit wild-type Rev function in trans. It seems possible that this trans-dominant inhibition results from competition between wild-type and mutant Rev proteins for binding to the viral RRE, whereupon the bound mutant Rev proteins are unable to interact with cellular factors involved in RNA transport from the nucleus (Malim *et al.*, 1989a). These observations suggest that the study of Rev function may lead to significant insights into the regulation of RNA export, an area which remains poorly understood.

VI. AUXILIARY PROTEINS OF HIV-1

In addition to Gag, Pol, and Env, HIV-1 encodes three additional late gene products, the Vpr, Vif, and Vpu proteins, which have been shown to affect the efficiency of viral spread (Fig. 4). The first of these, Vpr, shares with Nef the property of being dispensable for HIV-1 replication in culture. Indeed, several replication-competent proviruses derived from the HTLV-IIIB substrain of HIV-1 lack functional copies of both *vpr* and *nef*. Recent data suggest that viruses containing an intact *vpr* gene replicate slightly more rapidly in culture and may display enhanced cytopathicity (Ogawa *et al.*, 1989). The Vpr protein has recently been shown to be virion associated and thus may be considered a structural protein candidate (Cohen *et al.*, 1990a). Surprisingly, data have also been presented suggesting that the Vpr protein can transactivate HIV-1 LTR-specific gene expression by ~2- to 3-fold in transient expression assays (Cohen *et al.*, 1990b). Although this potential regulatory action of Vpr appears at first glance inconsistent with the late expression of Vpr in the replication cycle and its association with virion particles, a precedent for such a dual role has been reported from at least one other viral system (Post *et al.*, 1981).

The 23-kDa Vif protein is required for the efficient transmission of cell-free virus in culture (Fisher *et al.*, 1987; Strebel *et al.*, 1987). Proviruses lacking Vif appear to express fully normal levels of the other viral proteins and are able to release virus from infected cells with normal efficiency. However, these released virions, despite their apparently normal morphological appearance, have been found to be up to 1000-fold less infectious than virions released from Vif$^+$ provirus-infected cells (Fisher *et al.*, 1987; Strebel *et al.*, 1987). Interestingly, Vif does not appear to affect the efficiency of direct cell-to-cell spread of HIV-1. Vif has not been shown to be virion associated, and its mechanism of action is therefore uncertain. Current possibilities include roles of virion maturation and/or morphogenesis. It is, however, of interest that the Vif open reading frame is conserved among the primate lentiviruses and also appears to be retained in visna–maedi virus, a distantly related ungulate lentivirus.

In marked contrast to Vif, the HIV-1 Vpu protein appears to have no equivalent in other primate lentiviruses, including the other human representative, HIV-2. The Vpu protein is 81 amino acids in length, is phosphorylated *in vivo*, and is associated with the cytoplasmic membranes of expressing cells (Strebel *et al.*, 1989). Expression of Vpu is not essential for HIV-1 replication in culture. However, the loss of a functional Vpu gene product results in a 5- to 10-fold decrease in the production of progeny HIV-1 virions (Strebel *et al.*, 1988). This reduction does not appear to reflect a decrease in the biosynthesis of any of the other HIV-1 proteins, but instead results from a defect in virion release. Notably, lack of Vpu expression appears to result in an accumulation of cell-associated HIV-1 virions (Terwilliger *et al.*, 1989; Klimkait *et al.*, 1990). Again, the mechanism of action of this viral protein remains uncertain. It is possible that Vpu enhances budding of virions from the cell surface or, instead, functions to inhibit budding of virions through intracytoplasmic membranes. In either case it remains unclear why HIV-1 requires a Vpu gene product for efficient virion release while related viruses (e.g., HIV-2) appear to be able to replicate and bud efficiently without an equivalent protein.

VII. Conclusions

Although HIV-1 was discovered less than 10 years ago, the importance of this human pathogen has stimulated an intense research effort that has rapidly led to remarkable insights into the regulation of HIV-1 gene expression. The genetic complexity of the HIV-1 provirus proved to be only the first of several surprising results. Perhaps most remarkable was the discovery that the HIV-1 genome encodes two regulatory proteins that act through entirely novel mechanisms to influence both the quantity and quality of HIV-1 gene expression. The study of these novel regulatory pathways, which are mediated by specific protein–RNA interactions, appears likely to lead to new insights into eukaryotic gene regulation in general. Moreover, it is to be hoped that this research effort will also eventually lead to strategies to effectively combat the devastating pathogenic effects of this virus in humans.

References

Ahmad, N., and Venkatesan, S. (1988). *Science* **241,** 1481–1485.

Arya, S. K., Guo, C., Josephs, S. F., and Wong-Staal, F. (1985). *Science* **229,** 69–73.

Bachelerie, F., Alcami, J., Hazan, U., Israel, N., Goud, B., Arenzana-Seisdedos, F., and Virelizier, J.-L. (1990). *J. Virol.* **6,** 3059–3062.

Barry, P. A., Pratt-Lowe, E., Peterlin, B. M., and Luciw, P. A. (1990). *J. Virol.* **64,** 2932–2940.

Berkhout, B., and Jeang, K.-T. (1989). *J. Virol.* **63,** 5501–5504.

Berkhout, B., Silverman, R. H., and Jeang, K.-T. (1989). *Cell* **59,** 273–282.

Bohnlein, E., Lowenthal, J. W., Siekevitz, M., Ballard, D. W., Franza, B. R., and Greene, W. C. (1988). *Cell* **53,** 827–836.

Braddock, M., Chambers, A., Wilson, W., Esnouf, M. P., Adams, S. E., Kingsman, A. J., and Kingsman, S. M. (1989). *Cell* **58,** 269–279.

Chang, D. D., and Sharp, P. A. (1989). *Cell* **59,** 789–795.

Cohen, E. A., Dehni, G., Sodroski, J. G., and Haseltine, W. A. (1990a). *J. Virol.* **64,** 3097–3099.

Cohen, E. A., Terwilliger, E. F., Jalinoos, Y., Proulx, J., Sodroski, J. G., and Haseltine, W. A. (1990b). *J. Acquired Immune Defic. Syndromes* **3,** 11–18.

Cullen, B. R. (1986). *Cell* **46,** 973–982.

Cullen, B. R. (1990). *Cell* **63,** 655–657.

Cullen, B. R., and Greene, W. C. (1989). *Cell* **58**, 423–426.

Cullen, B. R., and Greene, W. C. (1990). *Virology* **178**, 1–5.

Cullen, B. R., and Malim, M. H. (1990). *Nucleic Acids Mol. Biol.* **4**, 176–184.

Daly, T. J., Cook, K. S., Gray, G. S., Maione, T. E., and Rusche, J. R. (1989). *Nature (London)* **342**, 816–819.

Dingwall, C., Ernberg, I., Gait, M. J., Green, S. M., Heaphy, S., Karn, J., Lowe, A. D., Singh, M., Skinner, M. A., and Valerio, R. (1989). *Proc. Natl. Acad. Sci. U.S.A.* **86**, 6925–6929.

Feinberg, M. B., Jarrett, R. F., Aldovini, A., Gallo, R. C., and Wong-Staal, F. (1986). *Cell* **46**, 807–817.

Felber, B. K., Hadzopoulou-Cladaras, M., Cladaras, C., Copeland, T., and Pavlakis, G. N. (1989). *Proc. Natl. Acad. Sci. U.S.A.* **86**, 1495–1499.

Feng, S., and Holland, E. C. (1988). *Nature (London)* **334**, 165–167.

Fisher, A. G., Feinberg, M. B., Josephs, S. F., Harper, M. E., Marselle, L. M., Reyes, G., Gonda, M. A., Aldovini, A., Debouk, C., Gallo, R. C., and Wong-Staal, F. (1986). *Nature (London)* **320**, 367–371.

Fisher, A. G., Ensoli, B., Ivanoff, L., Chamberlain, M., Petteway, S., Ratner, L., Gallo, R. C., and Wong-Staal, F. (1987). *Science* **237**, 888–893.

Frankel, A. D., Bredt, D. S., and Pabo, C. O. (1988). *Science* **240**, 70–73.

Garcia, J. A., Wu, F. K., Mitsuyasu, R., and Gaynor, R. B. (1987). *EMBO J.* **6**, 3761–3770.

Gimble, J. M., Duh, E., Ostrove, J. M., Gendelman, H. E., Max, E. E., and Rabson, A. B. (1988). *J. Virol.* **62**, 4104–4112.

Gojobori, T., Moriyama, E. N., Ina, Y., Ikeo, K., Miura, T., Tsujimoto, H., Hayami, M., and Yokoyama, S. (1990). *Proc. Natl. Acad. Sci. U.S.A.* **87**, 4108–4111.

Griffin, G. E., Leung, K., Folks, T. M., Kunkel, S., and Nabel, G. J. (1989). *Nature (London)* **339**, 70–73.

Guy, B., Kieny, M. P., Riviere, Y., Le Peuch, C., Dott, K., Girard, M., Montagnier, L., and Lecocq, J.-P. (1987). *Nature (London)* **330**, 266–269.

Hammes, S. R., Dixon, E. P., Malim, M. H., Cullen, B. R., and Greene, W. C. (1989). *Proc. Natl. Acad. Sci. U.S.A.* **86**, 9549–9553.

Hauber, J., and Cullen, B. R. (1988). *J. Virol.* **62**, 673–679.

Hauber, J., Perkins, A., Heimer, E. P., and Cullen, B. R. (1987). *Proc. Natl. Acad. Sci. U.S.A.* **84**, 6364–6368.

Hauber, J., Malim, M. H., and Cullen, B. R. (1989). *J. Virol.* **63**, 1181–1187.

Heaphy, S., Dingwall, C., Ernberg, I., Gait, M. J., Green, S. M., Karn, J., Lowe, A. D., Singh, M., and Skinner, M. A. (1990). *Cell* **60**, 685–693.

Jakobovits, A., Smith, D. H., Jakobovits, E. B., and Capon, D. J. (1988). *Mol. Cell. Biol.* **8**, 2555–2561.

Jones, K. A., Kadonaga, J. T., Luciw, P. A., and Tjian, R. (1986). *Science* **232**, 755–759.

Jones, K. A., Luciw, P. A., and Duchange, N. (1988). *Genes Dev.* **2**, 1101–1114.

Kaminchik, J., Bashan, N., Pinchasi, D., Amit, B., Sarver, N., Johnston, M. I., Fischer, M., Yavin, Z., Gorecki, M., and Panet, A. (1990). *J. Virol.* **64**, 3447–3454.

Kao, S.-Y., Calman, A. F., Luciw, P. A., and Peterlin, B. M. (1987). *Nature (London)* **330**, 489–493.

Katz, R. A., Kotler, M., and Skalka, A. M. (1988). *J. Virol.* **62**, 2686–2695.

Kestler, H. W., Ringler, D. J., Mori, K., Panicali, D. L., Sehgal, P. K., Daniel, M. D., and Desrosiers, R. C. (1991). *Cell* **65**, 651–662.

Kim, S., Byrn, R., Groopman, J., and Baltimore, D. (1989a). *J. Virol.* **63**, 3708–3713.

Kim, S., Ikeuchi, K., Byrn, R., Groopman, J., and Baltimore, D. (1989b). *Proc. Natl. Acad. Sci. U.S.A.* **86**, 9544–9548.

Klimkait, T., Strebel, K., Hoggan, M. D., Martin, M. A., and Orenstein, J. M. (1990). *J. Virol.* **64,** 621–629.

Laspia, M. F., Rice, A. P., and Mathews, M. B. (1989). *Cell* **59,** 283–292.

Lazinski, D., Grzadzielska, E., and Das, A. (1989). *Cell* **59,** 207–218.

Lenardo, M. J., and Baltimore, D. (1989). *Cell* **58,** 227–229.

Leonard, J., Parrott, C., Buckler-White, A. J., Turner, W., Ross, E. K., Martin, M. A., and Rabson, A. B. (1989). *J. Virol.* **63,** 4919–4924.

Lewin, B. (1990). *Cell* **61,** 1161–1164.

Lu, Y., Stenzel, M., Sodroski, J. G., and Haseltine, W. A. (1989). *J. Virol.* **63,** 4115–4119.

Lusso, P., Ensoli, B., Markham, P. D., Ablashi, D. V., Salahuddin, S. Z., Tschachler, E., Wong-Staal, F., and Gallo, R. C. (1989). *Nature (London)* **337,** 370–373.

Maddon, P. J., Dalgleish, A. G., McDougal, J. S., Clapham, P. R., Weiss, R. A., and Axel, R. (1986). *Cell* **47,** 333–348.

Malim, M. H., Hauber, J., Fenrick, R., and Cullen, B. R. (1988). *Nature (London)* **335,** 181–183.

Malim, M. H., Bohnlein, S., Hauber, J., and Cullen, B. R. (1989a). *Cell* **58,** 205–214.

Malim, M. H., Hauber, J., Le, S.-Y., Maizel, J. V., and Cullen, B. R. (1989b). *Nature (London)* **338,** 254–257.

Malim, M. H., Fenrick, R., Ballard, D. W., Hauber, J., Bohnlein, E., and Cullen, B. R. (1989c). *J. Virol.* **63,** 3213–3219.

Malim, M. H., Tiley, L. S., McCarn, D. F., Rusche, J. R., Hauber, J., and Cullen, B. R. (1990). *Cell* **60,** 675–683.

Marciniak, R. A., Garcia-Blanco, M. A., and Sharp, P. A. (1990). *Proc. Natl. Acad. Sci. U.S.A.* **87,** 3624–3628.

Mosca, J. D., Bednarik, D. P., Raj, N. B. K., Rosen, C. A., Sodroski, J. G., Haseltine, W. A., and Pitha, P. M. (1987). *Nature (London)* **325,** 67–70.

Muesing, M. A., Smith, D. H., Cabradilla, C. D., Benton, C. V., Lasky, L. A., and Capon, D. J. (1985). *Nature (London)* **313,** 450–458.

Nabel, G., and Baltimore, D. (1987). *Nature (London)* **326,** 711–713.

Nabel, G. J., Rice, S. A., Knipe, D. M., and Baltimore, D. (1988). *Science* **239,** 1299–1302.

Niederman, T. M. J., Thielan, B. J., and Ratner, L. (1989). *Proc. Natl. Acad. Sci. U.S.A.* **86,** 1128–1132.

Ogawa, K., Shibata, R., Kiyomasu, T., Higuchi, I., Kishida, Y., Ishimoto, A., and Adachi, A. (1989). *J. Virol.* **63,** 4110–4114.

Osborn, L., Kunkel, S., and Nabel, G. J. (1989). *Proc. Natl. Acad. Sci. U.S.A.* **86,** 2336–2340.

Pomerantz, R. J., Trono, D., Feinberg, M. B., and Baltimore, D. (1990). *Cell* **61,** 1271–1276.

Post, L. E., Mackem, S., and Roizman, B. (1981). *Cell* **24,** 555–565.

Rosen, C. A., Sodroski, J. G., and Haseltine, W. A. (1985). *Cell* **41,** 813–823.

Rosen, C. A., Sodroski, J. G., Goh, W. C., Dayton, A. I., Lippke, J., and Haseltine, W. A. (1986). *Nature (London)* **319,** 555–559.

Rosen, C. A., Terwilliger, E., Dayton, A., Sodroski, J. G., and Haseltine, W. A. (1988). *Proc. Natl. Acad. Sci. U.S.A.* **85,** 2071–2075.

Roy, S., Delling, U., Chen, C.-H., Rosen, C. A., and Sonnenberg, N. (1990). *Genes Dev.* **4,** 1365–1373.

Ruben, S., Perkins, A., Purcell, R., Joung, K., Sia, R., Burghoff, R., Haseltine, W. A., and Rosen, C. A. (1989). *J. Virol.* **63,** 1–8.

Schwartz, S., Felber, B. K., Benko, D. M., Fenyo, E.-M., and Pavlakis, G. N. (1990). *J. Virol.* **64,** 2519–2529.

Selby, M. J., Bain, E. S., Luciw, P. A., and Peterlin, B. M. (1989). *Genes Dev.* **3,** 547–558.

Seto, E., Yen, T. S. B., Peterlin, B. M., and Ou, J.-H. (1988). *Proc. Natl. Acad. Sci. U.S.A.* **85,** 8286–8290.

Sharp, P. A., and Marciniak, R. A. (1989). *Cell* **59,** 229–230.

Shaw, J.-P., Utz, P. J., Durand, D. B., Toole, J. J., Emmel, E. A., and Crabtree, G. R. (1988). *Science* **241,** 202–205.

Siekevitz, M., Josephs, S. F., Dukovich, M., Peffer, N., Wong-Staal, F., and Greene, W. C. (1987). *Science* **238,** 1575–1578.

Sodroski, J., Patarca, R., and Rosen, C. (1985). *Science* **229,** 74–77.

Southgate, C., Zapp, M. L., and Green, M. R. (1990). *Nature (London)* **345,** 640–642.

Strebel, K., Daugherty, D., Clouse, K., Cohen, D., Folks, T., and Martin, M. A. (1987). *Nature (London)* **328,** 728–730.

Strebel, K., Klimkait, T., and Martin, M. A. (1988). *Science* **241,** 1221–1223.

Strebel, K., Klimkait, T., Maldarelli, F., and Martin, M. A. (1989). *J. Virol.* **63,** 3784–3791.

Terwilliger, E. F., Cohen, E. A., Lu, Y., Sodroski, J. G., and Haseltine, W. A. (1989). *Proc. Natl. Acad. Sci. U.S.A.* **86,** 5163–5167.

Tong-Starksen, S. E., Luciw, P. A., and Peterlin, B. M. (1987). *Proc. Natl. Acad. Sci. U.S.A.* **84,** 6845–6849.

Varmus, H. (1988). *Science* **240,** 1427–1435.

Wright, C. M., Felber, B. K., Paskalis, H., and Pavlakis, G. N. (1986). *Science* **234,** 988–992.

Zapp, M. L., and Green, M. R. (1989). *Nature (London)* **342,** 714–716.

ADVANCES IN VIRUS RESEARCH, VOL. 40

IMMORTALIZING GENES OF EPSTEIN–BARR VIRUS

Tim Middleton, Toni A. Gahn, Jennifer M. Martin, and Bill Sugden

McArdle Laboratory for Cancer Research
University of Wisconsin
Madison, Wisconsin 53706

I. Introduction

Epstein–Barr virus (EBV) is a human pathogen that induces a striking phenotype in the B lymphocytes that it infects *in vitro*. These quiescent cells on infection can enter the cell cycle and continue to proliferate indefinitely. Our understanding of the involvement of

19

EBV in this phenotypic change in its host cell is reviewed in this chapter.

EBV is a herpesvirus that is causally associated with both benign and malignant B-cell lymphoproliferations and nasopharyngeal carcinoma in people. In particular, it causes infectious mononucleosis (Niederman *et al.*, 1968) and almost certainly contributes to the cause of Burkitt's lymphoma (BL) in the areas of the world in which this childhood tumor is endemic (de The *et al.*, 1978). The causal association of EBV with these diseases seems likely to be because EBV affects the state of growth of the infected cells profoundly. *In vitro*, this capacity is reflected in the ability of EBV to induce and maintain proliferation of the infected B cell. This induction and maintenance of proliferation is termed immortalization.

EBV has not been shown to infect human primary epithelial cells in culture. EBV DNA is, however, found in epithelial tumor cells of biopsies of nasopharyngeal carcinoma (Klein *et al.*, 1974). This finding indicates that the virus can infect certain epithelial cells *in vivo*. Our inability to infect primary epithelial cells with EBV has severely limited *in vitro* studies of the virus in this cell type, and few studies of established epithelial cells transfected with viral genes have been performed. In this chapter properties of the uninfected B cell are described briefly. Then changes that are wrought in this cell as a consequence of being immortalized by EBV are described. Finally, three genes of this virus that are likely to contribute to these cellular changes are discussed in detail. It is probable that genes of EBV in addition to these three also contribute to immortalization; the identity of these viral genes is, however, not known.

II. The Target Cell for Immortalization by EBV

Soon after it was appreciated that EBV could immortalize cells in culture (Pope *et al.*, 1968), classes of white blood cells were separated and surface immunoglobulin-positive (sIg$^+$) B lymphocytes were identified as targets for immortalization by the virus (Pattengale *et al.*, 1973). It seems likely that cells expressing any class of Ig are targets for immortalization by EBV. After infection by EBV, sIg$^+$ cells yield immortalized clones that express IgM, IgG, or IgA on their surfaces. More recent studies with lymphoid cells from the peripheral blood, bone marrow, or fetal liver of human donors have shown that on infection cells grow out that are immortalized and are either sIg$^+$ cells or precursors to such cells. Pro-B cells (cells that lack Ig gene rearrangements) have been shown to be immortalized by EBV (Katamine *et al.*,

1984), as have pre-B cells (cells that have rearrangements of Ig heavy-chain genes but fail to express Ig proteins on their surfaces) (Ernberg *et al.*, 1987). Although purified pro- and pre-B cells have not been shown directly to be targets for immortalization by EBV, the observation that such cells grow out to be immortalized on infection of bulk cultures with EBV indicates that it is likely that these cells are such targets.

Of these cells the one studied most as a target for immortalization by EBV is the sIg$^+$ resting B lymphocyte. These B cells are not cycling. They are small, having a diameter of about 8–10 μm, and do not incorporate [^{3}H]thymidine. They do not secrete Ig, as measured by an indirect Jerne plaque assay (Bird and Britton, 1979), nor do they efficiently express the blast antigen CD23, which is also the low-affinity receptor for IgE, for it is undetectable on the surface of most of these cells (Sugden and Metzenberg, 1983). Nor can RNAs encoded by the protooncogene c-*fgr* be detected in these cells (Cheah *et al.*, 1986; Knutson, 1990). Although resting B cells fail to express these cellular genes, they are not inert. Exposure of resting B lymphocytes to several mitogens (e.g., pokeweed mitogen in the presence of T cells or *Staphylococcus aureus* protein A in the absence of T cells) induces them to traverse at least one cell cycle and consequently to incorporate [^{3}H]thymidine. It is clear, however, that exposure of human B lymphocytes to such mitogens induces the cells to enter the cell cycle only transiently. There is as yet no way to induce continued proliferation of these cells *in vitro* other than by infecting them with EBV.

A. Infection of B Lymphocytes with EBV

It is informative to consider numerically the infection and immortalization of B lymphocytes by EBV. The numbers indicate that wild-type EBV can immortalize a wild-type B lymphocyte. This finding is quite different from the case of transformation of fibroblasts by simian virus 40 (SV40), in which transformation is inefficient and often involves viral mutants.

Both the infection and immortalization of primary B lymphocytes with EBV can be efficient. More than 50% of sIg$^+$ B cells isolated from the peripheral blood of an adult human can be infected with EBV, as evidenced by the expression of EBV-encoded antigens in cells of the infected population (Mark and Sugden, 1982; Robinson and Smith, 1981). The immortalization of these infected cells under optimal conditions is also efficient. When the infected cells are plated in agarose over a human fibroblast feeder layer, approximately 3% of the infected cells grow out to yield immortalized progeny (Sugden and Mark,

1977). This number does not take into consideration the cloning efficiency of these assay conditions. When recently immortalized cells are plated under identical conditions, they are found to clone with an efficiency of 3–10% (Sugden and Mark, 1977). If we assume that the cloning efficiency of the infected cells is equal to that of cells already immortalized by EBV, then the efficiency with which the infected cells are immortalized is 30–100%.

Several factors contribute to the optimal conditions for measuring immortalization. First, different strains of EBV immortalize infected B cells with different efficiencies. In particular, the laboratory strain B95–8 appears to be as efficient as any strain yet identified (Miller and Lipman, 1973). Second, the presence of a feeder layer of human fibroblasts increases the cloning efficiency of immortalized cells between 10- and 100-fold (Henderson *et al.*, 1977; Sugden and Mark, 1977). It is not known, however, what the fibroblasts contribute to increase this efficiency of cloning.

That EBV efficiently immortalizes primary human B lymphocytes has an important implication for the study of the contributions of EBV to immortalization. It is almost certainly a wild-type B lymphocyte that is immortalized by the virus and, therefore, it is not necessary to search for cellular mutations that are required for the cell to become immortalized on infection. It has also been shown that only one particle of EBV is required to immortalize a B lymphocyte and that 3–10% of the particles of EBV are competent to immortalize a cell (Henderson *et al.*, 1977; Sugden and Mark, 1977). These numbers indicate that it is likely that a wild-type strain of EBV is competent to immortalize a wild-type B cell. This likelihood has been confirmed by analyzing the EBV DNA within immortalized cells that have been infected by single particles of the virus. The viral information within these immortalized cells is capable of supporting productive replication of the virus, and the progeny virus is capable of immortalizing naive cells (Kintner and Sugden, 1981a; Sugden, 1984). In studying the contribution of EBV to the immortalization of a wild-type B lymphocyte, we therefore need to consider only the wild-type viral genome.

B. *Characteristics of EBV-Immortalized B Cells*

Cells immortalized by EBV are proliferating and therefore actively traversing the cell cycle. Those cells recently immortalized by the virus are large, with approximate diameters between 10 and 14 μm. However, individual cells within clonal populations often have varied morphologies and are heterogeneous in size. They incorporate [^{3}H]thymidine, but their cell cycles can be of varying lengths. Soon after immortalization clonal populations of cells have doubling times that range from 30 to 60 hours.

Cells recently immortalized by EBV do secrete Ig (Bird and Britton, 1979). It has not yet been shown that each cell in a population, as measured, for example, by an indirect Jerne plaque assay, secretes Ig, but it has been shown that among many clones almost all populations secrete it detectably (Brown and Miller, 1982). In some instances, B lymphocytes with a given antigen specificity have been isolated, then infected and immortalized with EBV, and shown to secrete Ig with that antigen specificity (Kozbor et al., 1979). In general the level of Ig secretion by EBV-immortalized B cells is less than or similar to that measured for murine plasmacytomas in cell culture.

In those rare instances in which EBV has been shown to have immortalized a pro-B cell, it has been found that Ig rearrangements do occur (Altiok et al., 1989; Otsu et al., 1987). The frequency of the rearrangement of the Ig genes in these cells has not been measured. Only those rearrangement events that presumably provide the cell a selective advantage in order to grow out of a population have been detected. However, the existence of these rearrangements indicates that immortalization by EBV does not alter the differentiated state of the pro-B cells, at least to the extent that the rearrangements can continue to occur.

The EBV-immortalized B lymphoblast expresses cellular genes not found to be expressed efficiently in the resting B cell. In particular, the blast antigen CD23 is expressed at high levels on cells recently immortalized by EBV in culture (Kintner and Sugden, 1981b).[1] Between 10^5 and 10^6 molecules of CD23 are found on the surface of these cells (Sugden and Metzenberg, 1983), and a portion of the CD23 molecule can be cleaved and then released into the medium from these cells (Swendeman and Thorley-Lawson, 1987). A role for this released moiety has not been identified (Uchibayashi et al., 1989). RNA for the protooncogene c-fgr is expressed detectably in these cells (Cheah et al., 1986; Knutson, 1990). Interleukins IL-5 and IL-6 have also been found to be secreted by them (Paul et al., 1990; Tosato et al., 1990; Yokoi et al., 1990). Which of these expressed cellular genes is central to the immortalized state and which are merely characteristic of the differentiated B lymphoblast are not now known. We can expect, however, that cellular genes central to immortalization will be expressed not only in the immortalized sIg$^+$ cells, but also in the immortalized pro- and pre-B cells.

Immortalized B lymphoblasts clone efficiently in semisolid medium

[1] In these early studies CD23 was termed EBVCS.

in the presence of a fibroblast feeder layer. The ability to clone in semisolid medium has been used as an assay to identify transformants of a variety of cell types, including fibroblasts. It is clear, however, that immortalized lymphoblasts are derivatives of nonadherent cells and that their ability to clone in semisolid medium cannot be considered to be a transformed phenotype, as can the growth of some derivatives of fibroblasts in semisolid medium. Although B lymphoblasts recently immortalized by EBV do clone efficiently in such medium, they do not grow as tumors in nude mice (Nilsson and Klein, 1982). They also have been found to have normal karyotypes (Nilsson and Klein, 1982). In these respects B lymphoblasts recently immortalized by EBV are distinct from B-cell lines derived from human malignancies, which do proliferate in nude mice to kill them.

III. The Context for Asking, "What Does the Virus Do?"

From this brief description of the target cell for immortalization prior to infection and of its EBV-immortalized progeny, there emerges a question: "What does the virus do to induce the changes characteristic of immortalization?" This question is addressed in the remainder of this chapter, once the structure of the viral genome and the expression of its genes in the infected B cell are outlined in order to provide a context for an answer.

The Genome of EBV and Its Expression in B Cells

The genome of EBV within its virion is a linear duplex DNA molecule of approximately 172,000 bp (Baer *et al.*, 1984). The viral DNA is usually maintained as a covalently closed circular molecule in immortalized B cells (Lindahl *et al.*, 1976) (Fig. 1). The ends of the linear virion DNA contain direct repeats, described as the terminal repeats, which are joined together on circularization of the molecule in the infected cell (Given *et al.*, 1979; Kintner and Sugden, 1979).

It is not known whether the virion of EBV contains proteins that affect viral transcription on infection of a resting B cell. It does seem likely, however, that a single viral promoter, the *Bam*W promoter (W_p), is used first to express precursors to viral mRNAs (Woisetschlaeger *et al.*, 1990) (Fig. 1). These precursor RNAs are up to 100,000 bases in length and must be highly spliced to yield mature mRNA species (Bodescot and Perricaudet, 1986). Two proteins, Epstein–Barr nuclear antigen leader protein (EBNA-LP) and Epstein–Barr nuclear antigen (EBNA-2), are synthesized first from transcripts

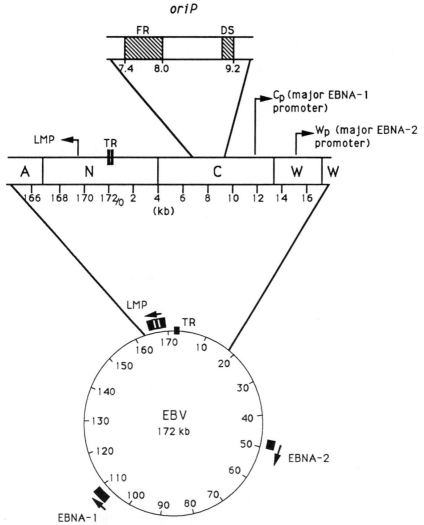

FIG. 1. The genome of the B95–8 strain of EBV as a plasmid. In infected B cells the EBV genome is maintained as a double-stranded circular DNA molecule of approximately 172 kilobase pairs (kb). The numbers inside the circle indicate the nucleotide number in kilobase pairs. The terminal repeats that are joined to form the circular plasmid are indicated (TR). The positions and directions of the open reading frames encoding the EBNA-1, EBNA-2, and LMP genes are indicated by the boxes and arrows outside the circular genome. Above the circular genome is an expanded diagram of the region extending from positions 165,000 to 18,000. The terminal repeats are indicated. The *Bam*HI segments that compose this portion of the genome are indicated by the letters A, N, C, and W. C_p, *Bam*C promoter; W_p, *Bam*W promoter. The arrows indicate the locations and directions of the promoters for the LMP, EBNA-1, and EBNA-2 genes. At the top is an expanded diagram of the region containing EBV's plasmid origin of replication (*oriP*). The family of repeats (FR) and the dyad symmetry element (DS) that compose oriP are indicated.

from W_p, apparently on alternatively spliced RNAs (Rogers *et al.*, 1990). Either one or both of these viral proteins are required to permit the expression of additional viral genes in the infected B cell. This conclusion has been derived from studies in which a mutant of EBV deleted for the last two exons of EBNA-LP and all of EBNA-2 was used to infect B lymphocytes. The truncated form of EBNA-LP, but no other viral gene product, was detected in the cells (Rooney *et al.*, 1989). The expression of EBNA-LP and EBNA-2 apparently is followed by transcription from three additional promoters; this additional transcription leads to the synthesis of eight viral proteins (reviewed by Speck and Strominger, 1989). Three of these are found in the plasma membrane. Their mRNAs are derived from transcripts that initiate from two promoters with the *Bam*N fragment of EBV (see Fig. 1 for their positions). The other five viral proteins now known usually to be expressed in B cells immortalized *in vitro* are nuclear antigens. Their mRNAs are derived from transcripts that initiate from either of two promoters that map in the *Bam*W (W_p) and *Bam*C (C_p) fragments (see Fig. 1). After the immortalized state has been established C_p apparently is used preferentially (Woisetschlaeger *et al.*, 1990).

The 10 open reading frames noted above that have been found to be expressed in B cells immortalized *in vitro* compose only 10% of the coding capacity of EBV (Baer *et al.*, 1984). The other 90% is expressed only when the immortalized cell, which is termed "latently infected," begins to support the lytic phase of the life cycle of EBV. The lytic phase of the viral life cycle ends with cell lysis and the release of EBV. Only a small fraction (usually 0.1–0.001% of the cells in different clonal populations) of immortalized cells support the lytic phase of the life cycle per cell generation (Sugden, 1984). This small fraction of lytically infected cells, however, may contain many copies of viral genes expressed only during the lytic cycle; these viral genes may be detected in the bulk population of immortalized cells and can confuse the assignment of a viral gene as "latent" or not.

At least three viral genes expressed in immortalized B cells are likely to be required for immortalization; the products of these genes are diagrammed in Fig. 2. EBNA-2 affects viral and cellular gene expression, the latent membrane protein (LMP) affects cellular growth control, and EBNA-1 mediates viral plasmid DNA replication. Each of these genes is discussed in detail. It is important to emphasize that it seems likely that additional viral genes are also involved in immortalizing the infected cell. Currently, however, there is little circumstantial evidence to identify them, so they are not discussed here.

EBNA-2

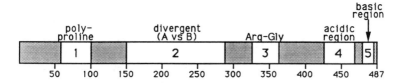

LMP

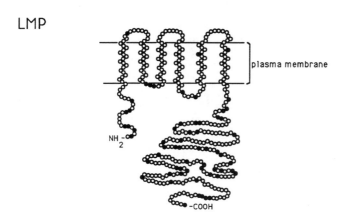

EBNA-1

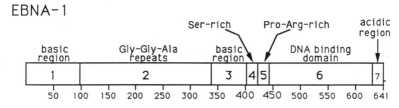

Fig. 2. The EBNA-2, LMP, and EBNA-1 proteins. (Top) The EBNA-2 protein encoded by the type A allele is shown. Small numbers representing amino acids in the open reading frame of EBNA-2 are shown beneath the bar and begin with the first amino-terminal residue. The large numbers within each box depict major domains and are described above each domain. Domain 1 consists of a polyproline sequence of 42 residues, and domain 2 represents EBNA-2 sequences that are divergent between type A and B alleles. The third domain contains alternating arginine–glycine residues, and the fourth and fifth domains are regions of acidic and basic residues, respectively. (Center) The predicted structure of the LMP protein is shown. LMP is predicted to span the membrane six times, with amino and carboxy termini located within the cytoplasm. The location of the first of the three extracellular loops, as well as that of the carboxy terminus, has been confirmed experimentally (see text). Solid circles represent the locations of charged residues in the protein. (Bottom) The structure of the EBNA-1 protein is shown. Small numbers beneath the bar represent the amino acids of EBNA-1 and begin with the first amino-terminal residue. There are two basic regions in EBNA-1 (domains 1 and 3) and an acidic region at the carboxy terminus (domain 7). Domains 4 and 5 are serine- and proline–arginine-rich regions, respectively. There is a large region of glycine–glycine–alanine repeats (domain 2) and another large domain defined experimentally (see text) as the DNA binding domain (domain 6).

IV. EBNA-2

Only one gene of EBV has been demonstrated to be required for immortalization of B lymphocytes thus far: that encoding EBNA-2. Although the function of EBNA-2 in immortalization remains unknown, several activities have been ascribed to the protein. After a description of the structure of the protein encoded by the EBNA-2 gene, each of these activities is considered. Among these activities are those that affect the expression of other viral genes and cellular genes. Potential roles for EBNA-2 in immortalization can then be formulated from these activities of the protein.

A. EBNA-2 Is Required for the Immortalization of B Lymphocytes

EBNA-2 is required for the immortalization of B lymphocytes. This requirement was demonstrated with an immortalization-defective deletion mutant of EBV termed P3HR1. P3HR1 arose during passage in tissue culture from the BL cell line Jijoye (Hinuma *et al.*, 1967). The deleted region in P3HR1 includes the entire EBNA-2 gene and a portion of EBNA-LP (Rowe *et al.*, 1985). While P3HR1 is capable of undergoing lytic replication, the virus does not immortalize B lymphocytes (Rabson *et al.*, 1982). Curing of the immortalization defect by either complementation or recombination with a vector encoding EBNA-2 provides proof that EBNA-2 is required for immortalization (Cohen *et al.*, 1989; Hammerschmidt and Sugden, 1989). EBNA-LP is not necessary for immortalization, although cells immortalized with an EBNA-LP⁻ virus initially grow more slowly than do those infected with wild-type EBV (Hammerschmidt and Sugden, 1989).

B. The EBNA-2 Protein

EBNA-2 exists in two allelic forms (Dambaugh *et al.*, 1984). Both alleles can function in the immortalization of B lymphocytes (Cohen *et al.*, 1989). Type A (or 1) consists of 487 amino acids, while type B (or 2) consists of 455 amino acids (Fig. 2). The two alleles share sequence similarity in both their amino- and carboxy-terminal one-thirds, whereas the central one-third of the two proteins is divergent. An unusual feature of both alleles is the presence of a polyproline sequence spanning 42 residues in type A and 14 residues in type B. The overall content of proline in type A EBNA-2 is approximately 28%. A class of transcriptional activators has been described in which the transcriptional activity resides in a domain of high proline content (Mermod *et al.*, 1989). Several lines of evidence indicate that EBNA-2 may function to activate transcription and thus may qualify as a member of this class of protein.

EBNA-2 exists in multiple isoelectric forms, indicating that the protein probably undergoes posttranslational modification (Petti *et al.*, 1990). One modification that has been demonstrated is phosphorylation (Petti *et al.*, 1990), which occurs on serine and threonine residues, but not on tyrosines (N. Muller-Lantzsch, personal communication). Since the half-life of the phosphate residues is only 6–9 hours, while that of EBNA-2 is approximately 24 hours (N. Muller-Lantzsch, personal communication), phosphorylation may play a role in modulating the activity of EBNA-2.

Phenotypic differences are observed when B lymphoblasts are immortalized with EBV that expresses type A, compared with type B, EBNA-2 (Rickinson *et al.*, 1987). The initial rates of colony formation are not significantly different between the two types, but on passage in culture several differences become apparent. Type A-immortalized cells yield cell lines more readily than do cells immortalized with type B virus. The cell lines immortalized with type B virus have a slower growth rate and a lower saturation density than do type A cell lines (Rickinson *et al.*, 1987). The basis for these differences, particularly the involvement of EBNA-2 in these phenotypes, is unclear. Recently, allelic differences have been observed among at least two additional viral genes expressed in latently infected cells (Arrand *et al.*, 1989; Rowe *et al.*, 1989; Sample *et al.*, 1990). There is a correlation between the presence of type A or B EBNA-2 alleles and particular alleles of these additional genes. Thus, there may be distinct strains of EBV that differ in at least three latently expressed genes, which makes identification of the genetic source of phenotypic differences in immortalized cells complex.

C. Activities of EBNA-2

The role of EBNA-2 in immortalization of B lymphocytes by EBV is unknown. However, several activities of EBNA-2 have been identified, most of which involve an increase in expression of a viral or cellular RNA or protein. Two viral proteins whose expression can be increased by EBNA-2 are LMP (Abbot *et al.*, 1990; Wang *et al.*, 1990b) and another membrane protein that is expressed in latently infected cells, called terminal protein (Zimber-Strobl *et al.*, 1991). Cellular proteins include the B-cell activation antigens CD21 (Cordier *et al.*, 1990) and CD23 (Cordier *et al.*, 1990; Wang *et al.*, 1990a) and the protooncogene c-*fgr* (Knutson, 1990).

1. Increase in Expression of LMP in Certain Cells

The relationship between the expression of EBNA-2 and LMP was first noted by comparing EBV-negative BL cell lines converted with either the B95–8 (EBNA-2$^+$) or P3HR1 (EBNA-2$^-$) strain of EBV.[2] In each case examined B95–8-converted BL lines express LMP, whereas P3HR1-converted lines do not (Murray *et al.*, 1988). Introduction of an EBNA-2 expression vector into several P3HR1-converted BL lines results in expression of LMP (Abbot *et al.*, 1990; Wang *et al.*, 1990b). Both types A and B of EBNA-2 are capable of inducing LMP (Wang *et al.*, 1990b). No other EBV genes are required for this effect because cotransfection of a vector expressing EBNA-2 with the LMP gene into an EBV-negative BL line results in increased LMP expression compared with transfection of the LMP gene alone (Wang *et al.*, 1990b). The effect of EBNA-2 on LMP expression appears to be at the level of accumulation of RNA, but it remains to be determined whether EBNA-2 affects the rate of transcription or acts posttranscriptionally.

The sequence upstream of the LMP gene has been analyzed in an attempt to identify an EBNA-2-responsive element (Fahraeus *et al.*, 1990a; Ghosh and Kieff, 1990). The EBV sequence extending 634 bp upstream of the LMP gene transcriptional start site is sufficient to confer responsiveness to EBNA-2 on reporter plasmids. Deletion analysis reveals a promoter element within 54 bp of the start site that is constitutively active in the absence of further upstream sequences. The presence of further upstream sequences results in the loss of LMP expression unless EBNA-2 is present. Thus, it has been concluded that EBNA-2 activates the LMP promoter by modulating the activity of a negative regulatory element (Fahraeus *et al.*, 1990a).

EBNA-2 is not effective in inducing LMP expression in all cell lines, however. In BL41 cells converted with P3HR1 virus, expression of EBNA-2 does not lead to the induction of LMP (Cordier *et al.*, 1990). In addition, hybrid cell lines generated by fusing EBV-negative Ramos cells with Raji cells (EBNA-2$^+$ and LMP$^+$) result in the loss of LMP expression even though expression of EBNA-2 is maintained. In contrast, a BJAB–Raji hybrid (BJAB is an EBV-negative BL-derived cell line) is permissive for both EBNA-2 and LMP expression (Contreras-Salazar *et al.*, 1989). Thus, some host cell environments are permissive for the EBNA-2 induction of LMP expression, while others are not. Whether this permissivity reflects an alteration of the activity of EBNA-2 or the presence or absence of dominant cellular factors remains to be determined.

[2] Converted cell lines are derived by infection of EBV-negative BL cells with EBV, and unspecified selection for maintenance of the EBV genome.

2. Increase in Expression of CD23

EBNA-2 and LMP both can increase expression of a B-cell activation antigen, CD23, the low-affinity Igε Fc receptor. Expression of either type A EBNA-2 or LMP in the EBV-negative BL lines Louckes or BJAB results in an increase in the levels of both CD23 mRNA and CD23 protein at the cell surface (Wang et al., 1987, 1990a). Interestingly, type B EBNA-2 does not increase the expression of CD23 in those cells that have been tested (Wang et al., 1990a).

There are two classes of CD23 mRNA: FcεRIIa and FcεRIIb. The former is restricted to B lymphocytes, in which it is constitutively expressed. FcεRIIb is expressed in monocytes and eosinophils and is induced in B lymphocytes by IL-4 (Yokota et al., 1988). EBNA-2 and LMP seem to have differential effects on the expression of the two classes of CD23 mRNA (Cordier et al., 1990; Wang et al., 1990a). In some cell lines, such as Daudi (an EBV-positive BL-derived cell line) and BL41 (an EBV-negative BL-derived cell line), expression of LMP has no detectable effect on CD23. In BJAB cells, however, expression of LMP preferentially induces FcεRIIb, thus altering the IIa/IIb ratio. Expression of EBNA-2 alone in BJAB increases the level of FcεRIIa, the class normally expressed in BJAB cells. The expression of EBNA-2 and LMP together in BJAB cells results in more than an additive effect on the level of CD23, while the IIa/IIb ratio remains that of cells expressing LMP alone. Thus, it appears that EBNA-2 and LMP can work cooperatively in inducing CD23. In those cells in which it has an effect, LMP preferentially induces FcεRIIb, while EBNA-2 provides a general stimulation of whichever CD23 is being expressed (Cordier et al., 1990; Wang et al., 1990a).

The effect of EBNA-2 on the expression of CD23 is observed even in a cell line in which the expression of EBNA-2 does not stimulate expression of LMP (Cordier et al., 1990). This finding indicates that the mechanisms by which EBNA-2 mediates these two activities either are distinct or involve host cell factors that are differentially expressed among cell lines.

3. Increase in Expression of c-fgr

Another cellular gene whose expression may be increased by EBNA-2 is the protooncogene c-fgr (Cheah et al., 1986; Knutson, 1990). c-fgr is a member of the src family of cytoplasmic tyrosine kinases and is normally expressed in peripheral blood granulocytes and monocytes and in tissue macrophages (Ley et al., 1989). c-fgr transcripts are also detected in EBV-positive BL-derived cell lines. When EBV-negative BL lines are infected with EBV, a 10- to 50-fold increase in the level of c-fgr mRNA is observed (Cheah et al., 1986; Knutson, 1990).

It has been observed that two EBV-negative BL cells transfected with and selected to express only EBNA-2 have an increased level of c-*fgr* mRNA (Knutson, 1990). The increase in expression of c-*fgr* by EBNA-2 is observed in a cell line (Ramos) that is nonpermissive for the induction of LMP by EBNA-2. Thus, these two activities of EBNA-2 either involve different host cell factors or are mediated by distinct mechanisms.

A comparison of c-*fgr* cDNAs from normal human monocytes with those from an EBV-infected BL cell line has revealed a 5'-untranslated exon found only in the BL c-*fgr* mRNA (Gutkind et al., 1991). One model proposed for the generation of the novel transcript is that a promoter is used to initiate the BL transcript different from that used for the transcripts found in monocytes (Gutkind et al., 1991). The interpretation that EBV in general and EBNA-2 in particular increases the expression of c-*fgr* in B cells is still being tested. Several groups have observed high levels of c-*fgr* RNA in an EBV-positive cell line derived from a BL, in converted cell lines, and in B-cell lines 48 hours after infection with EBV (Cheah et al., 1986; Klein et al., 1988; Knutson, 1990). Klein et al. (1988) did not, however, detect an increase in expression of c-*fgr* RNA in two B-cell lines derived by immortalization by EBV. The apparent differences in these studies may result from technical differences (the probe used to detect c-*fgr* RNA is derived from a region of the gene encoding its tyrosine kinase moiety and may cross-hybridize with other *src*-related RNAs) or may mean that EBV and EBNA-2 affect the expression of c-*fgr* in only certain B cells.

Each of the three activities of EBNA-2 discussed—increase in expression of LMP, CD23, and c-*fgr*—is consistent with the idea that the protein functions as an activator of transcription. But are any of these activities integral to the ability of EBNA-2 to participate in immortalization? Characteristics of the induced gene products indicate that they could be. LMP can function as an oncogene in rodent cells (Wang et al., 1985) and is likely to be required for immortalization by EBV. The protooncogene c-*fgr* has the potential to be activated so that it can transform cells (Kawakami et al., 1986), and its normal function is likely to be involved in the regulation of cell growth and differentiation (Ley et al., 1989). The function of CD23 is unknown, but its expression is associated with the activation of B lymphocytes to dividing lymphoblasts (Crow et al., 1986; Kintner and Sugden, 1981b; Thorley-Lawson et al., 1985). Thus, each of the gene products induced by EBNA-2 may be associated with functions that affect the proliferation of the cell.

4. Abrogation of the Antiproliferative Effects of α-Interferon

Another activity of EBNA-2 that may provide EBV a selective advantage *in vivo* is its abrogation of the antiproliferative effects of α-interferon (α-IFN) (Aman and von Gabain, 1990). Cells infected with mutant strains of EBV lacking EBNA-2 are 100-fold more sensitive to the antiproliferative effects of α-IFN than are those infected with wild-type EBV. The proliferation of several EBV-negative BL cell lines can be slowed by exposure to α-IFN. These same cell lines expressing a transfected EBNA-2 gene acquire the ability to proliferate in the presence of α-IFN even though α-IFN binds to the cells (Aman and von Gabain, 1990). The neutralization of the effects of α-IFN may promote the survival of latently infected B lymphoblasts *in vivo*. While EBNA-2 is involved in limiting the effect of α-IFN, the mechanism by which it does so is unknown.

D. Timing of EBNA-2 Expression

A study of the timing of EBV gene expression on infection of B lymphocytes has shown that EBNA-2 is the first gene product to be detected, appearing 24–40 hours after infection (Rooney *et al.*, 1989). Two small EBV-encoded RNAs (EBERs) appear 36–60 hours after infection, whereas EBNA-1 and -LP are not detected until about 70 hours post infection. This same order of expression of viral genes, but with an accelerated curve, is found when prolymphocytic leukemia cells are infected with the B95–8 strain of EBV (Allday *et al.*, 1989). The pattern of expression is quite different when B lymphocytes are infected with P3HR1. Because the EBNA-2 open reading frame is deleted in this virus, EBNA-2 is not expressed. EBNA-LP is detectable very early (i.e., 12 hours) after infection. Surprisingly, EBNA-1 is not expressed. In addition, only one of the EBERs (EBER-1) is detected and is seen only at a low level very late (i.e., 132 hours) after infection (Rooney *et al.*, 1989).

Expression of EBNA-2 is not necessarily required for expression of EBNA-1 or the EBERs, because infection of EBV-negative BL cells with P3HR1 results in expression of all these gene products within 12–24 hours after infection (Rooney *et al.*, 1989). Nonetheless, EBNA-2 may influence the expression of the other latent EBV gene products by inducing resting B lymphocytes to produce cellular factors for efficient transcription or translation of these genes. This proposed activity could be an important part of the role of EBNA-2 in immortalization.

There are two promoters from which the EBNA genes are known to be expressed: W_p and C_p (Speck and Strominger, 1989). W_p and C_p usage is mutually exclusive (Woisetschlaeger et al., 1989). W_p appears to be used first on infection of B lymphocytes by EBV, with a switch to C_p occurring approximately 3 days after infection (Woisetschlaeger et al., 1990). What regulates this switch and, in particular, whether or not EBNA-2 plays a role are unknown.

A substantial body of evidence indicates that EBNA-2 modulates the expression of both viral and cellular genes. Apparently, EBNA-2 acts to affect the level of RNA. However, it is not known whether EBNA-2 affects promoter usage, transcription initiation rates, alternative splicing, poly(A) site selection, message stability, or other processes that control RNA levels. Whether EBNA-2 acts directly in any of these processes or indirectly by affecting the activity of other viral or cellular proteins remains to be elucidated.

V. LMP

Although EBNA-2 is thus far the only EBV gene shown to be required for immortalization, LMP (previously referred to as BNLF-1) is also likely to be required for immortalization. This contention is supported here by outlining the effects of LMP on the growth of established cell lines *in vitro* and on cells *in vivo*. In order to consider these effects mechanistically, the known biochemical properties of LMP and their structure–function relationship are described next. Then, possible biochemical activities of LMP are explored, on the basis of the known structural and biochemical properties of the protein.

A. LMP Alters the Growth Properties of Rodent Cells

Although it has not been shown that LMP is required for the immortalization of B cells by EBV, circumstantial evidence indicates that LMP is likely to be essential for this process. Rat-1 fibroblasts selected to express the LMP gene from the metallothionein promoter require reduced serum levels for growth, grow in an anchorage-independent manner in soft agar, and can grow as tumors in nude mice (Wang et al., 1985). Similarly, expression of LMP from the SV40 early promoter–enhancer in BALB/c 3T3 murine cells induces these cells to grow in an anchorage-independent manner, and anchorage-independent clones grow as tumors in nude mice. The ability of LMP to transform BALB/c 3T3 cells is independent of drug selection, since cells transfected with an LMP expression vector alone are able to induce anchorage-independent growth (Baichwal and Sugden, 1988). These results indicate that LMP functions as an oncogene in rodent cell lines. It seems likely that the biochemical activity of LMP that results in tumorigenicity in rodent cells will underlie any ability of LMP to contribute to the immortalization of B cells by EBV.

B. LMP Alters the Growth Properties of Epithelial Cells

Expression of LMP can alter the growth properties of established human epithelial cell lines. Expression of LMP in cell lines established from an immortalized nontumorigenic variant of the human squamous cell carcinoma cell line SCC12 has no effect on the growth rate of subconfluent cultures (Dawson et al., 1990). However, cells expressing LMP do not stratify as do dense cultures of SCC12 cells, nor do they form cross-linked envelopes in response to the calcium ionophore A23187; such stratification reflects the capacity of epithelial cells to differentiate terminally in vitro. Furthermore, the ability of LMP-expressing cells to form organized epithelial structures on collagen rafts, a reflection of epithelial cell differentiation in vivo, is severely impaired. LMP-positive cells form a thicker, yet unorganized, epithelium that lacks terminal differentiation markers (e.g., involucrin). Epithelial cells transfected with the ras oncogene show similar effects (Dawson et al., 1990). Expression of LMP in another immortalized, yet nontumorigenic, epithelial cell line, RHEK-1, alters the morphology of these cells, converting them from a keratinocyte-like morphology to that of a fibroblast (Fahraeus et al., 1990b). This change is accompanied by the down-regulation of cytokeratin proteins. Together, these results indicate that the expression of LMP alters the ability of established epithelial cells to differentiate.

LMP may affect epithelial cell differentiation in vivo as well. Transgenic mice harboring an expressed LMP transgene exhibit the phenotype of epidermal hyperplasia that progresses to chronic dermatitis with the increasing age of the mice (Wilson et al., 1990). In mice that are transgenic for the LMP gene driven by the polyoma early promoter–enhancer, expression of the transgene at the level of RNA is limited to the skin (i.e., epidermis) and the tongue. In the one mouse tested for expression at the level of protein, LMP was also expressed in the skin and the tongue. The expression of LMP RNA in the epidermis correlates with an induction of the expression of keratin 6, a protein normally expressed only in proliferating epidermis and hair follicles (keratin 6 is expressed aberrantly in hyperplastic and malignant epidermis). These observations indicate that the expression of LMP alters the growth properties and differentiated state of epithelial cells *in vitro* and *in vivo*. These results support a possible role for LMP in EBV-induced transformation of nasopharyngeal epithelium to undifferentiated nasopharyngeal carcinoma. It is interesting to note that epithelial cells expressing LMP do not form tumors in nude mice (Dawson *et al.*, 1990), nor do mice that are transgenic for LMP exhibit tumors of any type (Wilson *et al.*, 1990). This failure indicates that events in addition to the expression of LMP are required to convert EBV-infected nasopharyngeal epithelium to nasopharyngeal carcinoma.

C. LMP Affects the Phenotype of B-Lymphoblastoid Cells

LMP alters the expression of surface antigens in EBV-negative lymphoblastoid cells, much as EBV does on infecting resting B lymphocytes. This observation supports a role for LMP in the immortalization of B cells by EBV. Louckes cells, an EBV-negative BL-derived cell line, typically proliferate in suspension as single cells. Expression of LMP in these cells results in their proliferating as clumps of cells. Louckes cells expressing a nonfunctional mutant of LMP are indistinguishable from parental cells in that they grow as single cells (Wang *et al.*, 1988). The increased aggregation of LMP-expressing Louckes cells is accompanied by a modest (i.e., 2- to 3-fold) induction of expression of the cell adhesion molecules LFA-1 and ICAM-1. Binding of LFA-1 to ICAM-1 results in homotypic cellular adhesion (Springer, 1990). It is possible that the induction of LFA-1 and ICAM-1 expression by LMP results in the observed cell–cell aggregation. In addition, LMP induces the expression of another cell adhesion molecule, LFA-3, in Louckes cells. LFA-3 is involved in B-cell–T-cell interactions by virtue of its ability to bind to the T-cell antigen CD2 (Springer, 1990). Infection of B cells by EBV results in the induction of expression of the B-cell activation markers CD23 and transferrin receptor (Gordon *et al.*, 1986; Kintner and Sugden, 1981b). Louckes cells selected to express LMP also exhibit increases in the expression of CD23 and the transferrin receptor.

The mechanism by which LMP expression in these cells results in an increase in cellular adhesion properties and an increase in the expression of surface antigens is unknown. It is not known whether these changes are a direct effect of the expression of LMP or an indirect result of the selection of the cells to express LMP. However, this mimicking by LMP in EBV-negative lymphoblastoid cells of some of the effects of EBV infection of resting B cells is consistent with the contribution of LMP to the immortalization of B lymphocytes by EBV. Direct evidence for such a role for LMP must be provided genetically, as has been done for EBNA-2 (Cohen *et al.*, 1989; Hammerschmidt and Sugden, 1989).

D. Biochemical Properties of LMP

Although no biochemical activity has been assigned to LMP, much is known about the biochemical properties of LMP. These properties are distinct from those of other latently expressed EBV proteins. LMP is a 62-kDa membrane protein (Bankier *et al.*, 1983) located in the plasma membranes of immortalized B cells (Hennessy *et al.*, 1984; Mann *et al.*, 1985) and rodent cells transformed by the LMP gene (Baichwal and Sugden, 1988; Wang *et al.*, 1985). The LMP protein is predicted, based on analysis of its open reading frame, to span the plasma membrane six times, with both the amino-terminal 25 amino acids and the carboxy-terminal 200 amino acids located in the cytoplasm (Bankier *et al.*, 1983; Liebowitz *et al.*, 1986) (Fig. 2). Immunological (Hennessy *et al.*, 1984; Mann *et al.*, 1985) and proteolytic analyses of LMP in intact cells (Liebowitz *et al.*, 1986) are consistent with this predicted positioning of LMP. The tertiary structure of LMP resembles that of the rhodopsin family of cell surface receptors (Hanley and Jackson, 1987), certain ion channels (Baumann *et al.*, 1988; Tempel *et al.*, 1987), and transport proteins. However, LMP shares no sequence similarity with known proteins.

LMP is localized to discrete patches in the plasma membranes of infected cells and rodent cells transformed by LMP (Liebowitz *et al.*, 1986; Mann *et al.*, 1985), is associated with the detergent-insoluble cytoskeleton (Liebowitz *et al.*, 1987; Mann and Thorley-Lawson, 1987), and has a short half-life (Baichwal and Sugden, 1987; Mann and Thorley-Lawson, 1987). These properties of LMP correlate with its ability to transform rodent cells. Nontransforming mutants of LMP do not turn over rapidly, are not patched in the plasma membrane, and are not associated with the cytoskeleton (Martin and Sugden, 1991b). In addition, the rapid turnover of LMP requires ongoing protein synthesis and is probably preceded by its internalization from the cell surface (Martin and Sugden, 1991b). The 25-kDa acidic carboxy terminus of LMP is cleaved at a specific site from the protein and accumulates in the cytoplasm, where it is stable (Moorthy and Thorley-Lawson, 1990). Mutants of LMP that are deleted for their acidic carboxy termini transform rodent cells in a fashion equivalent to wild-type LMP (Baichwal and Sugden, 1989; Martin and Sugden, 1991a).

LMP is a phosphoprotein and is phosphorylated on serine and threonine residues at the carboxy terminus (Baichwal and Sugden, 1987; Mann and Thorley-Lawson, 1987; Moorthy and Thorley-Lawson,

1990). The functional significance of the phosphorylation of LMP is not known.

E. Mutational Analysis of LMP

The LMP protein can be divided into three domains, based on its amino acid sequence (Bankier *et al.*, 1983; Fennewald *et al.*, 1984): an amino-terminal charged domain of 25 amino acids, a hydrophobic transmembrane domain of 160 amino acids, and an acidic carboxy-terminal domain of 200 amino acids (Fig. 2). The ability of LMP to function as an oncogene in rodent cells has been used to assay domains of LMP required for function (Baichwal and Sugden, 1989). A series of deletion mutants of LMP, in which these three domains are altered, has been assayed for the ability to induce anchorage-independent growth in BALB/c 3T3 cells (Baichwal and Sugden, 1989). These experiments reveal that the carboxy terminus of LMP is dispensable for transformation, but that the transmembrane domain is required. The amino terminus of LMP, while not sufficient, is clearly necessary for function, as demonstrated in experiments with a mutant of LMP lacking nine of the 25 amino-terminal amino acids (Hammerschmidt *et al.*, 1989). In addition, these experiments demonstrate that localization of the amino terminus at the plasma membrane is not sufficient to confer LMP function; this finding suggests that the hydrophobic transmembrane domain of LMP has a more complex function than simply to anchor the amino terminus at the membrane. These results are consistent with those obtained with a truncated form of LMP lacking approximately 128 amino acids from the amino terminus (including the entire amino terminus and four of the six transmembrane domains) that is expressed in certain EBV-positive lymphoblastoid cell lines. This deleted form of LMP does not function as an oncogene in rodent cells and does not alter the adhesive properties of EBV-negative lymphoblastoid cells (Wang *et al.*, 1988).

The expression of LMP at high levels from the cytomegalovirus immediate–early promoter–enhancer is toxic to all cell lines tested, which include fibroblastic, epithelial, erythroid, and lymphoblastoid cell lines (Hammerschmidt *et al.*, 1989). In contrast, a moderate level of expression of LMP from the SV40 early promoter–enhancer induces anchorage-independent growth in BALB/c 3T3 cells and does not exhibit the cytotoxic effects seen with the expression of LMP at higher levels (i.e., from the human cytomegalovirus immediate–early promoter–enhancer). The transforming and toxic activities of LMP are not separable by mutational analysis; mutants that do not transform cells also are not toxic, and vice versa (Hammerschmidt *et al.*, 1989).

Therefore, a high level of expression, specifically of the transforming domain of LMP, is cytotoxic. These findings imply that the activity of LMP that mediates transformation of certain rodent cells when expressed at a moderate level may be toxic in various cells when expressed at high levels. Thus, this biochemical activity of LMP is likely to function in a variety of cell types.

F. Possible Biochemical Activities of LMP

Although the tertiary structure of LMP resembles that of certain ion channels, the rapid turnover of LMP and the lack of conserved amino acid residues characteristic of ion channel proteins (Baumann *et al.*, 1988; Tempel *et al.*, 1987) are not consistent with LMP functioning as an ion channel; known ion channels are stable proteins (Schmidt and Catterall, 1986). The structure of LMP also resembles that of transport proteins such as mdr and STE6 (Kuchler *et al.*, 1989). These proteins have consensus sequences associated with the binding of ATP, which is required for their function; LMP lacks these consensus sequences. Thus, if LMP functions as a transport protein, it belongs to a class of transporters not yet identified.

Although much is known about the biochemical properties of LMP, no biochemical activity has been assigned to it that explains the phenotypes it induces in cells. The structure of LMP also resembles that of the rhodopsin family of cell surface receptors. Several properties of LMP are consistent with a function as a cell surface receptor. It is transported rapidly (i.e., in less than 20 minutes) to the plasma membrane following synthesis (Martin and Sugden, 1991a) and therefore presumably acts there. The localization of LMP in patches in the membrane is also consistent with a receptorlike function; addition of a ligand such as the epidermal growth factor (EGF) to intact cells results in the aggregation of EGF receptors into discrete patches (Boni-Schnetzler and Pilch, 1987; Glenney *et al.*, 1988; Yarden and Schlessinger, 1987).

The rapid turnover of LMP is consistent with a receptor function. Most receptors undergo down-regulation on binding ligands in which receptors are removed from the cell surface and degraded. Thus, in the presence of ligand, the half-lives of some receptors are quite short (i.e., 1–2 hours), and in its absence, quite long (i.e., >24 hours). Mutant receptors that are active in the absence of ligand can exhibit constitutively short half-lives (Stern *et al.*, 1988). If the turnover of LMP were to function as a negative regulatory mechanism to terminate a persistent intracellular signal, then its function should correlate with its rapid turnover. In fact, mutants of LMP that are transforming have short half-lives (i.e., 2–3 hours), while nontransforming mutants are stable (i.e., have half-lives of >24 hours) (Martin and Sugden, 1991b). In addition, the turnover of LMP resembles those of EGF (Gross et al., 1983) and insulin (Knutson et al., 1985) receptors in that the degradation of LMP requires ongoing protein synthesis and is probably preceded by internalization from the cell surface.

As with LMP, the degradation of the activated neu oncogene (a membrane protein tyrosine kinase similar in sequence to the EGF receptor) correlates with cellular transformation. The normal nontransforming form of neu has a half-life of 8 hours, which is about 5-fold longer than that of the activated transforming neu (Stern et al., 1988). In addition, the activated form of neu has the property of forming aggregates (i.e., patches) in the membrane (Weiner et al., 1989) similar to those seen for the EGF receptor in the presence of ligand and for both wild-type LMP and functional LMP mutants (Martin and Sugden, 1991b). Nontransforming neu and nontransforming LMP mutants do not form these aggregates in the membrane. Thus, the turnover and membrane localization (i.e., patching) of both LMP and neu correlate with cellular transformation. It is notable, however, that the turnover and biological activity of LMP are independent of serum factors (Martin and Sugden, 1991b).

These observations, together with the known properties of LMP, are consistent with a model in which LMP functions as a cell surface ligand-dependent or -independent receptor or as a protein that modifies the activity of such a receptor. If this hypothesis were accurate, LMP should trigger detectable changes in intracellular signal transduction pathways. An extension of this hypothesis is that it is possible that LMP functions in immortalization by interacting with a tyrosine kinase transducer such as the c-*fgr* protooncogene. The transforming/tyrosine kinase activity of several cytoplasmic tyrosine kinases of the src family has been shown to be regulated by plasma membrane-bound receptor proteins such as CD4 and CD8 (Veillette et al., 1988). A plasma membrane-bound receptorlike protein could function in an analogous manner.

VI. EBNA-1

EBNA-2 affects the expression of viral and cellular genes, and LMP affects the growth characteristics of cells. EBNA-1 indirectly supports each of these functions in that it is required to initiate plasmid replication of the viral genome in immortalized cells. This indirect role, and a potential direct role, of EBNA-1 in immortalization are examined first. Then, functions of EBNA-1 are considered in detail. EBNA-1 is the only one of the three genes described for which a biochemical activity of its encoded protein has been identified: It is a site-specific DNA-binding protein that binds to EBV's plasmid origin of replication (*oriP*) and is required for plasmid replication.

A. Is EBNA-1 Necessary for Immortalization of B Lymphocytes by EBV?

It is likely that the EBNA-1 gene of EBV is required to support early stages of proliferation of infected B cells, because its gene product is required at least initially for viral plasmid DNA replication (Lupton and Levine, 1985; Yates *et al.*, 1985). It is not obvious, for two reasons, that EBNA-1 is required to maintain indefinite proliferation. It has not been shown that the continued expression of EBNA-1 is required for continued plasmid DNA replication, nor has it been shown that EBV itself is required to maintain the immortalized state once it is initiated. Circumstantial evidence does, however, indicate both that EBNA-1 is likely to be required to maintain plasmid DNA replication and that EBV is likely to be required to maintain immortalization. Cells that maintain EBV DNA as plasmids have been found consistently to express EBNA-1, and cells immortalized by EBV have not been cured of their EBV, yet continue to proliferate.

Biopsies of EBV-positive BLs often express only EBNA-1 detectably (Gregory *et al.*, 1990; Klein, 1989). This finding indicates that, if EBV contributes anything to maintain proliferation of these tumor cells, this contribution is likely to be mediated directly by EBNA-1. Study of the function of EBNA-1 in these tumor cells, therefore, may reveal a direct role for EBNA-1 in maintaining immortalization.

B. Functions of EBNA-1

EBNA-1 was the first EBV-encoded protein for which functions were identified. These functions are to activate transcription from heterologous promoters and from a promoter in EBV that is used for synthesis of mRNAs encoding up to six of the 10 known latently synthesized proteins (Reisman and Sugden, 1986; Sugden and Warren, 1989) and to act as a required component for the replication of EBV DNA (Lupton and Levine, 1985; Yates *et al.*, 1985). Both of these functions are mediated via binding of EBNA-1 specifically to the two regions of DNA that make up *oriP* (Rawlins *et al.*, 1985). For understanding the significance of these studies, it is useful to examine some of the structural characteristics of *oriP* and EBNA-1.

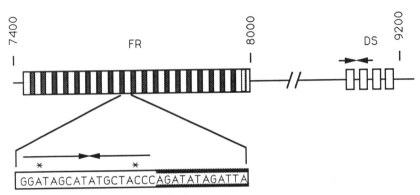

FIG. 3. Structure of the EBV latent origin of replication (*oriP*). The EBV origin of DNA replication consists of two elements, both of which are composed of binding sites for the protein EBNA-1. The first element, FR, contains 21 imperfect copies of a 30-bp repeat to which EBNA-1 binds. The second element, DS, has four related copies, two of which form a structure having dyad symmetry. The position of *oriP* in EBV is given in base pairs according to the numbering scheme of Baer *et al.* (1984). The sequence of a consensus repeat unit is given at the bottom, with the arrows indicating an inverted repeat within this unit, and the asterisks indicating a mismatch within that inverted repeat.

oriP (Fig. 3) contains two elements, both composed of multiple degenerate copies of a 30-bp segment to which EBNA-1 binds. The two elements are required for replication. One of these elements, as has been noted with several other viral DNA replicons, can affect the synthesis of RNA (reviewed by DePamphilis, 1988). A distinctive feature of the structure of *oriP* is that the function of these two elements does not depend on their relative orientations or the distance separating them. This flexibility contrasts with the positional requirements of the well-studied SV40 and adenovirus replicons. The positional flexibility of the EBV replication elements appears to be analogous to that of a generic transcriptional promoter–enhancer.

One of the elements, referred to here as DS (for dyad symmetry), contains four copies of the EBNA-1 binding site, arranged as two inverted repeats. One of these pairs is part of a region of dyad symmetry covering 65 bp. It is close to or at DS that DNA synthesis originates (Gahn and Schildkraut, 1989). The other element, termed the family of repeats (FR), contains 21 copies of binding sites arranged as direct repeats. In addition to its role in supporting DNA replication, this element can activate transcription when bound by EBNA-1 (Reisman and Sugden, 1986; Sugden and Warren, 1989).

C. The EBNA-1 Protein

EBNA-1 from the B95–8 strain is a protein of 641 amino acids. It can be divided into several domains (see Fig. 2). Roughly one-third of the protein is composed of a polymer of glycine and alanine. This region is bounded by two regions that are relatively rich in basic amino acids. Following the downstream basic region are two small segments rich in serine and proline–arginine, respectively. The carboxy-terminal 20 amino acids are highly acidic.

The function of one of these domains is known; it is required for DNA binding and composes roughly the carboxy-terminal one-third of the protein. This region alone binds to the EBNA-1-responsive element from *oriP* and has been used to define the DNA sequence to which EBNA-1 binds. In a DNase I protection assay, this protein fragment protects about 25 of the 30 bases that compose the repeated unit in *oriP* (Milman and Hwang, 1987; Rawlins *et al.*, 1985). The DNase footprint resulting from binding of this fragment of EBNA-1 to a consensus EBNA-1-responsive element is nearly identical to that resulting from binding of EBNA-1 from an extract of Raji cells (Jones *et al.*, 1989). Hydroxyl radical footprinting and methylation interference studies have shown that binding is centered over a palindromic sequence in the binding element (Kimball *et al.*, 1989). Binding of this EBNA-1 fragment to mutated binding sites has yielded a consensus sequence for binding: G(A/G)TAGCNNNNGCTA(T/C)C (Ambinder *et al.*, 1990). However, there is considerable tolerance for substitutions in this sequence, as might be expected from the variability in the sequence of sites that compose *oriP*.

D. EBNA-1's Support of DNA Replication

Replication of EBV DNA in the immortalized cell resembles replication of the host chromosomal DNA. EBV DNA is replicated once per cycle during the S phase of the cell cycle (Adams, 1987; Adams *et al.*, 1989; Hampar *et al.*, 1974). The machinery other than EBNA-1 used to replicate viral DNA is provided by host cellular proteins. A striking difference between the replication of viral and host chromosomal DNA is that *oriP* and EBNA-1 are sufficient to allow heterologous DNA to be replicated as a plasmid in mammalian cells. When circular fragments of host chromosomal DNA that contain origins of replication are introduced into cells, the chromosomal origins are not sufficient to maintain the DNA as plasmids (Burhans *et al.*, 1990). EBNA-1 must therefore provide two functions to maintain viral DNA. One is to adapt the host replication proteins to initiate synthesis at the viral origin. The other is to support the replication of newly introduced DNA as a plasmid in the nucleus of the cell. DNA replication originates from the region containing the DS element (Gahn and Schildkraut, 1989). The role of FR in replication is unknown. One copy of DS alone is insufficient to initiate replication. Either the FR enhancer or multiple copies of DS support replication transiently (Wysokenski and Yates, 1989), although there may be a requirement for FR in order to maintain the DNA for many generations (Chittenden et al., 1989). Whether this latter finding represents a requirement of the FR enhancer for replication or for proper segregation of the plasmid has not been determined.

Unlike viral lytic origins of DNA replication, oriP not only must initiate DNA replication, but must do so in synchrony with host replication. The mechanism of this timing of replication is an unstudied facet of EBV DNA synthesis. It is not clear whether the virus plays an active role in this regulation, or whether the viral origin simply comes under the same controls as do chromosomal origins. Mechanisms for repressing multiple rounds of bovine papillomavirus DNA synthesis have been proposed that can limit even the synthesis of a chimeric DNA molecule that uses an SV40 origin of replication (Roberts and Weintraub, 1986, 1988). Similar experiments with oriP have shown no such dominant repression of multiple rounds of replication (Heinzel et al., 1988). Whether the limitation of one round of DNA replication per cell cycle is a feature of oriP, EBNA-1, or the cellular machinery remains to be determined.

E. Activation of Transcription

Most of the viral transcription that occurs in latently infected cells originates from four promoters clustered in a region of the genome surrounding the FR enhancer of oriP (Bodescot and Perricaudet, 1986; Ghosh and Kieff, 1990; Rogers et al., 1990; Sample et al., 1986; Sample and Kieff, 1990; Speck et al., 1986; Speck and Strominger, 1985). [One of these promoters is present in up to 11 copies as part of the BamW repeated region (Fig. 1).] At least three of these promoters are close enough to FR to be affected by this enhancer. One of them, C_p, has been shown to be activated by the enhancer (Sugden and Warren, 1989). This enhancement requires the presence of EBNA-1 and works only in cells in which C_p is active (T. Middleton, unpublished observations), indicating that the enhancer functions through this promoter.

Expression of the EBNA genes is also directed by W_p. The effect of EBNA-1 bound to the FR enhancer on the activity of W_p has not been tested. RNA synthesis in this region occurs early after infection from W_p (Woisetschlaeger et $al.$, 1990), with EBNA-2 being translated first. By the time EBNA-1 is expressed, synthesis has shifted to C_p. Either EBNA-1 or any of the other nuclear proteins could mediate this switch, either by positively affecting C_p or by repressing synthesis from W_p.

One of the proteins synthesized from C_p is EBNA-1. Thus, EBNA-1 may regulate its own synthesis. This possibility has not been tested, but the level of EBNA-1 appears to be controlled in cells. The range in the average number of DNA molecules per cell is much greater than the range in concentrations of EBNA-1 among different cell lines (Sternas et $al.$, 1990). In fact, attempts to express higher levels of EBNA-1 in mammalian cells for extended periods have generally been unsuccessful, making it likely that levels of EBNA-1 much higher than those found in most lymphoblastoid cell lines are not tolerated (see, e.g., Vidal et $al.$, 1990).

F. Interactions between the Two Elements of oriP

The presence of a relatively large number of binding sites for EBNA-1 in both elements of $oriP$ indicates that EBNA-1 might bind them cooperatively, but such cooperativity has not been detected (Ambinder et $al.$, 1990). There is, however, an apparent cooperativity for support of transcription and replication by $oriP$. Maximal activity of the FR element requires six to eight copies of binding sites for EBNA-1 (Chittenden et $al.$, 1989; Wysokenski and Yates, 1989). When FR elements with increasing numbers up to eight are tested, it is found that transcription and replication in short-term assays increase exponentially (Wysokenski and Yates, 1989). That eight copies of the binding site yield both maximal transcriptional and replication activity, while 20 copies are maintained in the virus, may indicate that this element has an additional role not detected in the assays used in these studies.

Two experimental rearrangements of $oriP$ provide insight into its function. The first is that the organization of the enhancer element is not critical. Wysokenski and Yates (1989) have shown that the FR element can be substituted by multiple copies of the DS element and the modified $oriP$ still replicates. Its capacity to replicate is diminished relative to wild-type $oriP$, although this decrease may reflect the lower binding affinity of EBNA-1 for DS than for FR (Ambinder et $al.$, 1990; Jones et $al.$, 1989). The converse rearrangement is not functional; multiple copies of FR do not substitute for DS in supporting replication (Reisman et $al.$, 1985).

The other rearrangement is one in which the EBNA-1 binding sites are widely distributed throughout the DNA. This structure also replicates (Chittenden et al., 1989). Examination of one plasmid recovered from cells into which DNA containing two copies of the binding site in FR (and an intact DS) had been transfected showed that the plasmid recovered was a tetramer of the input DNA. The recovered plasmid thus contained eight copies of the binding element from FR, distributed uniformly at four points on the DNA, as well as four evenly distributed copies of DS. Transfection of a plasmid containing only DS did not give rise to recombinants that were maintained extrachromosomally. These rearrangements of oriP indicate that DS both provides a function distinct from that of FR and can substitute for FR. They also indicate that multiple copies of EBNA-1 binding sites need not be contiguous to support replication, but are needed for it.

G. Is There a Connection between Transcriptional Activation and Support of DNA Replication?

Studies of the elements of oriP and EBNA-1 link them both to transcription and replication. There is little constraint on the relative orientation or the distance separating the two elements of oriP for it to support replication (Reisman et al., 1985). This positional flexibility is reminiscent of that of cis-acting transcriptional elements. Mutational analyses of EBNA-1 (Polvino-Bodnar et al., 1988; Yates and Camiolo, 1988) have delineated several domains of the protein that affect its role in transcriptional activation and replication. However, no mutants have been found that affect its role only in transcription or only in replication. This failure indicates that EBNA-1 may contribute the same activities to transcription and replication. It is possible that viral DNA replication may require transcription per se or the assembly of a complex of proteins needed for transcription; EBNA-1 in activating transcription would meet either of these requirements.

Two observations argue against this possibility. First, the substitution of the FR enhancer in oriP by either an enhancer from SV40 (Reisman and Sugden, 1986) or the immediate–early enhancer from human cytomegalovirus (T. Middleton, unpublished observations) does not reconstitute an origin of replication in the presence of EBNA-1. Second, a chimeric protein that contains the EBNA-1 binding domain within the estrogen receptor binds to both elements of oriP, and activates transcription, but does not support replication (Middleton and Sugden, 1991). These observations indicate that a function of EBNA-1 in addition to those needed to support transcription is required for DNA replication to occur, although such a function has not been identified in mutational studies.

A factor that may provide a barrier to the discovery of this putative replication function of EBNA-1 is that the region of the protein needed for binding to DNA is a large portion of the total protein. Small deletions spread over about one-third of the protein destroy DNA binding capacity (Yates and Camiolo, 1988). The loss of DNA binding is likely to mask other lost activities. This region of the protein is, therefore, a strong candidate for providing interactions with proteins involved in replication originating from *oriP*.

VII. SUMMARY

EBV immortalizes human B lymphocytes efficiently. Ten of its approximately 100 genes are expressed in these proliferating lymphoblasts and are candidates for mediating the changes central to the immortalization of the cell. Enough has been learned now about three of these viral genes to indicate that they are likely to be required for immortalization. As more is learned, additional genes of EBV will probably be found to support the process of immortalization of the host cell.

EBNA-2 has been shown genetically to be required for EBV to immortalize an infected B lymphocyte. The biochemical activities of EBNA-2 that constitute this requirement have not been identified. Many experiments indicate that EBNA-2 affects the accumulation of specific viral and cellular RNAs. These effects, however, can be detected only in certain EBV-negative B-lymphoblastoid cells. It is, therefore, not clear that the known effects of EBNA-2 adequately explain its ubiquitous requirement in the immortalization of primary human B lymphocytes.

LMP is likely to be required for immortalization because it can affect the growth properties of established human lymphoid and epithelial cells and can transform at least two established rodent cells to proliferate in an anchorage-independent manner. The structure of this viral protein, its position in the plasma membrane, many of its biochemical properties, as well as studies of its mutant derivatives are consistent with its acting as a growth factor receptor or affecting the activity of such a receptor. However, no biochemical activity has been assigned directly to LMP, and both its mechanism of action and its possible contribution to immortalization by EBV remain enigmatic.

EBNA-1 presumably is required for EBV to immortalize a B lymphocyte because it is essential for the initiation of plasmid DNA replication by EBV. Circumstantial observations indicate also that EBNA-1 is probably necessary for sustaining viral DNA replication in the proliferating cell population. EBNA-1 may well affect the regulation of transcription of viral genes that themselves are required for immortalization. These roles of EBNA-1 are performed in part by its site-specific binding to the elements of *oriP* required in cis for the replication of EBV plasmid DNAs. It is probable that EBNA-1 also binds both to a set of cellular proteins that function in transcription and to a nonidentical set of cellular proteins that function in replication.

EBV effects a fascinating phenotypic change in B lymphocytes it infects. It does so by using several viral genes that alter the physiology of the cell by different means. We are now beginning to understand these genes and their activities.

ACKNOWLEDGMENTS

We thank Jeff Jones, Ilse Riegel, and Bayar Thimmapaya for reviewing the manuscript, and our colleagues who provided us with unpublished findings. Work was supported by U.S. Public Health Service grants CA-22443 and CA-07175 from the National Cancer Institute (to B.S.) and grant AI-29988 from the National Institute of Allergy and Infectious Diseases (to B.S.), American Cancer Society grant IN-35-31-19 (to T.M.), and fellowships from the Cancer Research Institute (to T.A.G. and T.M.) and the Leukemia Society of America (to J.M.M.).

REFERENCES

Abbot, S. D., Rowe, M., Cadwallader, K., Ricksten, A., Gordon, J., Wang, F., Rymo, L., and Rickinson, A. B. (1990). Epstein–Barr virus nuclear antigen 2 induces expression of the virus-encoded latent membrane protein. *J. Virol.* **64,** 2126–2134.

Adams, A. (1987). Replication of latent Epstein–Barr virus genomes in Raji cells. *J. Virol.* **61,** 1743–1746.

Adams, A., Pozos, T. C., and Purvey, H. V. (1989). Replication of latent Epstein–Barr virus genomes in normal and malignant lymphoid cells. *Int. J. Cancer* **44,** 560–564.

Allday, M. J., Crawford, D. H., and Griffin, B. E. (1989). Epstein–Barr virus latent gene expression during the initiation of B-cell immortalization. *J. Gen. Virol.* **70,** 1755–1764.

Altiok, E., Klein, G., Zech, L., Uno, M., Henriksson, B. E., Battat, S., Ono, Y., and Ernberg, I. (1989). Epstein–Barr virus-transformed pro-B-cells are prone to illegitimate recombination between the switch region of the μ chain gene and other chromosomes. *Proc. Natl. Acad. Sci. U.S.A.* **86,** 6333–6337.

Aman, P., and von Gabain, A. (1990). An Epstein–Barr virus immortalization associated gene segment interferes specifically with the IFN-induced anti-proliferative response in human B-lymphoid cell lines. *EMBO J.* **9,** 147–152.

Ambinder, R. F., Shah, W. A., Rawlins, D. R., Hayward, G. S., and Hayward, S. D. (1990). Definition of the sequence requirements for binding of the EBNA-1 protein to its palindromic target sites in Epstein–Barr virus DNA. *J. Virol.* **64,** 2369–2379.

Arrand, J. R., Young, L. S., and Tugwood, J. D. (1989). Two families of sequences in the small RNA-encoding region of Epstein–Barr virus (EBV) correlate with EBV types A and B. *J. Virol.* **63,** 983–986.

Baer, R., Bankier, A. T., Biggin, M. D., Deininger, P. L., Farrell, P. J., Gibson, T. J., Hatfull, G., Hudson, G. S., Satchwell, S. C., Seguin, C., Tuffnell, P. S., and Barrell, B. G. (1984). DNA sequence and expression of the B95–8 Epstein–Barr virus genome. *Nature (London)* **310**, 207–211.

Baichwal, V. R., and Sugden, B. (1987). Posttranslational processing of an Epstein–Barr virus-encoded membrane protein expressed in cells transformed by Epstein–Barr virus. *J. Virol.* **61**, 866–875.

Baichwal, V. R., and Sugden, B. (1988). Transformation of BALB/c 3T3 cells by the BNLF-1 gene of Epstein–Barr virus. *Oncogene* **2**, 461–467.

Baichwal, V. R., and Sugden, B. (1989). The multiple membrane-spanning segments of the BNLF-1 oncogene from Epstein–Barr virus are required for transformation. *Oncogene* **4**, 67–74.

Bankier, A. T., Deininger, P. L., Satchwell, S. C., Baer, R., Farrell, P. J., and Barrell, B. G. (1983). DNA sequence analysis of the EcoRI Dhet fragment of B95-8 Epstein–Barr virus containing the terminal repeat sequences. *Mol. Biol. Med.* **1**, 425–445.

Baumann, A., Grupe, A., Ackermann, A., and Pongs, O. (1988). Structure of the voltage-dependent potassium channel is highly conserved from *Drosophila* to vertebrate central nervous systems. *EMBO J.* **7**, 2457–2463.

Bird, A. G., and Britton, S. (1979). A new approach to the study of human B-lymphocyte function using an indirect plaque assay and a direct B-cell activator. *Immunol. Rev.* **45**, 41–67.

Bodescot, M., and Perricaudet, M. (1986). Epstein–Barr virus mRNAs produced by alternative splicing. *Nucleic Acids Res.* **14**, 7103–7114.

Boni-Schnetzler, M., and Pilch, P. F. (1987). Mechanism of epidermal growth factor receptor autophosphorylation and high-affinity binding. *Proc. Natl. Acad. Sci. U.S.A.* **84**, 7832–7836.

Brown, N. A., and Miller, G. (1982). Immunoglobulin expression by human B-lymphocytes clonally transformed by Epstein–Barr virus. *J. Immunol.* **128**, 24–29.

Burhans, W. C., Vassilev, L. T., Caddle, M. S., Heintz, N. H., and DePamphilis, M. L. (1990). Identification of an origin of bidirectional DNA replication in mammalian chromosomes. *Cell* **62**, 955–965.

Cheah, M. S. C., Ley, T. J., Tronick, S. R., and Robbins, K. C. (1986). *fgr* proto-oncogene mRNA induced in B-lymphocytes by Epstein–Barr virus infection. *Nature (London)* **319**, 238–240.

Chittenden, T., Lupton, S., and Levine, A. J. (1989). Functional limits of *oriP*, the Epstein–Barr virus plasmid origin of replication. *J. Virol.* **63**, 3016–3025.

Cohen, J. I., Wang, F., Mannick, J., and Kieff, E. (1989). Epstein–Barr virus nuclear protein 2 is a key determinant of lymphocyte transformation. *Proc. Natl. Acad. Sci. U.S.A.* **86**, 9558–9562.

Contreras-Salazar, B., Klein, G., and Masucci, M. G. (1989). Host cell-dependent regulation of growth transformation-associated Epstein–Barr virus antigens in somatic cell hybrids. *J. Virol.* **63**, 2768–2772.

Cordier, M., Calender, A., Billaud, M., Zimber, U., Rousselet, G., Pavlish, O., Banchereau, J., Tursz, T., Bornkamm, G., and Lenoir, G. M. (1990). Stable transfection of Epstein–Barr virus (EBV) nuclear antigen 2 in lymphoma cells containing the EBV P3HR1 genome induces expression of B-cell activation molecules CD21 and CD23. *J. Virol.* **64**, 1002–1013.

Crow, M. K., Jover, J. A., and Friedman, S. M. (1986). Direct T helper–B-cell interactions induce an early B-cell activation antigen. *J. Exp. Med.* **164**, 1760–1772.

Dambaugh, T., Hennessy, K., Chamnankit, L., and Kieff, E. (1984). U2 region of Epstein–Barr virus DNA may encode Epstein–Barr nuclear antigen 2. *Proc. Natl. Acad. Sci. U.S.A.* **81**, 7632–7636.

Dawson, C. W., Rickinson, A. B., and Young, L. S. (1990). Epstein–Barr virus latent membrane protein inhibits human epithelial cell differentiation. *Nature (London)* **344**, 777–780.

DePamphilis, M. L. (1988). Transcriptional elements as components of eukaryotic origins of DNA replication. *Cell* **52**, 635–638.

de The, G., Geser, A., Day, N. E., Tukei, P. M., Williams, E. H., Beri, D. P., Smith, P. G., Dean, A. G., Bornkamm, G. W., Feorino, P., and Henle, W. (1978). Epidemiological evidence for causal relationship between Epstein–Barr virus and Burkitt's lymphoma from Ugandan prospective study. *Nature (London)* **274**, 756–761.

Ernberg, I., Falk, K., and Hansson, M. (1987). Progenitor and pre-B-lymphocytes transformed by Epstein–Barr virus. *Int. J. Cancer* **39**, 190–197.

Fahraeus, R., Jansson, A., Ricksten, A., Sjoblom, A., and Rymo, L. (1990a). Epstein–Barr virus-encoded nuclear antigen 2 activates the viral latent membrane protein promoter by modulating the activity of a negative regulatory element. *Proc. Natl. Acad. Sci. U.S.A.* **87**, 7390–7394.

Fahraeus, R., Rymo, L., Rhim, J. S., and Klein, G. (1990b). Morphological transformation of human keratinocytes expressing the LMP gene of Epstein–Barr virus. *Nature (London)* **345**, 447–449.

Fennewald, S., van Santen, V., and Kieff, E. (1984). Nucleotide sequence of an mRNA transcribed in latent growth-transforming virus infection indicates that it may encode a membrane protein. *J. Virol.* **51**, 411–419.

Gahn, T. A., and Schildkraut, C. L. (1989). The Epstein–Barr virus origin of plasmid replication, *oriP*, contains both the initiation and termination sites of DNA replication. *Cell* **58**, 527–535.

Ghosh, D., and Kieff, E. (1990). cis-Acting regulatory elements near the Epstein–Barr virus latent-infection membrane protein transcriptional start site. *J. Virol.* **64**, 1855–1858.

Given, D., Yee, O., Griem, K., and Kieff, E. (1979). DNA of Epstein–Barr virus: V. Direct repeats of the ends of Epstein–Barr virus DNA. *J. Virol.* **30**, 852–862.

Glenney, J. J., Chen, W. S., Lazar, C. S., Walton, G. M., Zokas, L. M., Rosenfeld, M. G., and Gill, G. N. (1988). Ligand-induced endocytosis of the EGF receptor is blocked by mutational inactivation and by microinjection of anti-phosphotyrosine antibodies. *Cell* **52**, 675–684.

Gordon, J., Walker, L., Guy, G., Brown, G., Rowe, M., and Rickinson, A. (1986). Control of human B-lymphocyte replication. II. Transforming Epstein–Barr virus exploits three distinct viral signals to undermine three separate control points in B-cell growth. *Immunology* **58**, 591–595.

Gregory, C. D., Rowe, M., and Rickinson, A. B. (1990). Different Epstein–Barr virus–B-cell interactions in phenotypically distinct clones of a Burkitt's lymphoma cell line. *J. Gen. Virol.* **71**, 1481–1495.

Gross, J. L., Krupp, M. N., Rifkin, D. B., and Lane, M. D. (1983). Down-regulation of epidermal growth factor receptor correlates with plasminogen activator activity in human A431 epidermoid carcinoma cells. *Proc. Natl. Acad. Sci. U.S.A.* **80**, 2276–2280.

Gutkind, J. S., Link, D. C., Katamine, S., Lacal, P., Miki, T., Ley, T., and Robbins, K. C. (1991). A novel c-*fgr* exon utilized in Epstein–Barr virus-infected B lymphocytes but not in normal monocytes. *Mol. Cell. Biol.* **11**, 1500–1507.

Hammerschmidt, W., and Sugden, B. (1989). Genetic analysis of immortalizing functions of Epstein–Barr virus in human B-lymphocytes. *Nature (London)* **340**, 393–397.

Hammerschmidt, W., Sugden, B., and Baichwal, V. R. (1989). The transforming domain alone of the latent membrane protein of Epstein–Barr virus is toxic to cells when expressed at high levels. *J. Virol.* **63**, 2469–2475.

Hampar, B., Tanaka, A., Nonoyama, M., and Derge, J. G. (1974). Replication of the resident repressed Epstein–Barr virus genome during the early S phase (S-1 period) of nonproducer Raji cells. *Proc. Natl. Acad. Sci. U.S.A.* **71,** 631–633.

Hanley, M. R., and Jackson, T. (1987). Substance K receptor: Return of the magnificent seven. *Nature (London)* **329,** 766–767.

Heinzel, S. S., Krysan, P. J., Calos, M. P., and DuBridge, R. B. (1988). Use of simian virus 40 replication to amplify Epstein–Barr virus shuttle vectors in human cells. *J. Virol.* **62,** 3738–3746.

Henderson, E., Miller, G., Robinson, J., and Heston, L. (1977). Efficiency of transformation of lymphocytes by Epstein–Barr virus. *Virology* **76,** 152–163.

Hennessy, K., Fennewald, S., Hummel, M., Cole, T., and Kieff, E. (1984). A membrane protein encoded by Epstein–Barr virus in latent growth-transforming infection. *Proc. Natl. Acad. Sci. U.S.A.* **81,** 7207–7211.

Hinuma, Y., Konn, M., Yamaguchi, J., Wudarski, D. J., Blakeslee, J. R., Jr., and Grace, J. J. (1967). Immunofluorescence and herpes-type virus particles in the P3HR-1 Burkitt lymphoma cell line. *J. Virol.* **1,** 1045–1051.

Jones, C. H., Hayward, S. D., and Rawlins, D. R. (1989). Interaction of the lymphocyte-derived Epstein–Barr virus nuclear antigen (EBNA-1) with its DNA-binding sites. *J. Virol.* **63,** 101–110.

Katamine, S., Otsu, M., Tada, K., Tsuchiya, S., Sato, T., Ishida, N., Honjo, T., and Ono, Y. (1984). Epstein–Barr virus transforms precursor B-cells even before immunoglobulin gene rearrangements. *Nature (London)* **309,** 369–372.

Kawakami, T., Pennington, C. Y., and Robbins, K. C. (1986). Isolation and oncogenic potential of a novel human *src*-like gene. *Mol. Cell. Biol.* **6,** 4195–4201.

Kimball, A. S., Milman, G., and Tullius, T. D. (1989). High-resolution footprints of the DNA-binding domain of Epstein–Barr virus nuclear antigen 1. *Mol. Cell. Biol.* **9,** 2738–2742.

Kintner, C. R., and Sugden, B. (1979). The structure of the termini of the DNA of Epstein–Barr virus. *Cell* **17,** 661–671.

Kintner, C. R., and Sugden, B. (1981a). Conservation and progressive methylation of Epstein–Barr viral DNA sequences in transformed cells. *J. Virol.* **38,** 305–316.

Kintner, C. R., and Sugden, B. (1981b). Identification of antigenic determinants unique to the surfaces of cells transformed by Epstein–Barr virus. *Nature (London)* **294,** 458–460.

Klein, G. (1989). Viral latency and transformation: The strategy of Epstein–Barr virus. *Cell* **58,** 5–8.

Klein, C., Busson, P., Tursz, T., Young, L. S., and Raab-Traub, N. (1988). Expression of the c-*fgr* related transcripts in Epstein–Barr virus-associated malignancies. *Int. J. Cancer* **42,** 29–35.

Klein, G., Giovanella, B. C., Lindahl, T., Fialkow, P. J., Singh, S., and Stehlin, J. S. (1974). Direct evidence for the presence of Epstein–Barr virus DNA and nuclear antigen in malignant epithelial cells from patients with poorly differentiated carcinoma of the nasopharynx. *Proc. Natl. Acad. Sci. U.S.A.* **71,** 4737–4741.

Knutson, J. C. (1990). The level of c-*fgr* RNA is increased by EBNA-2, an Epstein–Barr virus gene required for B-cell immortalization. *J. Virol.* **64,** 2530–2536.

Knutson, V. P., Ronnett, G. V., and Lane, M. D. (1985). The effects of cycloheximide and chloroquine on insulin receptor metabolism. Differential effects on receptor recycling and inactivation and insulin degradation. *J. Biol. Chem.* **260,** 14180–14188.

Kozbor, D., Steinitz, M., Klein, G., Koskimies, S., and Makela, O. (1979). Establishment of anti-TNP antibody-producing human lymphoid lines by preselection for hapten binding followed by EBV transformation. *Scand. J. Immunol.* **10,** 187–194.

Kuchler, K., Sterne, R. E., and Thorner, J. (1989). *Saccharomyces cerevisiae STE6* gene product: A novel pathway for protein export in eukaryotic cells. *EMBO J.* **8,** 3973–3984.

Ley, T. J., Connolly, N. L., Katamine, S., Cheah, M. S., Senior, R. M., and Robbins, K. C. (1989). Tissue-specific expression and developmental regulation of the human *fgr* proto-oncogene. *Mol. Cell. Biol.* **9**, 92–99.

Liebowitz, D., Wang, D., and Kieff, E. (1986). Orientation and patching of the latent infection membrane protein encoded by Epstein–Barr virus. *J. Virol.* **58**, 233–237.

Liebowitz, D., Kopan, R., Fuchs, E., Sample, J., and Kieff, E. (1987). An Epstein–Barr virus transforming protein associates with vimentin in lymphocytes. *Mol. Cell. Biol.* **7**, 2299–2308.

Lindahl, T., Adams, A., Bjursell, G., Bornkamm, G. W., Kaschka-Dierich, C., and Jehn, U. (1976). Covalently closed circular duplex DNA of Epstein–Barr virus in a human lymphoid cell line. *J. Mol. Biol.* **102**, 511–530.

Lupton, S., and Levine, A. J. (1985). Mapping genetic elements of Epstein–Barr virus that facilitate extrachromosomal persistence of Epstein–Barr virus-derived plasmids in human cells. *Mol. Cell. Biol.* **5**, 2533–2542.

Mann, K. P., and Thorley-Lawson, D. A. (1987). Posttranslational processing of the Epstein–Barr virus-encoded p63/LMP protein. *J. Virol.* **61**, 2100–2108.

Mann, K. P., Staunton, D., and Thorley-Lawson, D. A. (1985). Epstein–Barr virus-encoded protein found in plasma membranes of transformed cells. *J. Virol.* **55**, 710–720.

Mark, W., and Sugden, B. (1982). Transformation of lymphocytes by Epstein–Barr virus requires only one-fourth of the viral genome. *Virology* **122**, 431–443.

Martin, J. M., and Sugden, B. (1991a). The LMP onco-protein resembles activated receptors in its properties of turnover. Submitted for publication.

Martin, J. M., and Sugden, B. (1991b). Transformation by the LMP onco-protein correlates with its rapid turnover, membrane localization, and cytoskeletal association. Submitted for publication.

Mermod, N., O'Neill, E. A., Kelly, T. J., and Tjian, R. (1989). The proline-rich transcriptional activator of CTF/NF-I is distinct from the replication and DNA binding domain. *Cell* **58**, 741–753.

Middleton, T., and Sugden, B. (1991). Uncoupling transcriptional activation from DNA replication for a mammalian replicon. Submitted for publication.

Miller, G., and Lipman, M. (1973). Release of infectious Epstein–Barr virus by transformed marmoset leukocytes. *Proc. Natl. Acad. Sci. U.S.A.* **70**, 190–194.

Milman, G., and Hwang, E. S. (1987). Epstein–Barr virus nuclear antigen forms a complex that binds with high concentration dependence to a single DNA-binding site. *J. Virol.* **61**, 465–471.

Moorthy, R., and Thorley-Lawson, D. (1990). Processing of the Epstein–Barr virus-encoded latent membrane protein p63/LMP. *J. Virol.* **64**, 829–837.

Murray, R. J., Young, L. S., Calender, A., Gregory, C. D., Rowe, M., Lenoir, G. M., and Rickinson, A. B. (1988). Different patterns of Epstein–Barr virus gene expression and of cytotoxic T-cell recognition in B-cell lines infected with transforming (B95-98) or nontransforming (P3HR1) virus strains. *J. Virol.* **62**, 894–901.

Niederman, J. C., McCollum, R. W., Henle, G., and Henle, W. (1968). Infectious mononucleosis. Clinical manifestations in relation to EB virus antibodies. *JAMA, J. Am. Med. Assoc.* **203**, 205–209.

Nilsson, K., and Klein, G. (1982). Phenotypic and cytogenetic characteristics of human B-lymphoid cell lines and their relevance for the etiology of Burkitt's lymphoma. *Adv. Cancer Res.* **37**, 319–380.

Otsu, M., Katamine, S., Uno, M., Yamaki, M., Ono, Y., Klein, G., Sasaki, M. S., Yaoita, Y., and Honjo, T. (1987). Molecular characterization of novel reciprocal translocation t(6;14) in an Epstein–Barr virus-transformed B-cell precursor. *Mol. Cell Biol.* **7**, 708·

Pattengale, P. K., Smith, R. W., and Gerber, P. (1973). Selective transformation of B-lymphocytes by Epstein–Barr virus. *Lancet* **2**, 93–94.

Paul, C. C., Keller, J. R., Armpriester, J. M., and Baumann, M. A. (1990). Epstein–Barr virus transformed B-lymphocytes produce interleukin-5. *Blood* **75**, 1400–1403.

Petti, L., Sample, C., and Kieff, E. (1990). Subnuclear localization and phosphorylation of Epstein–Barr virus latent infection nuclear proteins. *Virology* **176**, 563–574.

Polvino-Bodnar, M., Kiso, J., and Schaffer, P. A. (1988). Mutational analysis of Epstein–Barr virus nuclear antigen 1 (EBNA-1). *Nucleic Acids Res.* **16**, 3415–3435.

Pope, J. H., Horne, M. K., and Scott, W. (1968). Transformation of foetal human leukocytes *in vitro* by filtrates of a human leukaemic cell line containing herpes-like virus. *Int. J. Cancer* **3**, 857–866.

Rabson, M., Gradoville, L., Heston, L., and Miller, G. (1982). Non-immortalizing P3J-HR-1 Epstein–Barr virus: A deletion mutant of its transforming parent, Jijoye. *J. Virol.* **44**, 834–844.

Rawlins, D. R., Milman, G., Hayward, S. D., and Hayward, G. S. (1985). Sequence-specific DNA binding of the Epstein–Barr virus nuclear antigen (EBNA-1) to clustered sites in the plasmid maintenance region. *Cell* **42**, 859–868.

Reisman, D., and Sugden, B. (1986). Transactivation of an Epstein–Barr viral transcriptional enhancer by the Epstein–Barr viral nuclear antigen 1. *Mol. Cell. Biol.* **6**, 3838–3846.

Reisman, D., Yates, J., and Sugden, B. (1985). A putative origin of replication of plasmids derived from Epstein–Barr virus is composed of two cis-acting components. *Mol. Cell. Biol.* **5**, 1822–1832.

Rickinson, A. B., Young, L. S., and Rowe, M. (1987). Influence of the Epstein–Barr virus nuclear antigen EBNA-2 on the growth phenotype of virus-transformed B-cells. *J. Virol.* **61**, 1310–1317.

Roberts, J. M., and Weintraub, H. (1986). Negative control of DNA replication in composite SV40–bovine papilloma virus plasmids. *Cell* **46**, 741–752.

Roberts, J. M., and Weintraub, H. (1988). cis-Acting negative control of DNA replication in eukaryotic cells. *Cell* **52**, 397–404.

Robinson, J., and Smith, D. (1981). Infection of human B-lymphocytes with high multiplicities of Epstein–Barr virus: Kinetics of EBNA expression, cellular DNA synthesis, and mitosis. *Virology* **109**, 336–343.

Rogers, R. P. Woisetschlaeger, M., and Speck, S. H. (1990). Alternative splicing dictates translational start in Epstein–Barr virus transcripts. *EMBO J.* **9**, 2273–2277.

Rooney, C., Howe, J. G., Speck, S. H., and Miller, G. (1989). Influences of Burkitt's lymphoma and primary B-cells on latent gene expression by the nonimmortalizing P3J-HR-1 strain of Epstein–Barr virus. *J. Virol.* **63**, 1531–1539.

Rowe, D., Heston, L., Metlay, J., and Miller, G. (1985). Identification and expression of a nuclear antigen from the genomic region of the Jijoye strain of Epstein–Barr virus that is missing in its nonimmortalizing deletion mutant, P3HR-1. *Proc. Natl. Acad. Sci. U.S.A.* **82**, 7429–7433.

Rowe, M., Young, L. S., Cadwallader, K., Petti, L., Kieff, E., and Rickinson, A. B. (1989). Distinction between Epstein–Barr virus type A (EBNA 2A) and type B (EBNA 2B) isolates extends to the EBNA 3 family of nuclear proteins. *J. Virol.* **63**, 1031–1039.

Sample, J., and Kieff, E. (1990). Transcription of the Epstein–Barr virus genome during latency in growth-transformed lymphocytes. *J. Virol.* **64**, 1667–1674.

Sample, J., Hummel, M., Braun, D., Birkenbach, M., and Kieff, E. (1986). Nucleotide sequences of mRNAs encoding Epstein–Barr virus nuclear proteins: A probable transcriptional initiation site. *Proc. Natl. Acad. Sci. U.S.A.* **83**, 5096–5100.

Sample, J., Young, L., Martin, B., Chatman, T., Kieff, E., Rickinson, A., and Kieff, E.

54   TIM MIDDLETON et al.

(1990). Epstein–Barr virus types 1 and 2 differ in their EBNA-3A, EBNA-3B, and EBNA-3C genes. *J. Virol.* **64**, 4084–4092.

Schmidt, J. W., and Catterall, W. A. (1986). Biosynthesis and processing of the α subunit of the voltage-sensitive sodium channel in rat brain neurons. *Cell* **46**, 437–445.

Speck, S. H., and Strominger, J. L. (1985). Analysis of the transcript encoding the latent Epstein–Barr virus nuclear antigen I: A potentially polycistronic message generated by long-range splicing of several exons. *Proc. Natl. Acad. Sci. U.S.A.* **82**, 8305–8309.

Speck, S. H., and Strominger, J. L. (1989). Transcription of Epstein–Barr virus in latently infected, growth-transformed lymphocytes. *Adv. Viral Oncol.* **8**, 133–150.

Speck, S. H., Pfitzner, A., and Strominger, J. L. (1986). An Epstein–Barr virus transcript from a latently infected, growth-transformed B-cell line encodes a highly repetitive polypeptide. *Proc. Natl. Acad. Sci. U.S.A.* **83**, 9298–9302.

Springer, T. A. (1990). Adhesion receptors of the immune system. *Nature (London)* **346**, 425–434.

Stern, D. F., Kamps, M. P., and Cao, H. (1988). Oncogenic activation of p185[neu] stimulates tyrosine phosphorylation *in vivo. Mol. Cell. Biol.* **8**, 3969–3973.

Sternas, L., Middleton, T., and Sugden, B. (1990). The average number of molecules of Epstein–Barr nuclear antigen 1 per cell does not correlate with the average number of Epstein–Barr virus (EBV) DNA molecules per cell among different clones of EBV-immortalized cells. *J. Virol.* **64**, 2407–2410.

Sugden, B. (1984). Expression of virus-associated functions in cells transformed in vitro by Epstein–Barr virus cell surface antigen and virus release from transformed cells. *In* "Immune Deficiency and Cancer" (D. T. Purtillo, ed.), pp. 165–178. Plenum, New York.

Sugden, B., and Mark, W. (1977). Clonal transformation of adult human leukocytes by Epstein–Barr virus. *J. Virol.* **23**, 503–508.

Sugden, B., and Metzenberg, S. (1983). Characterization of an antigen whose cell surface expression is induced by infection with Epstein–Barr virus. *J. Virol.* **46**, 800–807.

Sugden, B., and Warren, N. (1989). A promoter of Epstein–Barr virus that can function during latent infection can be transactivated by EBNA-1, a viral protein required for viral DNA replication during latent infection. *J. Virol.* **63**, 2644–2649.

Swendeman, S., and Thorley-Lawson, D. (1987). The activation antigen BLAST-2, when shed, is an autocrine BCGF for normal and transformed B-cells. *EMBO J.* **6**, 1637–1642.

Tempel, B. L., Papazian, D. M., Schwarz, T. L., Jan, Y. N., and Jan, L. Y. (1987). Sequence of a probable potassium channel component encoded at *shaker* locus of *Drosophila. Science* **237**, 770–775.

Thorley-Lawson, D. A., Nadler, L. M., Bhan, A. K., and Schooley, R. T. (1985). BLAST-2 [EBVCS], an early cell surface marker of human B-cell activation, is superinduced by Epstein Barr virus. *J. Immunol.* **134**, 3007–2012.

Tosato, G., Tanner, J., Jones, K. D., Revel, M., and Pike, S. E. (1990). Identification of interleukin-6 as an autocrine growth factor for Epstein–Barr virus-immortalized cells. *J. Virol.* **64**, 3033–3041.

Uchibayashi, N., Kikutani, H., Barsumian, E. L., Hauptmann, R., Schneider, F. J., Schwendenwein, R., Sommergruber, W., Spevak, W., Maurer-Fogy, I., Suemura, M., and Kishimoto, T. (1989). Recombinant soluble Fe epsilon receptor II (Fc epsilon RII/CD23) has IgE binding activity but no B-cell growth promoting activity. *J. Immunol.* **142**, 3901–3908.

Veillette, A., Bookman, M. A., Horak, E. M., and Bolen, J. B. (1988). The CD4 and CD8 T cell surface antigens are associated with the internal membrane tyrosine-protein kinase p56[lck]. *Cell* **55**, 301–308.

Vidal, M., Wrighton, C., Eccles, S., Burke, J., and Grosveld, F. (1990). Differences in human cell lines to support stable replication of Epstein–Barr virus-based vectors. *Biochim. Biophys. Acta* **1048,** 171–177.

Wang, D., Liebowitz, D., and Kieff, E. (1985). An EBV membrane protein expressed in immortalized lymphocytes transforms established rodent cells. *Cell* **43,** 831–840.

Wang, D., Liebowitz, D., Wang, F., Gregory, C., Rickinson, A., Larson, R., Springer, T., and Kieff, E. (1988). Epstein–Barr virus latent infection membrane protein alters the human B-lymphocyte phenotype: Deletion of the amino terminus abolishes activity. *J. Virol.* **62,** 4173–4184.

Wang, F., Gregory, C. D., Rowe, M., Rickinson, A. B., Wang, D., Birkenbach, M., Kikutani, H., Kishimoto, T., and Kieff, E. (1987). Epstein–Barr virus nuclear antigen 2 specifically induces expression of the B-cell activation antigen CD23. *Proc. Natl. Acad. Sci. U.S.A.* **84,** 3452–3456.

Wang, F., Gregory, C., Sample, C., Rowe, M., Liebowitz, D., Murray, R., Rickinson, A., and Kieff, E. (1990a). Epstein–Barr virus latent membrane protein (LMP1) and nuclear proteins 2 and 3C are effectors of phenotypic changes in B-lymphocytes: EBNA-2 and LMP1 cooperatively induce CD23. *J. Virol.* **64,** 2309–2318.

Wang, F., Tsang, S., Kurilla, M. G., Cohen, J., and Kieff, E. (1990b). Epstein–Barr virus nuclear antigen 2 transactivates latent membrane protein LMP1. *J. Virol.* **64,** 3407–3416.

Weiner, D. B., Liu, J., Cohen, J. A., Williams, W. V., and Greene, M. I. (1989). A point mutation in the *neu* oncogene mimics ligand induction of receptor aggregation. *Nature (London)* **339,** 230–231.

Wilson, J. B., Weinberg, W., Johnson, R., Yuspa, S., and Levine, A. J. (1990). Expression of the BNLF-1 oncogene of Epstein–Barr virus in the skin of transgenic mice induces hyperplasia and aberrant expression of keratin 6. *Cell* **61,** 1315–1327.

Woisetschlaeger, M., Strominger, J. L., and Speck, S. H. (1989). Mutually exclusive use of viral promoters in Epstein–Barr virus latently infected lymphocytes. *Proc. Natl. Acad. Sci. U.S.A.* **86,** 6498–6502.

Woisetschlaeger, M., Yandava, C. N., Furmanski, L. A., Strominger, J. L., and Speck, S. H. (1990). Promoter switching in Epstein–Barr virus during the initial stages of infection of B-lymphocytes. *Proc. Natl. Acad. Sci. U.S.A.* **87,** 1725–1729.

Wysokenski, D. A., and Yates, J. L. (1989). Multiple EBNA1-binding sites are required to form an EBNA1-dependent enhancer and to activate a minimal replicative origin within oriP of Epstein–Barr virus. *J. Virol.* **63,** 2657–2666.

Yarden, Y., and Schlessinger, J. (1987). Epidermal growth factor induces rapid, reversible aggregation of the purified epidermal growth factor receptor. *Biochemistry* **26,** 1443–1451.

Yates, J. L., and Camiolo, S. M. (1988). Dissection of DNA replication and enhancer activation functions of Epstein–Barr virus nuclear antigen 1. *Cancer Cells* **6,** 197–205.

Yates, J. L., Warren, N., and Sugden, B. (1985). Stable replication of plasmids derived from Epstein–Barr virus in various mammalian cells. *Nature (London)* **313,** 812–815.

Yokoi, T., Miyawaki, T., Yachie, A., Kato, K., Kasahara, Y., and Taniguchi, N. (1990). Epstein–Barr virus-immortalized B-cells produce IL-6 as an autocrine growth factor. *Immunology* **70,** 100–105.

Yokota, A., Kikutani, H., Tanaka, T., Sato, R., Barsumian, E. L., Suemura, M., and Tadamitsu, K. (1988). Two species of human Fcε receptor II (FcεRII/CD23): Tissue-specific and IL-4-specific regulation of gene expression. *Cell* **55,** 611–618.

Zimber-Strobl, U., Suentzenich, K.-O., Laux, G., Eick, D., Cordier, M., Calender, A., Billaud, M., Lenoir, G. M., and Bornkamm, G. W. (1991). Epstein–Barr virus nuclear antigen-2 activates transcription of the terminal protein gene. *J. Virol.* **65,** 415–423.

ADVANCES IN VIRUS RESEARCH, VOL. 40

MOLECULAR BIOLOGY OF NON-A, NON-B HEPATITIS AGENTS: HEPATITIS C AND HEPATITIS E VIRUSES

Gregory R. Reyes* and Bahige M. Baroudy†

*Genelabs, Inc.
Redwood City, California 94063
and
†James N. Gamble Institute of Medical Research
Cincinnati, Ohio 45219

I. INTRODUCTION

The discovery of the "Australia antigen" (Blumberg *et al.*, 1965) as a marker of hepatitis B infection and the visualization of the 27-nm hepatitis A viral particle (Feinstone *et al.*, 1973) made it possible to develop reliable diagnostic assays for the detection of hepatitis A virus (HAV) and hepatitis B virus (HBV) (Prince, 1968; Walsh *et al.*, 1970; Bradley *et al.*, 1977, 1979b). Nevertheless, the appearance of post-transfusion-associated hepatitis in patients negative for markers of either virus led researchers to believe that another new agent(s) existed. Such an agent(s) came to be known as non-A, non-B hepatitis (NANBH).

In the past 2 years major discoveries in the area of viral hepatitis have led to the identification of two new agents: hepatitis C virus (HCV), which is believed to cause the majority (at least 90%) of parenterally transmitted non-A, non-B hepatitis (PT-NANBH) (Choo *et al.*, 1989), and hepatitis E virus (HEV), responsible for enterically trans-

57

mitted non-A, non-B hepatitis (ET-NANBH) (Reyes *et al.*, 1990a). This brings to five the number of known hepatotropic agents, namely, hepatitis A, B, C, D, and E viruses. HAV and HEV are spread by the fecal/oral route, whereas HBV, HCV, and HDV are transmitted principally by parenteral routes (e.g., blood and blood-derived products). These viral agents, with the exception of HBV, have RNA genomes.

The successful development and application of molecular cloning strategies that had previously been applied to the elucidation of rare or low-abundance genes and gene transcripts have hastened the cloning and characterization of low-titer viruses that have not yet been successfully propagated *in vitro*. In this chapter we review the developments of the last 2 years which led to the successful identification of HCV and HEV.

II. HEPATITIS C VIRUS

A. Infectious Agent

1. Early Studies

Several reports describing posttransfusion-associated hepatitis in patients negative for serological markers of HAV and HBV appeared in the literature in the 1970s (Prince *et al.*, 1974; Feinstone *et al.*, 1975; Mosley, 1975; Alter *et al.*, 1975; Hoofnagle *et al.*, 1977; Mosley *et al.*, 1977). In the first study the courses of 204 cardiovascular surgery patients were followed biweekly for serological conversion to HBV. Thirty-six of 51 cases (71%) of posttransfusion hepatitis were negative for HBV markers (Prince *et al.*, 1974). Involvement of other viral agents, such as HAV or cytomegalovirus (CMV), was excluded in this study on the basis that the mean incubation period of this new agent(s) was longer than that of HAV and that there was an equal exposure in this group of patients to CMV with or without hepatitis. In another study carried out at the National Institutes of Health, 22 patients who developed posttransfusion hepatitis after corrective cardiac surgery tested negative for HAV and HBV (Feinstone *et al.*, 1975). A third study implicated a new agent(s) as the cause of hepatitis in transfused recipients after retesting sera that were collected in the early 1950s. Sera from six asymptomatic blood donors were inoculated into 10–20 volunteers each. In retrospect, two of these donors were shown to be positive for hepatitis B surface antigen (HBsAg) and transmitted HBV to all susceptible recipients. Sera from three HBsAg-negative donors induced icterus in 10–47% of the recipients. In addition, evidence of

TABLE I

EPIDEMIOLOGICAL FEATURES OF NANBH

Parenterally transmitted NANBH	
Etiological agent	Hepatitis C Virus
Transmission	Parenteral, sexual(?), perinatal(?)
Chronic liver disease	In at least 50%
Seroprevalence	0.5–1.5% worldwide
Other	Association with hepatocellular carcinoma(?)
Enterically Transmitted NANBH	
Etiological agent	Hepatitis E Virus
Transmission	Fecal/oral (contaminated water)
Endemic	Potential for large epidemics
Seroprevalence	Unknown
Susceptibility	Principally a disease of young adults (15–40 years)
Chronic disease	Not a recognized sequela
Mortality	High (10–20%) in pregnant women

chronic liver disease was present in one recipient who was HBsAg-positive and in two recipients who had developed hepatitis in the absence of serological markers for known viral infections (Hoofnagle *et al.*, 1977). These early reports supported the existence of a new agent(s) capable of causing hepatitis that was unrelated to HAV, HBV, or other viral agents (e.g., CMV and Epstein–Barr virus) that are capable of causing liver disease. With the current availability of an HCV diagnostic test, these early studies have been largely confirmed. It is now established that HCV accounts for at least 90% of PT-NANBH. The epidemiological findings in PT-NANBH are summarized in Table I.

2. Animal Model

Many attempts were made to develop serological markers for this new NANBH agent without apparent success. On the other hand, the successful development of the chimpanzee model for NANBH was instrumental in furthering the study of this virus and ultimately culminated in the genomic cloning of this previously elusive agent. Two reports that demonstrated the existence of a transmissible agent responsible for PT-NANBH appeared in 1978 (Alter *et al.*, 1978; Tabor *et al.*, 1978). In the first study five chimpanzees were inoculated with plasma or serum obtained from five humans. One was a blood donor implicated in two cases of PT-NANBH and the remaining four were patients with acute or chronic PT-NANBH. All five chimpanzees developed hepatitis based on biochemical and histological evidence, suggesting the presence of a transmissible agent for PT-NANBH (Alter *et al.*, 1978). In the second study four chimpanzees were inoculated either with a human chronic NANBH serum suspected of transmitting this disease to a nurse (following accidental needle-stick exposure) or with two blood donor serum samples whose HBsAg-negative blood appeared to transmit hepatitis. All four chimpanzees developed hepatitis, demonstrating the transmissible nature of this disease and once again implicating a new virus as the cause (Tabor *et al.*, 1978). These studies led to the acceptance of the chimp as the "gold standard" by which authentic transmissable NANBH was gauged.

There were other successful studies that supported the presence of a transmissible agent that caused NANBH in chimpanzees. The work carried out at the Centers for Disease Control (CDC) in Atlanta, and reviewed recently by Bradley (1990a), is worthy of special mention since RNA extracted from the plasma of one of the NANBH-infected chimpanzees (chimp 910) was used to successfully clone the genome of this virus. The source inoculum in these studies was a Factor VIII preparation suspected of transmitting NANBH (Bradley *et al.*, 1979a).

Electron-microscopic studies revealed ultrastructural changes in hepatocytes of humans and infected chimpanzees (Jackson *et al.*, 1979; Shimizu *et al.*, 1979; Bradley *et al.*, 1980; Pfeifer *et al.*, 1980; Tsiquaye *et al.*, 1981; Schaff *et al.*, 1985). Peculiar tubular structures referred to as a double unit membrane were observed in the nucleus or cytoplasm of hepatocytes from humans or chimpanzees infected with NANBH (Shimizu *et al.*, 1979). In another study as many as four different changes were visualized by electron microscopy (Pfeifer *et al.*, 1980). These studies did not speculate as to whether these tubular structures arose directly from a virus-specific event or resulted from a host immune response. The tubular structures were present in the hepatocytes of experimentally infected chimpanzees, not in the nuclei of normal-appearing hepatocytes (Bradley *et al.*, 1980). Although these tubular structures were in hepatocytes infected with NANBH, they were not observed in hepatocytes infected with HBV or HAV (Shimizu *et al.*, 1979). On the other hand, tubular structures were found in hepatocytes of chimpanzees infected with HDV (Shimizu *et al.*, 1979; Kamimura *et al.*, 1983). In the absence of a specific test for HCV, the presence of tubular structures in hepatocytes of infected chimpanzees was considered to be one of the few criteria that demonstrated the presence of NANBH in infected liver tissue.

The detection of a cytoplasmic antigen by immunofluorescent staining with a monoclonal antibody was another criterion used to identify liver specimens infected with NANBH. A chimpanzee lymphoblastoid cell line producing antibody reactive with NANBH tissue was established by *in vitro* transformation with Epstein–Barr virus (Shimizu *et al.*, 1985). This particular antibody was reactive with liver biopsy specimens infected with NANBH or HDV, but not with ones infected with HAV and HBV (Shimizu *et al.*, 1986). The antigen recognized by this antibody was identified in subsequent studies to be an endogenous host protein response to NANBH or HDV apparently induced by interferon (Shimizu and Purcell, 1989). These studies are indicative of the inherent difficulties encountered by investigators as they attempted to generate specific serological reagents for the PT-NANBH agent.

3. Biophysical Properties and Classification

The success of the transmission studies in chimpanzees permitted further insight into the nature of the NANBH agent. The progress of research in this area was at times slow as well as controversial; nevertheless, some major advances are noted here due to their critical role in the cloning of the viral genome.

The NANBH agent capable of inducing cytoplasmic tubular structures was sensitive to chloroform treatment in chimpanzee transmission studies and as such was presumed to contain a lipid structure (enveloped virus). The tubule-forming (chloroform-sensitive) agent was shown to be smaller than 80 nm in diameter and thought to be togavirus-like based on these properties (Bradley *et al.*, 1985). Another sizing study was performed using microfiltration and testing for the presence of viable virus by chimp inoculation. These latter studies confirmed that the size of the putative virus was in the range of 30–60 nm (He *et al.*, 1987). Recently, the chloroform-sensitive agent has been shown definitively to be HCV by analysis of seroreactivity to C100-3 (HCV recombinant antigen; see Section II,B) in cross-challenge studies performed with chloroform-sensitive and -resistant inocula in the chimp (Bradley *et al.*, 1990). It is noteworthy that a second PT-NANBH agent resistant to treatment with chloroform and also infectious in chimpanzees was identified (Feinstone *et al.*, 1983; Bradley *et al.*, 1983, 1990). A summary of the animal model and the biophysical characteristics of the virus is presented in Table II.

The virus density in sucrose gradients was investigated by Bradley *et al.* (1991). It was at first predicted that the particle density would approximate that of other known togavirus genomes ($\sim$1.2 g/cm^3). It was determined, however, that the highest titer of viral infectivity was concentrated in the 1.09–1.11 g/cm^3 fraction in a chimpanzee titration study conducted using sucrose gradient fractionated chimp 910 serum. The retained infectivity of that particular fraction amounted to all that was loaded onto the gradient.

The recent availability of HCV molecular probes prompted a retro-

TABLE II

PT-NANBH: ANIMAL MODEL AND BIOPHYSICAL CHARACTERIZATION[a]

Animal model	Chimpanzee—ALT elevation; acute phase with chronic relapsing disease; reports of hepatocellular carcinoma
Pathology	Characteristic tubular changes in chimpanzee liver
Particle density	Reported as 1.09–1.11 g/cm^3 (sucrose gradient)
Particle size	30–60 nm by filtration
Inactivation	Chloroform-sensitive agent; enveloped

[a] ALT, Alanine aminotransferase.

spective analysis of chimp 910, and other similarly fractionated human PT-NANBH sera, for the presence of an HCV-specific sequence (G. R. Reyes *et al.*, unpublished observations). To our surprise, we did not detect HCV by a combined reverse transcriptase/polymerase chain reaction (PCR) protocol. These studies may be flawed by some technical problem, however, since the HCV genome could be detected in all starting samples if the virions were collected by direct extraction, pelleting from undiluted sera or precipitation using polyethylene glycol (Kim *et al.*, 1991).

It is quite remarkable and must be noted, however, that exogenous sequences different from HCV were isolated from all these 1.09–1.11 g/cm^3 gradient fractions when probed by immunoscreening (G. R. Reyes *et al.*, unpublished observations). The finding that the same density fraction from different sources yielded similar exogenous sequences raises the intriguing possibility that a second viral cofactor might play some role in the development of PT-NANBH. This hypothetical situation would be similar to those observed with HBV and HDV in which the latter defective virus not only requires HBV for its propagation, but also acts as a cofactor in the development of severe chronic hepatitis, leading to end-stage disease with associated high mortality (Rizzetto, 1983). The significance of these and other (see Section II,B) exogenous non-HCV clones will require further experimentation, and their availability should provide a starting point to unraveling this interesting conundrum.

B. Cloning and Genomic Organization

The identification of a molecular clone and its validation as derived from a novel viral agent presents a unique biological puzzle in which similarities can be drawn to Koch's postulates. The objective is to state with certainty that a previously unidentified molecular clone is de-

TABLE III

Validation of a Molecular Clone as Virus Derived

Disease associated	Present in infected tissues and individuals
Exogenous	Not present in the uninfected person or animal
Viruslike	Genes or genomic organization similar to other viruses
Immunogenic	Encodes diagnostic/prognostic proteins
Infectious	Reconstruction of infectious genome *in vitro* or *in vivo*

rived from a novel uncharacterized agent responsible for a pathological state or condition. A defined set of criteria that a molecular clone must satisfy is listed in Table III. These criteria were first presented in the context of identifying and validating the ET1.1 clone as derived from the HEV genome (Reyes *et al.*, 1990b).

The putative viral clone should be reproducibly associated with specific tissues obtained from infected individuals. The isolated clone should also be similar among different infected individuals. Transmission of these tissues to a susceptible animal causes the same disease in that animal. The clone should not be present in uninfected individuals or with uninfected tissues from infected individuals. This last point has generally been referred to as the exogenousity test for cloned viral sequences and is a rigorous requirement for a transmissable agent. Cloned sequences should, in some manner, resemble sequences already described for other authenticated viruses or virus clones. One example of this latter point is the consensus sequences associated with the RNA-directed RNA polymerase (RDRP) encoded by all positive-stranded RNA viruses (Kamer and Argos, 1984). Segments of the viral genome also encode proteins that are immunogenic in the infected host. These particular proteins have diagnostic utility and perhaps prognostic value in determining the ultimate outcome of a viral infection. This last point is amply illustrated by the various HBV antigens in current use.

One final measure of a clone as derived from a virus genome is the ability to use the isolated sequence to acquire other virus-specific sequences and extend the known sequence into a set of overlapping contiguous clones representing the entire viral genome. This overlapping contiguous set of clones together satisfies all the validation criteria noted above. This accumulated set of clones representing the entire genome has been achieved for HEV (Tam *et al.*, 1991a,b) and HCV (see below). Once a full set of independent clones has been obtained, they may then be used in the reconstruction of a full-length viral genome.

The final confirmatory experiment is the attempt to demonstrate the infectivity of the cloned genome in a susceptible cellular or animal host. Experiments of this type have been performed for numerous viral pathogens, including HAV (Cohen *et al.*, 1988) and HBV (Sureau *et al.*, 1986). The successful execution of such an experiment is an irrefutable final confirmation of the virus-specific nature of the acquired clones.

We have alluded to the efforts of Daniel W. Bradley and colleagues in the successful transmission studies of human PT-NANBH in chimpanzees (Bradley *et al.*, 1979a, 1980, 1983, 1985, 1990). These studies were recently reviewed by Bradley (1990a). A pool of plasma samples obtained from a well-characterized chronically infected chimpanzee (chimp 910) was estimated to contain ~10^6 chimpanzee infectious doses per milliliter (CID/ml) (Bradley *et al.*, 1985). This material was used as a source of nucleic acid in the cloning efforts that led to the identification of the chloroform-sensitive cytoplasmic tubule-forming PT-NANBH agent (Choo *et al.*, 1989). It had previously been recognized that the majority of NANBH inocula contain very low titers (i.e., <10^2 CID/ml) of infectious virions. The high-titer sera obtained from chimp 910 thus represented an unusual and possibly unique biological specimen derived from the chimpanzee.

Single-stranded RNA and DNA were generated by denaturing total nucleic acids purified from a virus-containing plasma pool (90 ml) that was diluted prior to pelleting by centrifugation (Choo *et al.*, 1989; European Patent Office Publ. #318,216). The denaturation step was essential in the cloning strategy of this agent, since it was not known whether the genome was single or double stranded. It was also not known whether the virus genome was polyadenylated at its 3' end. It was therefore necessary to synthesize the cDNA utilizing random primers (oligonucleotide hexamers of random sequence) and reverse transcriptase. The cDNA was subsequently cloned into the λgt11 expression vector. This bacteriophage vector is capable of producing a polypeptide product by expressing the cloned cDNA as a fusion protein with the *Escherichia coli* β-galactosidase gene. The expression of a fusion protein promotes the stability of the recombinant protein within the bacterial host. Phage infection of bacterial monolayers leads to the lysis and subsequent release of the fusion protein product. The primary immunoscreening approach is outlined in Fig. 1A. Screening approximately 1 million recombinant clones with serum from a bona fide chronic NANBH patient yielded a 155-bp clone, 5-1-1, that expressed detectable amounts of an NANBH antigen (Choo *et al.*, 1989).

By using a synthetic oligonucleotide derived from 5-1-1 as a hybridization probe, three larger overlapping clones were identified. Single-

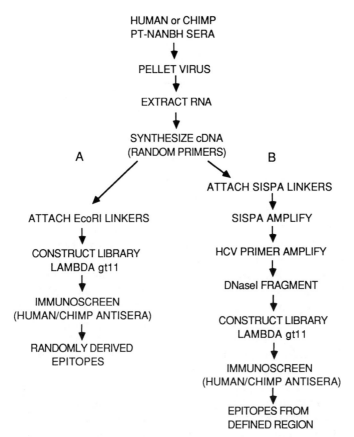

FIG. 1.   Identification of HCV molecular clones. The original procedure used to identify an HCV cDNA clone encoding an immunogenic region of the viral genome is depicted in the flow diagram (A). cDNA was synthesized from a serum/plasma source,
using random primers, and cloned into the λgt11 expression vector. This procedure was
successful in identifying, by primary immunoscreening, a number of immunogenic regions of the viral genome, all of which were derived from the nonstructural region.
Epitopes identified in this manner were "randomly derived" (i.e., isolated based on the
seroselectivity of the screening antisera). An alternative procedure for epitope identification has been described and is depicted in (B). Overlapping segments of the viral
genome were rescued by a sequential polymerase chain reaction protocol that required
the nonspecific preexpansion of the cDNA (by SISPA; see text) prior to amplification of
the desired segment using HCV sequence-specific oligonucleotides. The resulting segment of the genome was randomly degraded using DNase I prior to insertion in the
λgt11 expression vector for screening, using antisera from both chimp and human
sources. The latter procedure facilitated the execution of an epitope survey of the entire
nonstructural region and resulted in the identification of epitopes with greater diagnostic sensitivity than the C100-3 region.

TABLE IV

Genomic Characteristics of HCV

RNA genome	RNase-sensitive DNase-resistant nucleic acid
Single stranded	Oligonucleotide-specific hybridization
Positive sense	Oligonucleotide-specific hybridization; poly(A)(?)
Polyprotein	Single continuous open reading frame
Genomic organization	5' Structural; 3' nonstructural

stranded probes derived from one of these, clone 81, were used to show that the genome of NANBH was indeed a positive-stranded RNA molecule (Choo *et al.*, 1989). The size was estimated on a Northern blot of RNA extracted from liver to be nearly 10,000 nucleotides. An experiment reported in the original cloning report by Houghton and colleagues (Choo *et al.*, 1989) suggested that the HCV genome contained a poly(A) tract or tail, since the RNA was fractionated on the basis of binding to an oligo(dT) column. Although there has been no confirmation of the oligo(dT) binding specificity of HCV RNA, a short poly(A) tract at the 5' end of the genome was subsequently identified and could account for this observation (European Patent Office Publ. #388,232). A summary of the genomic characterization of HCV is presented in Table IV.

As noted above, the HCV-specific clone 5-1-1 was used to extend the HCV sequence into a contiguous set of overlapping clones. Three of these clones, 81, 36, and 32, were assembled to form a larger clone, C100, which, when fused to the human gene encoding superoxide dismutase (Steimer *et al.*, 1986), produced a new fusion polypeptide C100-3 that expressed 363 amino acids from the viral genome that amounted to nearly 4% of the total protein in a yeast expression system (Kuo *et al.*, 1989). Initially, the C100-3 polypeptide was used to coat wells of microtiter plates in order to capture antibodies to NANBH in blood samples. Using this assay, it was possible to identify six of seven human serum samples previously shown to transmit NANBH to chimpanzees (Alter *et al.*, 1982). Testing more sera led to the correct identification of the majority of clinically characterized NANBH samples and the recognition of this new agent, HCV, as the major cause of chronic NANBH (Choo *et al.*, 1989; Kuo *et al.*, 1989). Subsequently, Ortho Diagnostic Systems (Raritan, New Jersey) developed an enzyme-linked immunosorbent assay that has been used worldwide to detect the presence of HCV in human samples. Data collected thus far have confirmed the initial observation that HCV is present in the majority of NANBH cases (Kuo *et al.*, 1989; H. J. Alter *et al.*, 1989; van der Poel *et al.*, 1989; Mosley *et al.*, 1990; Esteban *et al.*, 1989, 1990) and also established a seropositivity rate of ~1% for volunteer blood donors (Stevens *et al.*, 1990). All but the last of the criteria listed in Table III have been satisfied for the confirmation of HCV as the etiological agent of PT-NANBH.

The nucleotide sequence information from this seminal cloning work on HCV was first published by the European Patent Office (Publ. #318,216). In this patent it was disclosed that the Chiron investigators had succeeded in extending the initial 5-1-1 clone bidirectionally to encompass a total of 7310 nucleotides in 24 overlapping clones. It was clear from the sequence of the 7310 nucleotides of the HCV genome that a single continuous translational open reading frame (ORF) existed that encoded an approximately 2440-amino-acid polypeptide, segments of which show some small homology with portions of the nonstructural proteins ns2a/2b and NS3 of dengue virus type 2 (European Patent Office Publ. #318,216) (see Fig. 2). In particular, the Gly–Asp–Asp tripeptide that forms part of the RDRP consensus motif was identified at the extreme 3′ end of the sequence, suggesting that the nonstructural portion of the viral genome was located there (Fig. 2). The available sequence information was used by other investigators to isolate defined regions from the nonstructural portion of the genome from other clinical isolates, using the PCR (Kubo *et al.*, 1989; Enomoto *et al.*, 1990).

Limited nucleotide and amino acid sequence comparisons were included in the Chiron patent application (European Patent Office Publ. #318,216). The most extensive sequence comparisons to date are those reported by Miller and Purcell (1990). The ns2a/2b–NS3 amino acid homology with dengue virus type 2 was confirmed and a second region with extensive global homology was identified. In this second identified region 20% protein sequence homology was shared with carnation mottle virus over 331 amino acid residues encompassing the putative RDRP in the ns5 region of the genome. More localized comparisons were made in a 190-amino-acid domain in the ns3 region of HCV. The greatest homology was found with two recently sequenced members of the pestivirus group: bovine viral diarrheal virus (Collett *et al.*, 1988a) and hog cholera virus (Meyers *et al.*, 1989). Homology was also shared with a member of the plant potyvirus group, tobacco vein mottling virus (Dormier *et al.*, 1986). All these alignments were statistically significant above those obtained with any members of the flavivirus family. This study indicates that HCV is related to both human and plant viruses and may therefore represent an evolutionary link between the two. It is also interesting to note the genomic similarities between HDV and the pathogenic viroids of plants (Branch *et al.*,

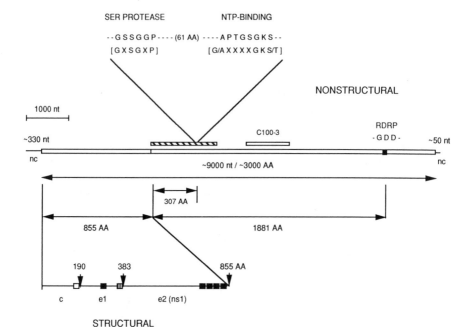

Fig. 2. Genetic organization of the HCV genome with the 5' end containing the structural genes and the 3' end containing the nonstructural gene components. The virus contains a single long open reading frame of approximately 9000 nt exclusive of the noncoding sequence (nc) at the 5' and 3' termini. The viral genes are presumed to be expressed as a polyprotein followed by co- or posttranslational processing to yield the individual gene products. Above the linear representation of the genome are the non-structural elements identified to date in the HCV genome. The hatched box indicates the location of an approximately 550-amino acid (AA) region that has ~20% homology with the ns2a/2b, NS3 region of dengue virus type 2. Within this region of global homology are conserved sequence motifs for the serine chymotrypsin-like protease (SER PRO-TEASE) and the NTP-binding motif of a putative helicase domain (NTP BINDING). Below the HCV amino acid sequences are the sequences of the consensus motifs in brackets (X represents any residue). The location of the RNA-directed RNA polymerase (RDRP) is also indicated toward the extreme 3' end of the nonstructural portion of the genome. Located in the structural gene region are a series of proteins that are recognized by their amino acid content, consensus signal sequences, transmembrane domains, and putative protease-sensitive cleavage sites. The first protein has a high basic amino acid content (pI = 12.53) and is presumed to encode the core (c). The c is followed by the putative envelope (e1) protein that begins with a signal peptide (open box) and is followed by a transmembrane domain (black box) and then a signal/transmembrane sequence (vertical hatching). The third possible structural gene has formerly been referred to as the ns1 gene product and is noted here as e2. The e2 utilizes the signal/transmembrane sequence at the e1/e2 junction for membrane targeting and ends with four predicted transmembrane domains (black boxes). All vertical arrows in the structural region mark the locations of putative consensus signal peptidase cleavages (Val–X–Ala). The multiple transmembrane segments at the COOH terminus of e2 possibly indicate it to be a polytopic membrane protein. The area of the genome identified by primary immunoscreening of cDNA libraries and utilized in the first HCV commercial diagnostic test is marked by the open box in the nonstructural region (C100-3). GDD, Gly–Asp–Asp.

1989). However, it is clear that there are no extensive homologies with any of the described RNA virus groups and that HCV may be the prototype member of an entirely new family of human pathogens.

Several potential enzymatic activities located in ns3 include the NTP-binding helicaselike domain (Lain *et al.*, 1989; Gorbalenya *et al.*, 1989b) and a serine protease domain toward the amino terminus of ns3 (Gorbalenya *et al.*, 1989a). Both domains are present in the NS3-like region of HCV within the region of global homology originally detected by Houghton and colleagues (European Patent Office Publ. #318,216) (see Fig. 2). The NS3 region of flaviviruses not only demonstrates the conservation of these motifs, but also displays similarity in the approximate distance between the two (~60 amino acids). The degree of conservation seen in the ns3 region among these diverse virus groups is an indication of the structural constraints and evolutionary relatedness of the encoded protein(s).

Mayumi's group was the first to present the HCV 5'-end sequence from two different isolates (chimp and human) (Okamoto *et al.*, 1990). It is interesting to note that their human isolate more closely resembled the chimp 910 sequence in the overlapping region. The chimp and human isolates were cloned by direct extension of the European Patent Office (Publ. #318,216) reported sequence of the chimp 910 isolate using a specifically primed cDNA library made from 1.8 ml of plasma and an oligonucleotide probe 5' to the primer site. The extreme 5' end clone from the human strain (HC-J1) was cloned from a PCR-generated product using a primer derived from the 5' end of the previously isolated chimp clone ϕ75. The HC-J1 (human) and HC-J4 (chimp) isolates were more distantly related than HC-J1 with the HCV(CDC) chimp isolate (see below). The sequence of the 1863 nucleotides corresponding to the 5'-terminal end of the HCV genome (Okamoto *et al.*, 1990) contained a region of 191 nucleotides at its 3'-terminal end which was 92.7% homologous with the first 191 nucleotides present at the 5'-terminal end of the patent sequence of HCV(CDC) (European Patent Office Publ. #318,216). It is of note that the sequence identifies a single major ORF coincident with that reported by the Chiron group. The AUG designating this ORF began at nucleotide 325. Prior to this initiation codon there are four other methionine codons. It has been speculated by some that the proper translational reading frame might be acquired through a ribosome-scanning mechanism.

Recently, the 5' end of the HCV(CDC) strain, as well as some additional 3'-end sequence, was disclosed by the Chiron investigators in a second European Patent Office publication (#388,232). They reported 319 nt of noncoding sequence at the 5' end of the virus genome, prior to the major ORF, and an additional 208 nt at the 3' end. A third group has also reported a sequence on another Japanese isolate cloned from a PT-NANBH-infected human (Takeuchi et al., 1990). The sequence information in this latter isolate extended only 190 bp into the noncoding region of the genome. The question immediately arises as to how far the genome extends at the 5'-noncoding end immediately preceding the single continuous ORF of the coding sequence and beyond the available sequence at the 3' end. A primer extension experiment similar to that performed for HEV (Tam et al., 1991a,b; see Section III,C) has not been reported. The propects for direct sequencing of the genomic RNA termini are not promising, due to the low HCV titer present in the majority of virus-containing samples and the absence of an in vitro system for high-titer virus propagation.

An ingenious procedure has been described by Riley et al. (1990) that might have utility in identifying and extending any sequence that might be present at the 5' or 3' end. The procedure relies on a novel "bubble" linker design that has a short noncomplementary segment (bubble). This bubble linker is ligated onto the ends of cDNA synthesized from an HCV-infected specimen. Primer extension from a fixed point in the HCV sequence into the bubble region synthesizes the template for the second primer for subsequent rounds of PCR. By this procedure a unique specific subset of sequences is amplified that extends in the 5' direction to the previously defined HCV sequence. With appropriately designed primers a similar procedure may be applied to the 3' end of the genome to confirm the extent of the 3' end. Experiments using this procedure are currently in progress.

One could conclude from the available sequencing data that the genome of HCV contains at least 324 nucleotides of potential noncoding sequence at its 5' terminus, followed by a continuous ORF of at least 2955 amino acid residues (European Patent Office Publ. #388,232; Okamoto et al., 1990). Very recently, data compiled from a series of cDNA clones (derived from nine individuals) indicated a single major ORF of 3010 amino acid residues for the so-called HCV-J (Japanese) "subtype" of HCV (Kato et al., 1990). This report was soon followed by the full-length sequence of a set of overlapping clones as determined by the Okayama and Houghton groups (Takamizawa et al., 1991; Choo et al., 1991b).

The flavi- and pestiviruses have similar genetic organization with the structural genes located at the 5' end and the nonstructural genes placed at the 3' end of the single large ORF. The expression strategy for the structural and nonstructural gene products would involve translational or cotranslational processing of a large polyprotein precursor molecule. It would therefore be expected from the similarities already noted in the nonstructural region that the 5′ end of the HCV genome would encode the structural genes. This prediction was borne out by an analysis of the translated nucleotide sequence (Fig. 2). The first 120 amino acids encode a protein with a theoretical isoelectric point of 12.53 and a high arginine and lysine content amounting to 20% of the first 150 residues. This high pI is characteristic of nucleocapsid (core) proteins and is believed to be due in part to their association with the viral genomic nucleic acid. The core protein has a predicted M_r of ~20,000 and, like the pestivirus p20, lacks any N-linked glycosylation sites (Collett et al., 1988b). A comparison of flavi- and pestiviruses by Collett et al. (1988c) equates the p20 of pestivirus with the C and prM proteins of flaviviruses. Accordingly, it would appear that the HCV core protein appears more similar to pestiviruses than to flaviviruses.

Following the core region of ~150 amino acids, there lies a 700-amino-acid stretch containing seven different hydrophobic domains, as predicted by the method of Klein et al. (1985). The similarity between signal and transmembrane consensus sequences (Boyd and Beckwith, 1990) and the absence of any reported amino-terminal sequencing data to firmly define the actual polyprotein cleavage points make the interpretation of this region difficult. There are also no clear similarities with either flavi- or pestiviruses, and, in particular, their hydropathicity profile upstream of ns2 "diverges from the flavivirus model" (European Patent Office Publ. #388,232). This dissimilarity in the structural gene region following c has led some to refer to it as the "x" region (Choo et al., 1990). The following discussion may therefore be considered a preliminary assessment based on the deduced amino acid sequence, computer-predicted signal and transmembrane segments, and other structural features already known for both flavi- and pestiviruses. The lower-case letter designations signify that these HCV protein products are only predicted and have not been experimentally verified (i.e., observed) either in vitro or in vivo.

This region probably encodes at least two different posttranslational products. The first of these has been referred to as the envelope (e) gene (Okamoto et al., 1990; Takeuchi et al., 1990) and is referred to here as e1. The e1 gene would appear to span at least 210 amino acids and is marked at its amino end by a signal peptide followed by a series of potential cleavage sites. At its carboxy terminus a transmembranelike sequence indicates its probable end (Fry et al., 1991b). Within this region there are five potential N-linked glycosylation sites as well as another predicted transmembrane sequence that possibly indicates e1 to be a polytopic transmembrane protein with two different membrane-spanning segments (Nilsson and von Heijne, 1990). Models have been developed for the membrane topology of two different flavivirus E proteins: the tick-borne encephalitis virus (Mandl *et al.*, 1989) and the West Nile fever virus (Nowak and Wengler, 1987). Both models predict two transmembrane segments toward the carboxy terminus, with the carboxy end out (extracellular). All flavivirus E proteins have 12 highly conserved cysteine residues forming six disulfide bridges (Mandl *et al.*, 1989). The cysteine content (10 residues) of this putative e protein nearly matches that of the flaviviruses. There are five potential N-linked glycosylation sites in the HCV e1 protein and all these are conserved among the four different isolates sequenced to date (see below). This differs from the relative lack of N-linked glycosylation (i.e., two or fewer sites) seen in the E proteins of flaviviruses (Rice *et al.*, 1986). The *e1* gene equivalent in the pestivirus bovine viral diarrheal virus is the *gp62*, which has 10 N-linked glycosylation sites (Collett *et al.*, 1988b). The predicted HCV e1 gene product is smaller than the equivalent E proteins expressed by either the flavi- or pestiviruses. A coupled *in vitro* transcription/translation experiment using microsomal membranes has indicated that this region might encode a glycosylated e1 gene product of M_r ~33,000 (gp33) that is reduced to M_r ~18,000 with endoglycosidase H treatment (Choo *et al.*, 1991a). The predicted size of the e1 gene product, as well as the cysteine content and abundant N-linked glycosylation sites, may mark HCV as a related but unique virus family.

The third gene roughly aligns (by hydropathic profile) with the flavivirus *NS1* gene and has therefore been referred to as *ns1* (European Patent Office Publ. #388,232). The *NS1* equivalent in pestiviruses is the *gp53*, which actually serves as a structural component of the virus (Collett *et al.*, 1988b). For the purposes of this discussion, we refer to ns1 as e2 to signify its possible role as a structural component in HCV. The start of the e2 is indicated by the consensus cleavage site for a putative signal peptidase [Val–X–Ala (von Heijne, 1985)]. This cleavage site also marks the carboxy terminus of the e1 transmembrane segment and is shared by a number of flaviviruses: West Nile fever, Yellow fever, Murray Valley encephalitis, Japanese encephalitis, St. Louis encephalitis, and dengue virus types 2 and 4 (Coia *et al.*, 1988; Speight *et al.*, 1988). The size of the putative HCV e2 gene product is estimated to be 460 amino acids and extends from the predicted signal peptidase cleavage point (after amino acid 383) through to the end of a highly hyrophobic stretch of four predicted transmembrane segments located at the carboxy terminus of the protein. This clustered group of transmembrane segments may indicate the presence of a polytopic transmembrane region at the carboxy terminus of e2. These four transmembrane segments are followed by another consensus peptidase cleavage site at amino acid 853 (Val–Glu–Ala). There are two to four N-linked glycosylation sites in the NS1 of flaviviruses and four potential sites in the gp53 of pestiviruses (Collett *et al.*, 1988c). The e2 of HCV would appear to be substantially different in containing 11 N-linked glycosylation sites. This level of glycosylation appears to be more related to that seen in the pestivirus gp62 (*E* gene in flaviviruses), which has 10 N-linked glycosylation sites (Collett *et al.*, 1988b).

The availability of the 5′-end sequence from several different isolates permits their comparison in the important structural gene-encoding region of the viral genome. The extent of diversity (genetic variation) in the structural region has important implications for future diagnostic and vaccine development. The homologies between the various sequenced HCV strains at both the nucleotide and amino acid levels have been investigated (Kremsdorf *et al.*, 1991; Fry *et al.*, 1990b; Weiner *et al.*, 1991). Although the strains are, for the most part, homologous, isolated pockets of heterogeneity located in e1 (at about residues 246–258) and the extreme amino terminus of e2 (polyprotein residues 384–410) are clearly evident. The sequence heterogeneity in these regions has recently been reported for four U.S. and two Italian isolates (Weiner *et al.*, 1991). These may be referred to as the variable and hypervariable regions of the *e1* and *e2* genes, respectively, and are perhaps analogous to those identified in the *e* gene of the human immunodeficiency virus (HIV). The significance of this sequence variation is not known. It is, however, tempting to speculate that these regions are in some way involved in viral pathogenesis. The extreme 5′-end capsid gene sequence is highly conserved among the different isolates and therefore has possible utility in the diagnosis of HCV infection.

The identification of the 5-1-1 clone may have been fortuitous, considering the experience of other laboratories. Although it was presumed that the most immunogenic region of the HCV genome would be identified by immunoscreening, a different antigen-encoding region of the genome was identified by all groups that pursued this procedure as their primary screening methodology. Also, a nonstructural, rather than structural, antigen was identified by all groups. One of our laboratories had used immunoscreening to identify an HCV-specific clone (Reyes *et al.*, 1991b). This procedure was undertaken in 1985 after evaluating several key assumptions. Foremost was the assumption that the agent contained a nucleic acid genome. This assumption would be invalid had HCV been a prionlike agent, as hypothesized for the etiological agent of scrapie (reviewed by Prusiner, 1989). Furthermore, it was assumed that certain virus-specific proteins encoded immunoreactive proteins in the infected host. This latter assumption was also contingent on the existence of an agent that did not mask its immunoreactive regions or evade immune surveillance by mutating immunogenic regions at an extremely high frequency. The final key assumption was that the procedures utilized to obtain and clone the nucleic acid were sufficient to the task when both the biophysical and biochemical characteristics of the etiological agent were considered.

We identified over 100 immunoreactive clones from various PT-NANBH libraries (Reyes et al., 1991b). One difference in our immunoscreening protocol compared to that of Choo et al. (1989) was that the second-step antibody was conjugated to alkaline phosphatase, rather than a radioiodinated anti-human immunoglobulin. Probably the most significant difference, however, was the utilization of a pelleting procedure to concentrate virus directly from undiluted serum or plasma. Of the ~111 immunoscreened clones identified from human and chimp sources, ~50% were excluded based on the exogenousity test (hybridization to human DNA on a whole-genome Southern blot). Of the remaining exogenous clones a significant number failed to reproducibly react with a larger panel of PT-NANBH-derived sera. Furthermore, sequence comparisons of certain of these clones with HCV indicated that they were clearly different. Apart from the identification of authentic HCV clones, other exogenous sequences were easily isolated by immunoscreening from these libraries. It is possible that these clones might represent other exogenous parenterally transmitted agents similar to those identified in the Bradley gradient fraction (see Section II,A,3). It should be noted that the preparation of anti-hemophiliac Factor VIII involves the pooling and purification of material from hundreds of individual units for each lot and therefore had the potential for infecting the chimp with multiple different pathogenic agents.

Shikata and colleagues also identified a molecular clone by an immunoscreening protocol (Maeno et al., 1990). They began with ~8 liters of pooled plasma collected from 148 Japanese donors selected by virtue of their elevated alanine aminotransferase (ALT) levels. Precipitation of the "viral fraction" was performed with 3.6% polyethylene glycol prior to partial purification on a sucrose step gradient. RNA extracted from the pelleted material was used for cDNA synthesis and λgt11 library construction. Only one of the three positive clones obtained, C8-2, was shown to be specific to NANBH. The 280-bp clone was again located in the nonstructural region of the genome (see Fig. 2).

Other reports describing the cloning and sequencing of "NANBH" by similar immunoscreening techniques have appeared. Of particular interest are the so-called Arima clones (Arima et al., 1989a,b, 1990). These were derived from nearly 100 liters of human sera collected from NANBH patients and pelleted to concentrate the putative viral agent. The sequences of 201 nucleotides corresponding to clone 18 (Arima et al., 1989a), 67 nucleotides of clone 2 (Arima et al., 1989b), and 115 nucleotides of clone 14 (Arima et al., 1990) showed no nucleic acid sequence homology with the original 7310 bp of HCV (K. E. Fry, unpublished observations). Before the disclosure of the 5'-end sequence, it remained a possibility that the sequence information contained in the Arima clones was present in the extreme 5' or 3' end of the genome. Experiments undertaken in one of our laboratories (G.R.R.) established that oligonucleotide primers derived from the Arima clones failed to prime the expected fragment from authentic HCV cDNA prepared from RNA extracted from the chimp 910 serum. Nor was it possible to "link" the Arima-cloned sequences with the established HCV sequence or with each other. The approach that was taken to establish linkage between the Arima clones (18 and 2) and HCV is outlined in Fig. 3. Control PCR reactions performed using HCV primer pairs established that the cDNA source and preparation were not responsible for the negative results obtained with the Arima primers.

It is possible that the Arima clones represent HCV variants with substantial sequence heterogeneity that would prevent successful annealing and priming in a PCR reaction. The recent availability of additional 5' and 3' HCV sequence information, however, confirms the lack of detectable nucleic acid homology with the Arima clones (K. E. Fry et al., unpublished observations). As already noted, we also identified exogenous clones from HCV material that did not have any genomic relationship with HCV. These cloned sequences are also unrelated to those of Arima. Both sets of exogenous immunoreactive clones are unrelated to the HCV sequence, but might represent an altogether different agent(s) of PT-NANBH. The relevance of these cloned sequences as human pathogens requires further investigation.

Immunoscreening failed to identify clones from the structural region of the genome. This may be related to comparatively low antibody titers to these gene products, the immunogenicity of these products, or the differential expression of nonstructural and structural gene products in a chronically infected host. These questions are relevant to the pathobiology of the virus. The lack of an effective high-titer neutralizing humoral immune response to HCV may be related, in part, to the mechanism of virus persistence (chronic infection). The development of a delayed high-titer antibody response to predominantly nonstruc-

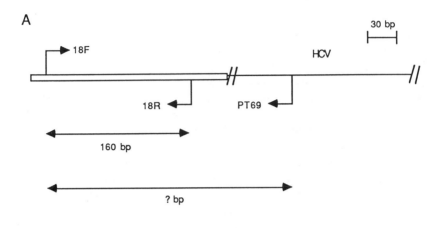

B   ARIMA SEQUENCE PRIMERS

18F: 5'-GGAGAAGCCAGCAATGGAG-3'
18R: 5'-CTGATATGCGTCCTTCCTTC-3'
18H: 5'-CGGCAACAAATAACCCAGGA-3'

2F: 5'-CCAACGCGTCGGCTTGGC-3'
2R: 5'-CACGGCCATCAGCGCGGG-3'
2H: 5'-CCGCGCCTTGGCCGCCGACC-3'

Fig. 3. Arima cDNA clones are unrelated to HCV. (A) Arima clone 18 is hypothetically presented as positioned 5' to the authentic HCV sequence (see text). Attempts were made either to detect the Arima sequence in an HCV-containing cDNA source or to link the Arima sequences with authentic HCV sequences. The HCV cDNA tested for the presence of Arima-type sequences was the 10^6 chimpanzee infectious doses (CID) titered material developed at the Centers for Disease Control (chimp 910) and used for the initial cloning work (Choo *et al.*, 1989). The sequences for the forward (F) and reverse (R) primers are shown in (B) for two Arima clones (clones 18 and 2). The H primer identifies the oligonucleotide situated between the F and R primers that was used for confirmatory hybridizations. HCV cDNA was made by either random priming (lanes 1, 2, 5, and 6) or oligo(dT) priming (lanes 3, 4, 7, and 8) before nonspecific amplification using SISPA (see text). The SISPA cDNA was confirmed to contain HCV-specific sequences by positive polymerase chain reaction analysis with nine different HCV primer pairs (Kim *et al.*, 1991) (data not shown). These same cDNAs were positive using the nested primer set derived from the 5'-end noncoding region (see Fig. 4). Shown under (C) is the result obtained using primers 18F and R (lanes 1–4) or primers 2F and R (lanes 5–8). All amplified fragments generated (see, e.g., lanes 5–8) were determined to be nonspecific by oligonucleotide hybridization. Attempts to link the Arima sequence with that of authentic HCV (e.g., primer PT69) were also unsuccessful (data not shown). Size markers (M) are fragments generated by *Hae*III cleavage of φX174 RF DNA.

C

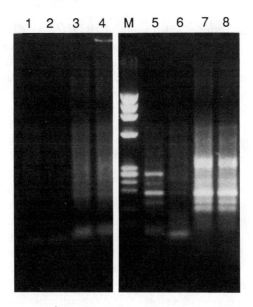

FIG. 3. (cont.)

tural antigens has utility in diagnosis, but is a serious deficiency in detecting early seroconversion and recent infection.

C. Viral Replication

A recent publication describing early events following transmission of HCV into two chimpanzees (1304 and 1313) is by far the most complete study undertaken to date (Shimizu et al., 1990). Several parameters were determined on individual samples in order to assess the course of infection. These included measurement of ALT levels, detection by electron microscopy of cytoplasmic tubular structures, detection by immunofluorescent (IF) staining of a cytoplasmic antigen, detection by cDNA/PCR of HCV genomic RNA, and detection by the C100-3 immunodiagnostic test for anti-HCV antibody. HCV genomic RNA was detected in serum as early as 3 days after inoculation and reached a maximum at 7 weeks as ALT values were on the rise. Cytoplasmic antigen and tubular structures appeared in hepatocytes 3 and 6 days, respectively, after the appearance of HCV RNA in serum. At a later stage these markers were still present in chimp 1304, but disappeared in chimp 1313, suggesting that the former became a chronic carrier, while the latter resolved his acute infection. This is the first report to describe early replication stages of HCV in chimpanzees.

The transient *in vitro* expression of HCV from primary hepatocytes established in culture from an infected chimp was recently reported (Jacob *et al.*, 1990). A previously described system using hormonally defined medium was developed and shown to be successful in the production and secretion of hepatocyte-specific markers into the culture medium (Lanford *et al.*, 1989). The transient expression of HBV from hepatocytes taken from an HBV chronically infected chimp (Jacob *et al.*, 1989) suggested that HCV might also be expressed from an HCV-infected animal. Hepatocytes established in culture from an NANBH animal were positive for an NANBH-associated antigen, as detected by immunochemical staining (Burk *et al.*, 1984). An enveloped viruslike particle (VLP) with diameters ranging between 39 and 46 nm was also observed by electron microscopy in concentrated culture supernatants. The infectivity of these supernatants containing VLPs was assessed by chimp inoculation. Although the animal developed elevated ALT values, these were not as pronounced as those obtained with the authentic plasma inocula used in control infections. The real test of the *in vitro* expression of HCV would be the development of chronic hepatitis in the inoculated chimp. Testing of these cultures with authentic molecular probes would, of course, provide a definitive result regarding the presence of virus in these cultures. The results of these experiments were not reported.

Another attempt to propagate HCV *in vitro* was through the generation of an immortalized hepatocytelike host (G. R. Reyes *et al.*, unpublished observations). Human fetal hepatocytes were fused to the SBC-H20 cell line that had originally been established for the purpose of immortalizing lymphocytes for human monoclonal antibody expression (Foung *et al.*, 1984). There was a transient expression of hepatocyte-specific markers such as albumin, C3 component of complement, and fibronectin into the culture medium of some of the established cell lines, including one named GL424. One of our early experiments to validate the susceptibility of GL424 to infection with hepatotropic viruses was performed with HBV. After a period of time, we were able to detect viral particles and filaments in concentrated media by electron microscopy, specific cell surface IF using anti-HBsAg reagents, and HBV core antigen comigrating on sucrose gradient fractions with an authentic HBV inoculum run in parallel. HBV expression was transient, however, in that the cultures eventually ceased production of these virus-specific markers.

GL424 cells, when exposed to chimp- or human-derived HCV inocula, demonstrated the expression of specific cytoplasmic and surface IF. The specific IF signal was detected only in the HCV-infected cultures and only when HCV sera from either chimps or humans was used to detect expressed antigen by indirect IF. These IF changes could be reproducibly subpassaged using cell-free lysates into new GL424 cultures with the development of the same IF changes. These experiments may represent, for both HBV and HCV, the first *de novo* infection experiments performed *in vitro*, and contrast with those experiments reported for HBV using transfection of a molecularly cloned viral genome to establish a producer cell line (Sureau *et al.*, 1986). The significance of these and other suggestive observations, performed before the availability of molecular probes, may now be definitively measured by verifying the presence of HCV genomic information or the expression of HCV-specific antigens using specific immunological reagents. These experiments are currently in progress. The utility of a cell line expressing HCV for studies of viral expression and replication cannot be overestimated.

D. Prevention and Control

The licensing and availability of diagnostic kits for the detection of antibodies to HCV in blood and blood products (Ortho Diagnostic Systems; Abbott Labs, North Chicago, Illinois) have decreased the risk of transfusion-transmitted hepatitis. Originally, the risk of posttransfusion hepatitis was estimated to be between 6.7 and 12.6% (Hoofnagle and Alter, 1984). Prior to the cloning and development of a specific serological test for past virus infection, a surrogate test based on elevated serum ALT and the presence of anti-HBV core antibody (anti-HBc) was adopted to screen donated blood (Stevens *et al.*, 1984). New studies using the specific HCV recombinant antigen test indicates that the risk of contracting PT-NANBH may now be considerably lower (Di Bisceglie and Hoofnagle, 1990); however, this may be due to the concomitant screening of donated units for HIV and the elimination of high-risk donors in the last 5 years. Nevertheless, a high percentage, about 50% of infected patients, estimated at 75,000 per year in the United States alone, develop chronic hepatitis, and about 20% of these chronically infected patients develop chronic active hepatitis or cirrhosis (Alter and Sampliner, 1989). As noted above, the test antigen was derived from a 363-amino-acid segment in the nonstructural portion of the genome. Seroconversion to the C100-3 antigen generally occurs 2–3 months after infection with the agent, but in certain cases may occur considerably later. As the early clinical phases of the disease may not be apparent, a significant period may exist during which the infected individual may transmit the virus. Due to this potential "window of infectivity," the surrogate marker test continues to be used (Dienstag, 1990).

The random screening approach was successful at epitope identification, but failed to identify the most highly immunogenic, and hence diagnostic, epitopes in two of three cases (Maeno et al., 1990; Reyes et al., 1991b). 5-1-1 may be considered exceptional in its broad pattern of immunoreactivity. Other antigenic immunodominant regions of the nonstructural portion of the genome have been identified, using an alternative procedure to random screening (Moeckli et al., 1991; Kim et al., 1990). This procedure was based in part on the sequence-independent single-primer amplification (SISPA) protocol and was taken once the partial sequence of the genome became available (Fig 1B). SISPA (Reyes et al., 1990a) is a technique that, when applied to DNA or cDNA populations, amplifies all nucleic acid molecules, irrespective of their specific sequence content. This is achieved by the directed end ligation onto all molecules of a specially designed linker/primer oligonucleotide. By the provision of a common end sequence, a single strand of the linker/primer oligo may be used as primer for subsequent Taq polymerase-mediated polymerization.

In brief, SISPA was used to expand the original set of randomly primed cDNA fragments to permit multiple HCV primer-specific amplifications and multiple attempts to recover longer cDNA sequences that might be present in the original starting population in reduced amounts. The full-length genome was obtained as a defined set of overlapping contiguous clones by sequence-specific PCR using the SISPA-amplified pool of cDNA fragments (Kim et al., 1991). The defined set of fragments was recovered for subcloning directly into an expression vector or for epitope library construction, using the DNase I procedure (Fig. 1B). In this way an epitope was identified that displayed equivalent reactivity to 5-1-1 on Western blot, but definite superiority in terms of earlier seroconversion and broader reactivity when tested in other formats (Kim et al., 1991). The identified regions have a high degree of reactivity with acute-phase sera derived from experimentally infected chimpanzees. Diagnostic tests based on these newer antigens should lead to improved diagnosis of HCV and further reduce the possibility of parenteral spread of HCV by blood and blood products.

Detection of virus in clinical samples may be attempted directly by use of the PCR. Several laboratories have identified HCV in clinical samples, using this technique (Kubo et al., 1989; Weiner et al., 1990; Kaneko et al., 1990; Garson et al., 1990). It would be preferable to identify and utilize a highly conserved portion of the viral genome for such an application. From the sequence data reported to date, it is clear that the 5'-noncoding region of the genome is highly conserved (Okamoto *et al.*, 1990; Takeuchi *et al.*, 1990; Fry *et al.*, 1991b). This region may be conserved due to its putative role in the regulation of replication and expression of the viral genome. This is suggested by the proposed complex secondary structure assumed by the 5' end of the viral genome having a negative free energy of -150 kcal (Fry *et al.*, 1991b). Two pairs of nested primers based on the highly conserved 5'-noncoding sequence were tested in such a detection assay (Fig. 4). With these primers we have detected a positive correlation between PCR results and serological assays in ~40% of cases. It should also be noted that a positive PCR result may also be obtained in the absence of serological conversion to an identified HCV antigen in some patients with chronic hepatitis (R. A. Moeckli *et al.*, unpublished observations). However, it has been observed that α-interferon treatment does result in the reduction/removal of PCR-detectable viral genome from sequential serum samples taken after treatment (A. M. Di Bisceglie, personal communication). The application of PCR in this manner provides a clear measure, where applicable, of the treatment efficacy of a particular therapeutic regimen.

Among the many approaches undertaken to treat chronically infected HBV patients are immunosuppression with corticosteroids, immunostimulation with bacille Calmette–Guérin vaccination or levamisole, use of plant extracts, or antiviral therapy with α-interferon. Two reports appeared in the past year describing the treatment efficacy of recombinant α-interferon in a cohort of patients with chronic HCV (Davis *et al.*, 1989; Di Bisceglie *et al.*, 1989). Both groups found that interferon therapy was beneficial to patients with chronic HCV; however, the response was often transient. Treatment efficacy was based on the normalization of elevated serum ALT levels and histological improvement. Although there has been some question as to the overall efficacy of such treatment (Koretz, 1990), as noted above, the PCR assay documents a decrease in circulating virus and is indicative of a reduced capacity of the HCV genome to replicate.

The risk of developing hepatocellular carcinoma (HCC) after infection with HBV is well established (Beasley, 1982). There have been reports indicating that PT-NANBH might have some role in the development of HCC (Resnick *et al.*, 1983; Kiyosawa *et al.*, 1984; Gilliam *et al.*, 1984). A review by Purcell (1989) underscores the fact that the molecular mechanisms behind the association of hepatitis viruses and HCC have not yet been elucidated. With the availability of a diagnostic kit, there is a growing body of evidence suggesting that HCV represents a significant risk for the development of HCC in chron-

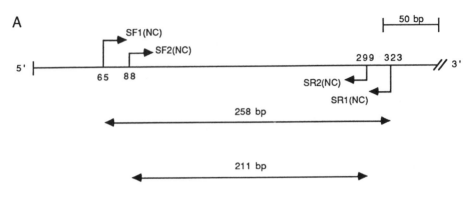

B   PRIMER SEQUENCES

SF1(NC): 5'-GCCATGGCGTTAGTATGAG-3'
SF2(NC): 5'-GTGCAGCCTCCAGGACCC-3'
SR1(NC): 5'-GCACGGTCTACGAGACCT-3'
SR2(NC): 5'-GGGCACTCGCAAGCACCC-3'

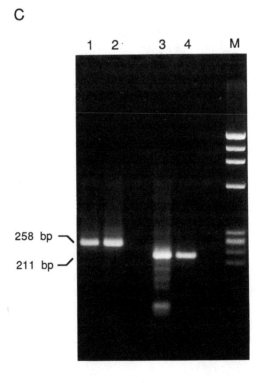

ically infected patients (Colombo *et al.*, 1989; Bruix *et al.*, 1989; Saito *et al.*, 1990; Kiyosawa *et al.*, 1990; Hasan *et al.*, 1990). These studies indicate that a significant percentage of HCC patients, ranging from 65 to 94.4%, have had prior exposure to HCV. Although a high rate of exposure to blood transfusion for this HCC/HCV-positive group was detected (30–42%), it does indicate that additional modes of transmission and exposure exist (Kiyosawa *et al.*, 1990; Saito *et al.*, 1990). The study by Kiyosawa *et al.* (1990) additionally lends support to the stepwise progression toward HCC dating from the initial exposure to HCV. In the documented cases of HCV contracted through blood transfusion, the mean interval to diagnosis of chronic hepatitis, cirrhosis, and HCC went from 10 to 21.1–29 years, respectively. These data support the progression of the chronically infected individual into cirrhosis and finally into HCC.

The high percentage (i.e., 50%) of infected individuals going on to chronic disease may be an underestimate. A prospective study of acute hepatitis cases indicated that at 12 months 80% of PT-NANBH individuals developed chronic hepatitis in association with seroconversion to HCV (Nishioka, 1990). This is of obvious concern due to the association of HCV with HCC. The incidence of deaths due to HBsAg-negative HCC in Japan has been increasing, while the death rate associated with HBsAg-positive HCC has remained nearly constant. Although the number of HBV carriers in Japan is estimated at 2.4 million versus 1.4 million HCV antibody-positive individuals, the incidence of HCC in the latter population is nearly four times that seen with HBV (Nishioka, 1990).

The prospects for vaccine prophylaxis of HCV are currently an area of intense interest. As noted above, the disease has a seroprevalence in the range of 0.5–1.5%. The high degree of seropositivity in the general population indicates that the virus is spread predominantly by mechanisms other than iatrogenic (i.e., transfusion of blood or blood products) or those involving intravenous substance abuse. Although the data should be considered preliminary based on seroconversion to a

---

FIG. 4.   Conserved nested primers in the HCV 5′-noncoding region (NC). The positioning of the oligonucleotide primer pairs at the 5′ end of the viral genome (within the highly conserved noncoding region) is indicated in (A). The actual sequence of the primers is shown under (B). The resulting 258-bp product from the first priming (SF1/SR1 primer set) and the second 211-bp product resulting from the second priming (SF2/SR2 set) is indicted in (C) after ethidium bromide staining. The samples tested were chimp 910 cDNA generated by oligo(dT) priming (lanes 1 and 3) or chimp 910 cDNA made by random priming (lanes 2 and 4). The expected products from primer set 1 and set 2 are indicated. Size markers (M) are fragments generated by *Hae*III cleavage of φX174 RF DNA.

single antigen (C100-3), sexual transmission does not appear to be as effective for HCV as it is for HBV (M. J. Alter *et al.*, 1989). The question of an endemic reservoir, as exists for certain flavi- and pestiviruses, has not yet been fully investigated.

The ultimate success of vaccine development will be determined by basic research investigation into how HCV is capable of establishing a persistent infection. In particular, how the virus escapes immune surveillance will be an area of critical importance. At present there are few answers and many questions. Does immune tolerance play a role, as, for example, in the antigenic mimicry of host antigens displayed by other viruses? Is there viral modulation of its antigenic components? Is the virus capable of maintaining itself in a quiescent state, as, for example, in the latency of herpesviruses? Does the virus undergo genetic drift, as seen in HIV, and does this lead to evasion of the immune response? As noted above, substantial genomic heterogeneity has already been detected in portions of the structural gene region outside of the capsid. The effect, if any, on vaccine development will depend on the positioning of neutralization epitopes in these gene products and also whether true neutralization serotypes exist, as seen with dengue virus (Halstead, 1988). As HCV has not yet been efficiently propagated *in vitro*, a recombinant subunit vaccine would seem to be the most viable approach.

The similarities between HCV and the flavi-/pestiviruses have already been alluded to and therefore indicate that the most suitable target for vaccine evaluation would be e1 and e2. Due to the high degree of glycosylation, recombinant expression of these genes in eukaryotic systems will be an important area of future study relevant to vaccine development. We have already acquired preliminary data on the expression of an ns3/ns4 junctional region by vaccinia virus (Kotwal *et al.*, 1991). Investigation of cell-mediated immunity will also be facilitated by the vaccinia-based expression of HCV.

III. Hepatitis E Virus

The designation "non-A, non-B hepatitis" (NANBH) was first used to define the clinical diagnosis of viral hepatitis in the absence of defined serological markers for infection with HAV or HBV. NANBH, in effect, represented a "disease" caused by multiple etiological agents not yet characterized. PT-NANBH, or what is now recognized as HCV infection, represented the major form of NANBH in the developed world. The recognition of the existence of a second distinct form of NANBH occurred through the accumulation of epidemiological evidence for a form of NANBH with fecal/oral transmission that was unrelated to HAV. These two very different routes of infection for NANBH implied that an altogether different virus with different physicochemical properties was responsible. This situation is similar to the distinction between the "infectious" and "serum" hepatitides which ultimately led to the discovery, characterization, and differentiation of HAV and HBV, respectively.

The first documented epidemic of ET-NANBH occurred in 1955 in Delhi, India (Viswanathan, 1957; Wong et al., 1980). Since then multiple epidemics with similar epidemiological and clinical features have occurred the world over. These epidemic outbreaks of ET-NANBH have been variously referred to as "epidemic" or "water-borne, non-A, non-B hepatitis" (reviewed by Bradley, 1990a,b). The recent molecular characterization of the virus responsible for ET-NANBH (Reyes et al., 1990a) has clearly demonstrated it to be distinct from the principal agent of PT-NANBH: HCV. As previously suggested (Purcell and Ticehurst, 1988), the etiological agent of ET-NANBH has been named HEV.

A. Infectious Agent

1. Clinical/Epidemiological Features

The key epidemiological features of ET-NANBH are listed in Table I. Epidemic outbreaks of ET-NANBH have occurred worldwide (Sreenivasan et al., 1978; Tandon et al., 1982; Khuroo et al., 1983; Kane et al., 1984; Bellabes et al., 1984; Myint et al., 1985). These epidemics are notable in that they tend to occur in the urban areas of developing countries and can generally be traced to point-source contamination of common water supplies. The magnitude of these epidemics can be quite large, involving tens of thousands of individuals (Viswanathan, 1957; Sergeev et al., 1957; Hillis et al., 1973; Shrestha and Malia, 1975). The disease is not generally considered in the differential diagnosis of hepatitis in the United States and other industrialized countries, although there are reports of cases imported into the United States by travelers returning from endemic areas (De Cock et al., 1987; Fortier et al., 1989). Of particular interest is whether there is a natural host in the wild that serves as an endemic virus reservoir. The mechanism of this endemicity requires clarification in order to take adequate measures to control the virus and its spread.

An examination of a limited set of sporadic hepatitis cases (i.e., those contracted with unknown etiology) was undertaken using a direct IF blocking assay (Krawczynski, 1989). Of 20 sporadic cases examined from the United States, none was found to be related to HEV. This situation was in contrast to that seen in developing countries, where a significant percentage of the sporadic cases were found to be caused by HEV (Krawczynski, 1989; Goldsmith et al., 1991). These preliminary observations will be more fully examined once a diagnostic test becomes available.

The typical presentation of ET-NANBH was described in the original report by Khuroo (1980) and is not unlike the clinical presentation of the other acute viral hepatitides. The course of the disease is generally uneventful and the vast majority of those infected recover without the chronic sequelae seen with HCV. One peculiar epidemiological feature of this disease, however, is the markedly high mortality rate observed in pregnant women, reported in numerous studies to be on the order of 10–20% (reviewed by Bradley, 1990a,b). This finding has been seen in a number of epidemiological studies, but at present remains unexplained. Whether this reflects viral pathogenicity, the lethal consequence of the interaction of virus and immunosuppressed (i.e., pregnant) host, or a reflection of the debilitated prenatal health of a susceptible malnourished population remains to be clarified.

2. Animal Model

The development of animal models for human pathogens permits the dissection and understanding of important pathobiological parameters in a controlled manner. The objective in an animal model is to reproduce the disease and all its manifestations as they might occur in the human host. The development of an animal model also provides important specimens that may not otherwise be available for molecular and biophysical studies of the pathogenic agent. The cloning of both HCV and HEV was greatly facilitated by the development of animal models and the availability of specimens containing sufficient quantities of virus for molecular cloning efforts. The studies describing the biochemical characterizations performed on HEV by Bradley et al. (1988) are an important illustration of this fact.

Various primate species have been tested for their susceptibility to infection with authentic HEV inocula (Balayan et al., 1983; Andjaparidze et al., 1986; Bradley et al., 1987; Arankalle et al., 1988; Uchida et al., 1990a; reviewed by Bradley, 1990a,b). By these studies it was determined that the cynomolgus macaque gave very reproducible liver enzyme elevations indicating infection and resultant hepatocellular dysfunction. The more sensitive indicator in this species was serum isocitrate dehydrogenase, rather than ALT. The biophysical characterization of HEV is summarized in Table V. These findings led to the supposition that the agent of ET-NANBH was an

TABLE V

ET-NANBH: Animal Model and Biophysical Characterization[a]

Animal model	Cynomolgus macaque—acute phase ALT/SICD elevation; chronic disease not observed
Particle size	27–34 nm by immunoelectron microscopy; possible particle degradation in gut
Viruslike particles	Present in feces and bile; identical to those present in clinical specimens
Particle density	183 S (potassium tartrate gradients)

[a] ALT, Alanine aminotransferase; SICD, serum isocitrate dehydrogenase.

RNA virus and possibly a member of the Caliciviridae (Bradley and Balayan, 1988).

B. Cloning and Genomic Organization

The absence of a cell culture system for virus propagation dictated the use of either cynomolgus macaque or human specimens as a source of virus for molecular cloning efforts. Virus was present in stool samples of clinical and cynomolgus specimens. However, it was necessary to screen through more than 2000 stool specimens by immunoelectron microscopy (IEM) (Bradley, 1990a) in order to identify a sample that had aggregated VLPs under IEM using convalescent cynomolgus and human sera. Unlike HCV, no titered HEV specimens were available for molecular cloning. These IEM-positive samples were used to fully develop the cynomolgus monkey as the HEV model at the CDC and to confirm the serological association of the 27 to 34-nm VLPs with ET-NANBH (Bradley et al., 1987).

An important finding by Bradley (1990a,b) and subsequently others (Uchida et al., 1990b) was that virus was present in the bile of infected cynomolgus monkeys. This meant that it should be possible to use a virus-containing sample in which the total nucleic acid complexity would be much lower than virus present in infected serum, liver, or stool. Cloning from bile would, in effect, reduce the background of irrelevant clones through which one would have to search in order to identify a virus-specific clone.

The molecular cloning of HEV was accomplished by two very different approaches (Fig. 5). The first successful identification of a molecular clone was based on the differential hybridization of putative HEV cDNA clones to heterogeneous cDNA from infected and uninfected cynomolgus bile (Reyes et al., 1990a,b). cDNAs from both sources were labeled to high specific activity with ^{32}P to identify a

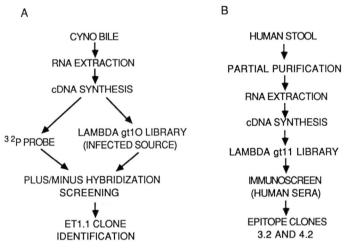

FIG. 5.   Identification of HEV cDNA clones. Outlined are the two approaches taken to identify genomic cDNA clones from the HEV[Burma (B)] strain (A) and the HEV-[Mexico (M)] strain (B). The HEV(B) cDNA clone ET1.1 was identified by a differential hybridization screening protocol applied to a cDNA library made from the bile of an experimentally infected cynomolgus macaque (CYNO). Clones identified by this procedure were confirmed as HEV specific as described in the text. The direct immunoscreening protocol was applied to a cDNA library made from RNA extracted from a semipurified virion preparation made from a human stool sample. Two epitope clones were identified from the 3′ half of the viral genome: 406.3.2 (3.2) and 406.4.2 (4.2).

clone that hybridized specifically to the infected source probe. A cynomolgus sample infected with HEV(Burma) was used in these first experiments. The sensitivity of this procedure is directly related to the relative abundance of the specific sequence compared with the overall background. In control experiments we found that identification of a target sequence may be obtained with as little as 1 specific part per 1000 background sequences (J. P. Kim, unpublished observations). A number of clones were identified by this procedure, using libraries and probes made from infected (Burma isolate) and control uninfected cynomolgus bile. The first extensively characterized clone of the 16 plaques purified by this protocol was ET1.1.

ET1.1 was first characterized as both derived from and unique to the infected source cDNA. Heterogeneous cDNA was amplified from both infected and uninfected sources using SISPA. The limited pool of cDNA made from Burma-infected cynomolgus bile could then be amplified enzymatically prior to cloning or hybridization, using putative HEV clones as probes. ET1.1 hybridized specifically to the original bile cDNA from the infected source. Further validation of this clone, as derived from the genome of HEV, was demonstrated by the similarity of the ET1.1 sequence and those present in SISPA cDNA prepared from five human stool samples collected from different ET-NANBH epidemics in Somalia, Tashkent, Borneo, Mexico, and Pakistan (Reyes *et al.*, 1990a). These molecular epidemiological studies established the isolated sequence as derived from the virus that represented the major cause of ET-NANBH worldwide.

The viral specificity of ET1.1 was further established by the finding that the clone hybridized specifically to RNA extracted from infected cynomolgus liver (Reyes *et al.*, 1990a). Hybridization analysis of polyadenylated RNA demonstrated a unique ~7.5-kb polyadenylated transcript not present in uninfected liver. The size of this transcript suggested that it represented the full-length viral genome. Strand-specific oligonucleotides were also used to probe viral genomic RNA extracted directly from semipurified virions prepared from human stool (Reyes *et al.*, 1991). The strand specificity was based on the RDRP ORF identified in ET1.1 (see below). Only the probe detecting the sense strand hybridized to the nucleic acid. The target nucleic acid was also shown to be RNase A sensitive and DNase I resistant. These studies characterized HEV as a plus-sense single-stranded genome. Strand-specific hybridization to RNA extracted from infected liver also established that the vast majority of intracellular transcript was positive sense. Barring any novel mechanism for virus expression and replication, the negative strand, although not detectable, would be present at a ratio of less than 1:100 when compared with the sense strand.

ET1.1 was documented as exogenous when tested by both Southern blot hybridization and PCR, using genomic DNAs derived from uninfected humans, infected and uninfected cynomolgus macaques, and also the genomic DNAs from *E. coli* and various bacteriophage sources (Reyes *et al.*, 1990a,b). The latter were tested in order to rule out trivial contamination with an exogenous sequence introduced during the numerous enzymatic manipulations performed during cDNA construction and amplification. It was also found that the nucleotide sequence of the ET1.1 clone was not homologous to any entries in the Genbank database. The translated ORF of the ET1.1 clone did, however, demonstrate limited homology with consensus amino acid residues consistent with an RDRP. In fact, the highest degree of homology was detected with beet necrotic yellow vein mosaic virus (Fry *et al.*, 1991a). The similarity between HEV and a multisegmented positive-stranded RNA plant virus is once again a possible indication of a common evolutionary origin for both plant and human RNA viruses. This consensus amino acid motif is shared among all positive-stranded RNA viruses (Kamer and Argos, 1984) and, as noted in Section II,B, was present at the 3' end of the HCV genome. The 1.3-kb clone was therefore presumed to be derived, at least in part, from the nonstructural portion of the viral genome. These experiments satisfied the first three criteria outlined in Table III.

A second altogether different approach was taken to isolate a clone from HEV(Mexico) (Reyes et al., 1991a; Yarbough et al., 1991a,b). cDNA libraries were made directly from a semipurified human stool specimen collected from the Telixtac outbreak (Velazquez et al., 1990). The recovery of cDNA and the construction of representative libraries was assured by the application of SISPA. A cDNA library constructed in λgt11 from such an amplified cDNA population was screened with serum considered to have "high"-titer anti-HEV antibodies, as assayed by direct IF on liver sections from infected cynomolgus monkeys (Krawczynski and Bradley, 1989). Two cDNA clones, denoted 406.3.2 and 406.4.2, were identified by this approach from a total of 60,000 recombinant phage that were screened. The sequence of these clones was subsequently localized to the 3' half of the viral genome by homology comparison to the HEV(Burma) sequence obtained from clones isolated by hybridization screening of libraries with the original ET1.1 clone.

These isolated cDNA epitopes, when used as hybridization probes on Northern blots of RNA extracted from infected cynomolgus liver, gave a somewhat different result when compared to the Northern blots obtained with the ET1.1 probe (Yarbough et al., 1991a). In addition to the single 7.5-kb transcript seen using ET1.1, two additional transcripts of ~3.7 and ~2.0 kb were identified, using either of these epitopes as hybridization probes. These polyadenylated transcripts were identified using the extreme 3'-end epitope clone (406.3.2) as probe and therefore established these transcripts as coterminal with the 3' end of the genome (see Section III,C). One of the epitope clones (406.4.2) was subsequently shown to react in a specific fashion with antisera collected from five geographic epidemics (Somalia, Burma, Mexico, Tashkent, and Pakistan). The 406.3.2 clone reacted with sera from four of these same five epidemics. Both clones reacted with only postinoculation antisera from infected cynomolgus monkeys. The latter experiment confirmed that seroconversion in experimentally infected cynomolgus macaques was related to the isolated exogenous cloned sequence.

Do the epitopes described above represent type-common antigens among all HEV strains? This question has only partially been addressed, using sera from a combination of infected human and animal sources. We have already determined that the immune response to a particular antigenic determinant may not necessarily be a reflection of the immunodominance or type commonality of that determinant among the different geographic variants. Here we must distinguish geographic variants from serotypes, since we cannot yet assert that actual neutralization serotypes exist. We have seen that the immune response between cynomolgus and human samples varies somewhat in that the cynomolgus samples may not recognize the same determinant seen by the infected human (Yarbough *et al.*, 1991a). This may be due to species-specific reactivity or immune tolerance to certain viral determinants, rather than the existence of distinct neutralization serotypes.

C. Expression Strategy

The identification and confirmation of HEV-specific clones led to the acquisition of a contiguous set of overlapping cDNA clones representing the entire viral genome (Tam *et al.*, 1991a,b). The 5′ extent of the viral genome was established by a primer extension experiment. The 3′ end was fixed by the identification of a clone containing a poly(A) tail. Nearly the complete genome was encompassed in five overlapping clones, with the exception of the extreme 5′-end clone that lacked 50 bp. This was cloned separately after specific priming for cDNA synthesis. Sequencing of the viral genome indicated that it was not a member of the picorna-, toga-, or flavivirus families, as it clearly had a discontinuous ORF (see below; Tam *et al.*, 1991a,b). These experiments were performed in the Burma strain. The nearly completed cloning of the Mexico strain demonstrates the same consensus RDRP sequence (Reyes *et al.*, 1991a), genomic organization, and expression strategy as that seen with HEV(Burma) (Huang *et al.*, 1991).

A hypothesis for the expression strategy of HEV was deduced from several lines of evidence, summarized below (Fig. 6). As already noted, consensus residues for the RDRP were identified in the HEV(Burma) strain clone ET1.1. Once a contiguous overlapping set of clones was accumulated, it became clear that the nonstructural elements containing the RDRP, as well as what were identified as consensus residues for the helicase domain, were located in the first large ORF (ORF1). ORF1 covers the 5′ half of the genome and begins at the first methionine, after the 27 bp of the apparent noncoding sequence, and then extends 5079 bp before reaching a termination codon. Beginning 37 bp downstream of the ORF1 stop codon in the plus-1 frame is the second major ORF (ORF2), extending 1980 bp and terminating 68 bp upstream of the point of poly(A) addition. The third forward ORF (in the plus-2 frame) is also utilized by HEV. ORF3 is only 370 bp in length and would not have been predicted to be utilized by the virus were it not for the identification of the immunoreactive cDNA clone 406.4.2

GENOMIC ORGANIZATION

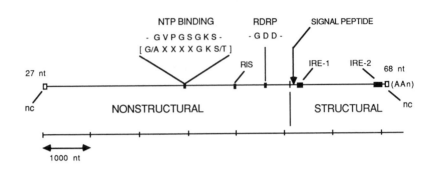

OPEN READING FRAMES

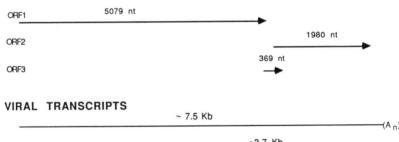

FIG. 6. A model for the genetic organization of HEV is presented. Noncoding regions (nc) of 27 and 68 nt flank the single-stranded positive-sense polyadenylated genome, which can be divided into regions encoding the nonstructural and structural genes. The delineation of both types is based in part on the recognition of sequence elements for two nonstructural gene products. A helicaselike domain (NTP BINDING) and an RNA-directed RNA polymerase domain (RDRP) were identified in the first major open reading frame (ORF1) that extends 5079 bp from the first AUG. The second major ORF (ORF2) begins with a signal peptide and is followed by an arginine/lysine-rich region. ORF2 is therefore similar to other high-pI viral capsid proteins. Also in the 3'-structural region are two immunoreactive epitopes (IRE), identified by a blind immunoscreening protocol (see Fig. 5B). The IRE-1 region is actually encoded in the 369-bp ORF3, which overlaps both ORF1 and ORF2. IRE-2 is located in ORF2. Three different viral transcripts have been detected in cynomolgus monkey-infected liver tissue by Northern blot hybridization. All three are polyadenylated and coterminal with the 3' end. A region has been identified having homology to an RNA transcription initiation site (RIS) described for Sindbis virus (Simmons and Strauss, 1972). GDD, Gly–Asp–Asp.

TABLE VI

GENOMIC CHARACTERISTICS OF HEV

RNA genome	RNase-sensitive DNase-resistant nucleic acid
Single stranded	Strand-specific probe of feces-derived virion RNA
Positive sense	Strand-specific probes and recognition of viral genes
Polyadenylated	Oligo(dT) selection of RNA from infected liver
Subgenomic messages	Two poly(A) messages coterminal with 3' end
Genomic organization	5' Nonstructural; 3' structural

from the Mexico SISPA cDNA library. This epitope confirmed the utilization of ORF3 by the virus, although the means by which this ORF is expressed have not yet been fully elucidated. If we assume that the first methionine is utilized, ORF3 overlaps ORF1 by 1 bp at its 5' end and overlaps ORF2 by 328 bp at its 3' end. ORF2 contains a signal sequence at its extreme 5' end and the broadly reactive 406.3.2 epitope at its 3' end. The first half of this ORF2 also has a high pI value (i.e., >10), similar to that seen with other virus capsid proteins. These data suggest that the ORF2 might be the predominant structural gene of HEV.

The existence of subgenomic transcripts prompted a set of experiments to determine whether these RNAs were produced by splicing from the 5' end of the genome. An analysis using subgenomic probes from throughout the genome, including the extreme 5' end, did not provide evidence for a spliced transcript (Tam et al., 1991a). However, it was discovered that a region of the genome displayed a high degree of homology with a 21-bp segment identified in Sindbis virus as a probable internal initiation site for RNA transcription and used in the production of its subgenomic messages (Simmons and Strauss, 1972). We are currently in the process of determining the exact start sites for the HEV subgenomic messages in order to verify the utilization of this putative internal transcription initiation site. A summary of the genomic characterization of HEV is presented in Table VI.

D. Prevention and Control

Measures to prevent and control HEV-induced ET-NANBH should at this point be aimed at improving the level of health and sanitation in areas known to be affected. The practicality of this solution, however, is an major issue for public health planners in view of the magnitude of the task in those areas principally affected.

A large number of sporadic cases of NANBH were recently recognized among a group of children that were diagnosed as having hepatitis of non-A, non-B origin (Goldsmith *et al.*, 1990). These children were confirmed as having HEV-induced hepatitis by their seroreactivity with the 406.3.2 and 406.4.2 recombinant epitopes. The occurrence of ET-NANBH among this group of patients indicates that overt disease is not limited to the young adult population, as surmised from previously available epidemiological data. It is likely that many of the concepts and accepted dogma, some of which are presented here, will undergo considerable revision as newer more sensitive diagnostic tests become available.

The efficacy of a possible vaccine for HEV is only speculative at this point. Studies into the pathogenesis of this agent will better define the potential for vaccine prophylaxis of ET-NANBH. Several factors in the pathogenesis of ET-NANBH will require elucidation before the ultimate efficacy of a vaccine program can be estimated. The young adult susceptibility to disease might be a reflection of inapparent or subclinical infection in children. This age distribution could also result from immune enhancement of infection, as recognized with dengue virus (Halstead, 1988), thereby increasing the severity of subsequent infection in a susceptible population. The occurrence of cyclical outbreaks/epidemics in what is normally considered an endemic disease might result from the introduction of highly virulent subtypes into the region and the absence of humoral surveillance against them. Alternatively, "escape" mutants of the normally endogenous HEV serotype might be generated and lead to an outbreak. It has been postulated by some that waning antibody titers could also lead to a reduced immune status and hence an increased susceptibility to viral infection. As with HEV, it has been postulated that Norwalk virus is a member of the Caliciviridae (reviewed by Greenberg *et al.*, 1990). It has been clearly demonstrated that, after Norwalk virus-induced gastroenteritis, immunity (in certain individuals) is not lifelong and reinfection is possible. The recent cloning of the Norwalk virus (Jiang *et al.*, 1990; Matsui *et al.*, 1991) will permit a comparison of the two viruses at the molecular level to determine the extent to which they are related.

IV. Conclusions

There has been rapid progress in recent years into elucidating the viral causation of hepatitis. The development and application of different cloning strategies have led to the identification of two different viruses that were formerly classified as NANBH agents. The molecular virology of hepatitis in general has now led to the identification and characterization of at least five different viruses that appear

TABLE VII

SUMMARY OF HCV AND HEV[a]

Disease	Virus	Genome	Size (Kb)[b]	Type
PT-NANBH	HCV	ss, +, RNA	~10	Flavi/pesti(?)
ET-NANBH	HEV	ss, +, pA, RNA	~7.5	Calici(?)

[a] PT-NANBH, Parenterally transmitted non-A, non-B hepatitis; ET-NANBH, enterically transmitted non-A, non-B hepatitis; HCV, hepatitis C virus; HEV, hepatitis E virus; ss, single strand; +, positive sense; pA, polyadenylated; flavi, Flaviviridae; pesti, Pestiviridae; calici, Caliciviridae.

[b] Size for HCV was determined by the original Northern blot hybridization of Choo et al. (1989); Size for HEV was determined by Northern blot (Reyes et al., 1990a).

to be primarily hepatotropic. The description of four of these agents represents the initial discovery of that particular agent as a human pathogen. The cloning and characterization of the HBV agent (Galibert et al., 1979) led to the elucidation of the prototype member of the Hepadnaviridae. The molecular characterization of HDV (Wang et al., 1986) also indicated the existence of a unique viroid/satellite-like agent of humans requiring HBV for its particle maturation and dissemination. Molecular cloning has now yielded two very different agents of NANBH.

The data available to date seem to indicate that the HCV genome, although related to both pesti- and flaviviruses, might, in fact, represent an entirely new, but related, subgroup to these positive-stranded RNA agents. The genetic organization and expression strategy of the HEV genome clearly indicate that it is a new representative of a positive-stranded RNA virus family that has not been previously described as a human pathogen. A summary of our current understanding of HCV and HEV is presented in Table VII.

The possibility of a different parenterally transmitted agent has been alluded to above. The biophysical characterization (chloroform resistance) and biological properties (parenteral transmission) of this new agent would indicate that it is possibly an entirely different type of NANBH. It is now possible to successfully isolate molecular clones of human pathogenic agents in the absence of an *in vitro* culture system (e.g., HEV and HCV) and even in the absence of an *in vitro* system or an animal model for viral propagation (e.g., Norwalk virus). The application of these techniques to the next NANBH or what may more appropriately be called the next non-A, -B, -C, -D, -E virus, should lead to even further insights into the nature of virus-induced hepatitis in humans.

ACKNOWLEDGMENTS

We thank Jung-suh P. Kim, K. E. Fry, and M. Lovett for critical review of this manuscript; K. E. Fry, A. W. Tam, and Qi-ming Sun for assistance in the nucleic acid sequence analysis; Frances F. McDonald and L. M. Young for assistance in manuscript preparation; and J. Fernandez and Genelabs Visual Arts Department for assistance in figure preparation.

REFERENCES

Alter, M.J., and Sampliner, R. E. (1989). *N. Engl. J. Med.* **321,** 1538–1540.
Alter, H. J., Purcell, R. H., Holland, P. V., Feinstone, S. M., Morrow, A. G., and Moritsugu, Y. (1975). *Lancet* **2,** 838–841.
Alter, H. J., Purcell, R. H., Holland, P. V., and Popper, H. (1978). *Lancet* **1,** 459–463.
Alter, H. J., Purcell, R. H. Feinstone, S. M., and Tegtmeier, G. E. (1982). *In* "Viral Hepatitis: 1981 International Symposium" (W. Szmuness, H. J. Alter, and J. E. Maynard, eds.), pp. 279–294. Franklin Inst. Press, Philadelphia, Pennsylvania.
Alter, H. J., Purcell, R. H., Shih, J. W., Melpolder, J. C., Houghton, M., Choo, Q.-L., and Kuo, G. (1989). *N. Engl. J. Med.* **321,** 1494–1500.
Alter, M. J., Coleman, P. J., Alexander, W. J., Kramer, E., Miller, J. K., Mandel, E., Hadler, S. C., and Margolis, H. S. (1989). *JAMA, J. Am. Med. Assoc.* **262,** 1201–1205.
Andjaparidze, A. G., Balayan, M. S., Savinov, A. P., Braginskiy, D. M., Poleschuk, V. F., and Zamyatina, N. A. (1986). *Vopr. Virusol.* **1,** 73–80.
Arankalle, V. A., Ticehurst, J., Sreenivasan, M. A., Kapikian, A. Z., Popper, H., Pavri, K. M., and Purcell, R. H. (1988). *Lancet* **2,** 550–554.
Arima, T., Nagashima, H., Murakami, S., Kaji, C., Fujita, J., Shimomura, H., and Tsuji, T. (1989a). *Gastroenterol. Jpn.* **24,** 540–544.
Arima, T., Takamizawa, A., Mori, C., Muramaki, S., Kaji, C., and Fujita, J. (1989b). *Gastroenterol. Jpn.* **25,** 545–548.
Arima, T., Mori, C., Takamizawa, A., Shimomura, H., and Tsuji, T. (1990). *Gastroenterol. Jpn.* **25,** 218–222.
Balayan, M. S., Andjaparidze, A. G., Savinskaya, S. S., Ketiladze, E. S., Braginsky, D. M., Savinov, A. P., and Poleschuk, V. F. (1983). *Intervirology* **20,** 23–31.
Beasley, R. P. (1982). *Hepatology* **2,** 21S–26S.
Bellabes, H., Benatallah, A., and Bourguermouh, A. (1984). *In* "Viral Hepatitis and Liver Disease" (G. N. Vyas, J. L. Dienstag, and J. H. Hoofnagle, eds.), p. 637. Grune & Stratton, Orlando, Florida.
Blumberg, B. S., Alter, H. J., and Visnich, S. (1965). *JAMA, J. Am. Med. Assoc.* **191,** 101–106.
Boyd, D., and Beckwith, J. (1990). *Cell* **62,** 1031–1033.
Bradley, D. W. (1990a). *Prog. Med. Virol.* **37,** 101–135.
Bradley, D. W. (1990b). *Br. Med. Bull.* **46,** 442–461.
Bradley, D. W., and Balayan, M. S. (1988). *Lancet* **1,** 819.
Bradley, D. W., Maynard, J. E., Hindman, S. H., Hornbeck, C. L., Fields, H. A., McCaustland, K. A., and Cook, E. H. (1977). *J. Clin. Microbiol.* **5,** 521–530.
Bradley, D. W., Cook, E. H., Maynard, J. E., McCaustland, K. A., Ebert, J. W., Dolana, G. H., Petzel, R. A., Kantor, R. J., Heilbrunn, A., Fields, H. A., and Murphy, B. L. (1979a). *J. Med. Virol.* **3,** 253–269.
Bradley, D. W., Fields, H. A., McCaustland, K. A., Maynard, J. E., Decker, R. H., Whittington, R., and Overby, L. R. (1979b). *J. Clin. Microbiol.* **9,** 120–127.
Bradley, D. W., Maynard, J. E., Cook, E. H., Ebert, J. W., Gravelle, C. R., Tsiquaye, K.

N., Kessler, H., Zuckerman, A. J., Miller, M. F., Ling, C.-M., and Overby, L. R. (1980). *J. Med. Virol.* **6,** 185–201.

Bradley, D. W., Maynard, J. E., Popper, H., Cook, E. H., Ebert, J. W., McCaustland, K. A., Schable, C. A., and Fields, H. A. (1983). *J. Infect. Dis.* **148,** 254–265.

Bradley, D. W., McCaustland, K. A., Cook, E. H., Schable, C. A., Ebert, J. W., and Maynard, J. E. (1985). *J. Infect. Dis.* **148,** 254–265.

Bradley, D. W., Krawczynski, K., Cook, E. H., Jr., McCaustland, K. A., Humphrey, C. D., Spelbring, J. E., Myint, H., and Maynard, J. E. (1987). *Proc. Natl. Acad. Sci. U.S.A.* **84,** 6277–6281.

Bradley, D. W., Andjaparidze, A., Cook, E. H., Jr., McCaustland, K., Balayan, M., Stetler, H., Velasquez, O., Robertson, B., Humphrey, C., and Kane, M. (1988). *J. Gen. Virol.* **69,** 731–738.

Bradley, D. W., Krawczynski, K., Ebert, J. W., McCaustland, K. A., Choo, Q.-L., Houghton, M., and Kuo, G. (1990). *Gastroenterology* **99,** 1054–1060.

Bradley, D. W., *et al.* (1991). Manuscript in preparation.

Branch, A. D., Berenfield, B. J., Baroudy, B. M., Wells, F. V., Gerin, J. L., and Robertson, H. D. (1989). *Science* **243,** 649–652.

Bruix, J., Calvet, X., Costa, J., Ventura, M., Bruguera, M., Castillo, R., Barrera, J. M., Ercilla, G., Sanchez-Tapias, J. M., Vall, M., Bru, C., and Rodes, J. (1989). *Lancet* **2,** 1004–1006.

Burk, K. H., Oefinger, P. E., and Dreesman, G. R. (1984). *Proc. Natl. Acad. Sci. U.S.A.* **81,** 3195–3199.

Choo, Q.-L., Kuo, G., Weiner, A. J., Overby, L. R., Bradley, D. W., and Houghton, M. (1989). *Science* **244,** 359–362.

Choo, Q.-L., Weiner, A. J., Overby, L. R., Kuo, G., Houghton, M., and Bradley, D. W. (1990). *Br. Med. Bull.* **46,** 423–441.

Choo, Q.-L., *et al.* (1991a). *In* "VII International Symposium on Viral Hepatitis and Liver Disease" (F. B. Hollinger, S. M. Lemon, and H. S. Margolis, eds.). Williams & Wilkins, Baltimore, Maryland. In press.

Choo, Q. L., Richman, K. H., Han, J. H., Berger, K., Lee, C., Dong, C., Gallegos, C., Coit, D., Medina-Selby, A., Barr, P. J., Weiner, A. J., Bradley, D. W., Kuo, G., and Houghton, M. (1991b). *Proc. Natl. Acad. Sci. U.S.A.* **88,** 2451–2455.

Cohen, J. I., Ticehurst, J. R., Feinstone, S. M., Rosenblum, B., and Purcell, R. H. (1988). *In* "Viral Hepatitis and Liver Disease" (A. J. Zuckerman, ed.), pp. 67–69. Liss, New York.

Coia, G., Parker, M. D., Speight, G., Byrne, M. E., and Westaway, E. G. (1988). *J. Gen. Virol.* **69,** 1–21.

Collett, M. S., Larson, R., Gold, C., Strick, D., Anderson, D. K., and Purchio, A. F. (1988a). *Virology* **165,** 191–199.

Collett, M. S., Larson, R., Belzer, S. K., and Retzel, E. (1988b). *Virology* **165,** 200–208.

Collett, M. S., Anderson, D. K., and Retzel, E. (1988c). *J. Gen. Virol.* **69,** 2637–2643.

Colombo, M., Choo, Q.-L., Del Ninno, E., Dioguardi, N., Kuo, G., Donato, M. F., Tommasini, M. A., and Houghton, M. (1989). *Lancet* **2,** 1006–1008.

Davis, G. L., Balart, L. A., Schiff, E. R., Lindsay, K., Bodenheimer, H. C., Perrillo, R. P., Carey, W., Jacobson, I. M., Payne, J., Dienstag, J. L., Van Thiel, D. H., Tamburro, C., Lefkowitch, J., Albrecht, J., Meschievitz, C., Ortego, T. J., Gibas, A., and the Hepatitis Intervention Therapy Group (1989). *N. Engl. J. Med.* **321,** 1501–1506.

De Cock, K. M., Bradley, D. W., Sandford, N. L., Govindarajan, S., Maynard, J. E., and Redeker, A. G. (1987). *Ann. Intern. Med.* **106,** 227–230.

Di Bisceglie, A. M., and Hoofnagle, J. H. (1990). *Am. J. Gastroenterol.* **85,** 650–654.

Di Bisceglie, A. M., Martin, P., Kassianides, C., Lisker-Melman, M., Murray, L., Waggoner, J., Goodman, Z., Banks, S. M., and Hoofnagle, J. H. (1989). *N. Engl. J. Med.* **321,** 1506–1510.

Dienstag, J. L. (1990). *Gastroenterology* **99**, 1177–1180.

Dormier, L. L., Franklin, K. M., Shahabuddin, M., Hellman, G. M., Overmeyer, J. H., Hiremath, S. T., Siaw, M. F. E., Lomonossoff, G. P., Shaw, J. G., and Rhoads, R. E. (1986). *Nucleic Acids Res.* **14**, 5417–5430.

Enomoto, N., Takada, A., Nakao, T., and Date, T. (1990). *Biochem. Biophys. Res. Commun.* **170**, 1021–1025.

Esteban, J. I., Esteban, R., Viladomiu, L., Lopez-Talavera, J. C., Gonzalez, A., Hernandez, J. M., Roget, M., Vargas, V., Genesca, J., Buti, M., Guardia, J., Houghton, M., Choo, Q.-L., and Kuo, G. (1989). *Lancet* **2**, 294–297.

Esteban, J. I., Gonzalez, A., Hernandez, J. M., Viladomiu, L., Sanchez, C., Lopez-Talavera, J. C., Lucea, D., Martin-Vega, C., Vidal, X. J., Esteban, R., and Guardia, J. (1990). *N. Engl. J. Med.* **323**, 1107–1112.

European Patent Office Publ. #318,216. Appl. #88310922.5, 5/31/89 Bulletin 89/22 (M. Houghton, Q.-L. Choo, and G. Kuo, inventors).

European Patent Office Publ. #388,232, Appl. #90302866.0, 9/19/90 Bulletin 90/38 (M. Houghton, Q.-L Choo, and G. Kuo, inventors).

Feinstone, S. M., Kapikian, A. Z., and Purcell, R. H. (1973). *Science* **182**, 1026–1028.

Feinstone, S. M., Kapikian, A. Z., Purcell, R. H., Alter, H. J., and Holland, P. V. (1975). *N. Engl. J. Med.* **292**, 767–770.

Feinstone, S. M., Mihalik, K. B., Kamimura, T., Alter, H. J., London, W. T., and Purcell, R. H. (1983). *Infect. Immun.* **41**, 816–821.

Fortier, D., Treadwell, T. L., and Koff, R. S. (1989). *N. Engl. J. Med.* **320**, 1281–1282.

Foung, S. K. H., Perkins, S., Raubitschek, A. A., Larrick, J., Lizak, G., Fisshwild, D., Engleman, E. G., and Grumet, C. (1984). *J. Immunol. Methods* **70**, 83–90.

Fry, K. E., Tam, A. W., Smith, M. M., Kim, J. P., Luk, K.-C., Young, L. M., Piatak, M., Feldman, R. A., Purdy, M. A., McCaustland, K. A., Bradley, D. W., and Reyes, G. R. (1991a). In press.

Fry, K. E., Sun, Q., Fernandez, J., Kim, J. P., and Reyes, G. R. (1991b). Manuscript in preparation.

Galibert, F., Mandart, E., Fitoussi, F., Tiollais, P., and Charnay, P. (1979). *Nature (London)* **281**, 646–650.

Garson, J. A., Tedder, R. S., Briggs, M., Tuke, P., Glazebrook, J. A., Trute, A., Parker, D., Barbara, J. A. J., Contreras, M., and Aloysius, S. (1990). *Lancet* **335**, 1419–1422.

Gilliam, J. H., Geisinger, K. R., and Richter, J. E. (1984). *Ann. Intern. Med.* **101**, 794–795.

Goldsmith, R., *et al.* (1991). Manuscript in preparation.

Gorbalenya, A. E., Donchenko, A. P., Koonin, E. V., and Blinov, V. M. (1989a). *Nucleic Acids Res.* **17**, 3889–3897.

Gorbalenya, A. E., Koonin, E. V., Donchenko, A. P., and Blinov, V. M. (1989b). *Nucleic Acids Res.* **17**, 4713–4730.

Greenberg, H. B., Skaar, M., and Monroe, S. S. (1990). In "Viral Diarrheas of Man and Animals" (L. J. Saif and K. W. Thiel, eds.), pp. 135–157. CRC Press, Boca Raton, Florida.

Halstead, S. B. (1988). *Science* **239**, 476–481.

Hasan, F., Jeffers, L. J., De Medina, M., Reddy, K. R., Parker, T., Schiff, E. R., Houghton, M., Choo, Q.-L., and Kuo, G. (1990). *Hepatology* **12**, 598–591.

He, L. F., Alling, D., Popkin, T., Alter, H. J., and Purcell, R. H. (1987). *J. Infect. Dis.* **156**, 636–640.

Hillis, A. Shrestha, S. M., and Saha, N. K. (1973). *J. Nepal Med. Assoc.* **11**, 145–151.

Hoofnagle, J., and Alter, H. J. (1984). In "Viral Hepatitis and Liver Disease" (G. N. Vyas, J. L. Dienstag, and J. H. Hoofnagle, eds.), pp. 97–113. Grune & Stratton, Orlando, Florida.

Hoofnagle, J. H., Gerety, R. J., Tabor, E., Feinstone, S. M., Barker, L. F., and Purcell, R. H. (1977). *Ann. Intern. Med.* **87,** 14–20.

Huang, C.-C., *et al.* (1990). Manuscript in preparation.

Jackson, D., Tabor, E., and Gerety, R. J. (1979). *Lancet* **1,** 1249–1250.

Jacob, J. R., Eichberg, J. W., and Lanford, R. E. (1989). *Hepatology* **10,** 921–927.

Jacob, J. R., Burk, K. H., Eichberg, J. W., Dreesman, G. R., and Lanford, R. E. (1990). *J. Infect. Dis.* **161,** 1121–1127.

Jiang, X., Graham, D. Y., Wang, K., and Estes, M. K. (1990). *Science* **250,** 1580 –1583.

Kamer, G., and Argos, P. (1984). *Nucleic Acids Res.* **12,** 7269.

Kamimura, T., Ponzetto, A., Bonino, F., Feinstone, S. M., Gerin, J. L., and Purcell, R. H. (1983). *Hepatology* **3,** 631–637.

Kane, M. A., Bradley, D. W., Shrestha, S. M., May (1984). *JAMA, J. Am. Med. Assoc.* **252,** 3140–3145.

Kaneko, S., Unoura, M., Kobayashi, K., Kuno, K., Murakami, S., and Hattori, N. (1990). *Lancet* **1,** 976.

Kato, N., Hijikata, M., Ootsuyama, Y., Nakagawa, M., Ohkoshi, S., Sugimura, T., and Shimotohno, K. (1990). *Proc. Natl. Acad. Sci. U.S.A.* **87,** 9524–9528.

Khuroo, M. S. (1980). *Am. J. Med.* **68,** 818–824.

Khuroo, M. S., Duermeyer, W., Zargar, S. A., Ahanger, M. A., and Shah, M. A. (1983). *Am. J. Epidemiol.* **118,** 360–364.

Kim. J. P., Moeckli, R. A., Warmerdam, M., Yun, K. M., Young, L. M., and Reyes, G. R. (1991). Manuscript in preparation.

Kiyosawa, K., Akahane, Y., Nagata, A., and Furuta, S. (1984). *Am. J. Gastroenterol.* **79,** 771–781.

Kiyosawa, K., Sodeyama, T., Tanaka, E., Gibo, E., Yoshizawa, K., Nakano, Y., Furuta, S., Akahane, Y., Nishioka, K., Purcell, R. H., and Alter, H. J. (1990). *Hepatology* **12,** 671–675.

Klein, P., Kanehisa, M., and DeLisi, C. (1985). *Biochim. Biophys. Acta* **815,** 468–476.

Koretz, R. L. (1990). *Hepatology* **12,** 613–615.

Kotwal, G., *et al.* (1991). Manuscript in preparation.

Krawczynski, K. (1989). *Liver Update* **3,** 5–6.

Krawczynski, K., and Bradley, D. W. (1989). *J. Infect. Dis.* **159,** 1042–1049.

Kremsdorf, D., Porchon, C., Kim, J. P., Reyes, G. R., and Brechot, C. (1991). In press.

Kubo, Y., Takeuchi, K., Boonmar, S., Katayama, T., Choo, Q.-L., Kuo, G., Weiner, A. J., Bradley, D. W., Houghton, M., Saito, I., and Miyamura, T. (1989). *Nucleic Acids Res.* **17,** 10367–10372.

Kuo, G., Choo, Q.-L., Alter, H. J., Gitnick, G. L., Redeker, A. G., Purcell, R. H., Miyamura, T., Dienstag, J. L., Alter, M. J., Stevens, C. E., Tegtmeier, G. E., Bonino, F., Colombo, M., Lee, W.-S., Kuo, C., Berger, K., Shuster, J. R., Overby, L. R., Bradley, D. W., and Houghton, M. (1989). *Science* **244,** 362–364.

Lain, S., Reichmann, J. L., Martin, M. T., and Garcia, J. A. (1989). *Gene* **82,** 357–362.

Lanford, R. E., Carey, K. D., Estlack, L. E., Smith, G. C., and Hay, R. V. (1989). *In Vitro Cell. Dev. Biol.* **25,** 174–182.

Maeno, M., Kaminaka, K., Sugimoto, H., Esumi, M., Hayashi, N., Komatsu, K., Abe, K., Sekiguchi, S., Yano, M., Mizumo, K., and Shikata, T. (1990). *Nucleic Acids Res.* **18,** 2685–2689.

Mandl, C. W., Guirakhoo, F., Holzmann, H., Heinz, F. X., and Kunz, C. (1989). *J. Virol.* **63,** 564–571.

Matsui, S. M., Kim, J. P., Greenberg, H. B., Su, W., Sun, Q., Johnson, P. C., DuPont, H. L., Oshiro, L., and Reyes, G. R. (1991). *J. Clin. Invest.* **87,** 1456–1461.

Meyers, G., Rumenapf, T., and Thiel, H.-J. (1989). *Virology* **171,** 555–567.

Miller, R. H., and Purcell, R. H. (1990). *Proc. Natl. Acad. Sci. U.S.A.* **87,** 2057–2061.

Moeckli, R. A., Warmerdam, M., Young, L. M., Sun, Q., Kim, J. P., and Reyes, G. R. (1991). *In* "VII International Symposium on Viral Hepatitis and Liver Disease" (F. B. Hollinger, S. M. Lemon, and H. S. Margolis, eds.). Williams & Wilkins, Baltimore, Maryland. In press.

Mosley, J. W. (1975). *JAMA, J. Am. Med. Assoc.* **233**, 967–969.

Mosley, J. W., Redeker, A. G., Feinstone, S. M., and Purcell, R. H. (1977). *N. Engl. J. Med.* **296**, 75–78.

Mosley, J. W., Aach, R. D., Hollinger, F. B., Stevens, C. E., Barbosa, L. H., Nemo, G. J., Holland, P. V., Bancroft, W. H., Zimmerman, H. J., Kuo, G., Choo, Q.-L., and Houghton, M. (1990). *JAMA, J. Am. Med. Assoc.* **263**, 77–78.

Myint, H., Soe, M. M., Khin, T., Myint, T. M., and Tin, K. M. (1985). *Am. J. Trop. Med. Hyg.* **34**, 1183–1189.

Nilsson, I. N., and von Heijne, G. (1990). *Cell* **62**, 1135–1141.

Nishioka, K. (1990). *Ortho Diagnostics Update* **4**, 2–3.

Nowak, T., and Wengler, G. (1987). *Virology* **156**, 127–137.

Okamoto, H., Okada, S., Sugiyama, Y., Yotsumoto, S., Tanaka, T., Yoshizawa, H., Tsuda, F., Miyakawa, Y., and Mayumi, M. (1990). *Jpn. J. Exp. Med.* **60**, 167–177.

Pfeifer, U., Thomssen, R., Legler, K., Boettcher, U., Gerlich, W., Weinman, E., and Klinge, O. (1980). *Virchows Arch. B* **33**, 233–243.

Prince, A. M. (1968). *Proc. Natl. Acad. Sci. U.S.A.* **60**, 814–821.

Prince, A. M., Brotman, B., Grady, G. F., Kuhns, W. J., Hazzi, C., Levine, R. W., and Millian, S. J. (1974). *Lancet* **2**, 241–246.

Prusiner, S. B. (1989). *Annu. Rev. Microbiol.* **43**, 345–374.

Purcell, R. H. (1989). *Cancer Detect. Prev.* **14**, 203–207.

Purcell, R. H., and Ticehurst, J. (1988). *In* "Viral Hepatitis and Liver Disease" (A. J. Zuckerman, ed.), pp. 131–137. Liss, New York.

Resnick, R. H., Stone, K., and Antonioli, D. (1983). *Dig. Dis. Sci.* **28**, 908–911.

Reyes, G. R., Purdy, M. A., Kim, J. P., Luk, K.-C., Young, L. M., Fry, K. E., and Bradley, D. W. (1990a). *Science* **247**, 1335–1339.

Reyes, G. R., Purdy, M. A., Kim, J. P., Young, L. M., Fry, K. E., Luk, K.-C., and Bradley, D. W. (1990b). *In* "Viral Hepatitis and Hepatocellular Carcinoma," (J.-L. Sung and D.-S. Chen, eds.), pp. 249–255. Excerpta Medica, Hong Kong.

Reyes, G. R., Huang, C.-C., Yarbough, P., Young, L. M., Tam, A. W., Moeckli, R. A., *et al.* (1991a). *In* "Viral Hepatitis C, D and E" (T. Shikata, R. H. Purcell, and T. Uchida, eds.), pp. 237–245. Elsevier, Amsterdam. In press.

Reyes, G. R., Young, L. M., Sharma, V., Fernandez, J., Fry, K. E., and Kim, J. P. (1991b). Manuscript in preparation.

Rice, C. M., Strauss, E. G., and Strauss, J. H. (1986). *In* "The Togaviridae and Flaviviridae" (S. Schlesinger and M. J. Schlesinger, eds.), p. 279. Plenum, New York.

Riley, J., Ogilvie, D., Finniear, R., Jenner, D., Powell, S., Anand, R., Smith, J. C., and Markham, A. F. (1990). *Nucleic Acids Res.* **18**, 2887–2890.

Rizzetto, M. (1983). *Hepatology* **3**, 729–737.

Saito, I., Miyamura, T., Ohbayashi, A., Harada, H., Katayama, T., Kikuchi, S., Watanabe, Y., Koi, S., Onji, M., Ohta, Y., Choo, Q.-L., Houghton, M., and Kuo, G. (1990). *Proc. Natl. Acad. Sci. U.S.A.* **87**, 6547–6549.

Schaff, A., Gerety, R. J., Grimley, P. M., Iwarsson, S. A., Jackson, D. R., and Tabor, E. (1985). *J. Exp. Pathol.* **2**, 25–36.

Sergeev, N. W., Paktoris, E. A., Ananev, W. A., Sinajko, G. A., Antinova, A. I., and Semenov, E. P. (1957). *Sov. Healthcare Kirgizii* **5**, 16–23.

Shimizu, Y. K., and Purcell, R. H. (1989). *Hepatology* **10**, 764–768.

Shimizu, Y. K., Feinstone, S. M., Purcell, R. H., Alter, H. J., and London, W. T. (1979). *Science* **205**, 197–200.

Shimizu, Y. K., Oomura, M., Abe, K., Uno, M., Yamada, E., Ono, Y., and Shikata, T. (1985). *Proc. Natl. Acad. Sci. U.S.A.* **82**, 2138–2142.

Shimizu, Y. K., Purcell, R. H., Gerin, J. L., Feinstone, S. M., Ono, Y., and Shikata, T. (1986). *Hepatology* **6**, 1329–1333.

Shimizu, Y. K., Weiner, A. J., Rosenblatt, J., Wong, D. C., Shapiro, M., Popkin, T., Houghton, M., Alter, H. J., and Purcell, R. H. (1990). *Proc. Natl. Acad. Sci. U.S.A.* **87**, 6441–6444.

Shrestha, S. M., and Malia, D. S. (1975). *J. Nepal Med. Assoc.* **13**, 58–69.

Simmons, D. T., and Strauss, J. H. (1972). *J. Mol. Biol.* **71**, 615–631.

Speight, G., Coia, G., Parker, M. D., and Westaway, E. G. (1988). *J. Gen. Virol.* **69**, 23–34.

Sreenivasan, M. A., Banerjee, K., Pandya, P. G., Kotak, R. R., Pandyr, P. M., Desai, N. J., and Vaghela, L. H. (1978). *Indian J. Med. Res.* **67**, 197–206.

Steimer, K., Higgin, K. W., Powers, M. A., Stephans, J. C., Gyenes, A., George-Nascimento, C., Lucin, P. A., Bar, P. J., Hallewell, R. A., and Sanchez-Pescador (1986). *J. Virol.* **58**, 9–16.

Stevens, C. E., Aach, R. D., Hollinger, F. B., Mosley, J. W., Szmuness, W., Kahn, R., Werch, J., and Edwards, V. (1984). *Ann. Intern. Med.* **101**, 733–738.

Stevens, C., Taylor, P. E., Pindyck, J., Choo, Q.-L., Bradley, D. W., Kuo, G., and Houghton, M. (1990). *JAMA, J. Am. Med. Assoc.* **263**, 49–53.

Sureau, C., Romet-Lemonne, J. L., Mullins, J. I., and Essex, M. (1986). *Cell* **47**, 37–47.

Tabor, E., Gerety, R. J., Drucker, J. A., Seeff, L. B., Hoofnagle, J. H., Jackson, D. R., April, M., Barker, L. F., and Pineda-Tamondong, G. (1978). *Lancet* **1**, 463–466.

Takamizawa, A., Mori, C., Fuke, I., Manabe, S., Murakami, S., Fujita, J., Onishi, E., Andoh, T., Yoshida, I., and Okayama, H. (1991). *J. Virol.* **65**, 1105–1113.

Takeuchi, K., Kubo, Y. Boonmar, S., Watanabe, Y., Katayama, T., Choo, Q.-L., Kuo, G., Houghton, M., Saito, I., and Miyamura, T. (1990). *Nucleic Acids Res.* **18**, 4626.

Tam, A. W., Smith, M. W., Guerra, M. E., Huang, C.-C., Bradley, D. W., Fry, K. E., and Reyes, G. R. (1991a). In press.

Tam, A. W., Smith, M. M., Guerra, M. E., Bradley, D. W., Fry, K. E., and Reyes, G. R. (1991b). *In* "VII International Symposium on Viral Hepatitis and Liver Disease" (F. B. Hollinger, S. M. Lemon, and H. S. Margolis, eds.). Williams & Wilkins, Baltimore, Maryland. In press.

Tandon, B. N., Joshi, Y. K., Jain, S. K., Gandhi, B. M., Mathiesen, L. R., and Tandon, H. D. (1982). **75**, 739–744.

Tsiquaye, K. N., Amini, S., Kessler, H., Bird, R. G., Tovey, G., and Zuckerman, A. J. (1981). *Br. J. Exp. Pathol.* **62**, 41–51.

Uchida, T., Win, K. M., Suzuki, K., Komatsu, K., Iida, F., Shikata, T., Rikihisa, T., Mizuno, K., Soe, S., Myint, H., Tin, K. M., and Nakane, K. (1990a). *Japan J. Exp. Med.* **60**, 13–21.

Uchida, T., Suzuki, K., Komatsu, K., Iida, F., Shikata, T., Rikihisa, T., Mizuno, K., Soe, S., Win, K. M., and Nakane, K. (1990b). *Jpn. J. Exp. Med.* **60**, 23–29.

van der Poel, C. L., Reesink, H. W., Lelie, P. N., Leentvaar-Kuypers, A., Choo, Q.-L., Kuo, G., and Houghton, M. (1989). *Lancet* **1**, 297–298.

Velazquez, O., Stetler, H. C., Avila, C., Ornelas, G., Alvarez, C., Hadler, S. C., Bradley, D. W., and Sepulveda, J. (1990). *JAMA, J. Am. Med. Assoc.* **263**, 3281–3285.

Viswanathan, R. (1957). *Indian J. Med. Res., Suppl.* **45**, 1–30.

von Heijne, G. (1985). *J. Mol. Biol.* **184**, 99–105.

Walsh, J. H., Yalow, R, and Berson, S. A. (1970). *J. Infect. Dis.* **121**, 550–554.

Wang, K., Choo, Q.-L., Weiner, A. J., Gerin, J. L., and Houghton, M. (1986). *Nature (London)* **323**, 508–514.

Weiner, A. J., Kuo, G., Bradley, D. W., Bonino, F., Saracco, G., Lee, C., Rosenblatt, J., Choo, Q.-L., and Houghton, M. (1990). *Lancet* **1,** 1–3.

Weiner, A. J., Braver, M. J., Rosenblatt, J., Richman, K. H., Tung, J., Crawford, K., Bonino, F., Saracco, G., Choo, Q.-L., Houghton, M., and Han, J. H. (1991). *Virology* **180,** 842–848.

Wong, D. C., Purcell, R. H., Sreenivasan, M. A., Prasad, S. R., and Pavri, K. M. (1980). *Lancet* **2,** 876–878.

Yarbough, P. O., Tam, A. W., Fry, K. E., Krawczynski, K., McCaustland, K. A., Bradley, D. W., and Reyes, G. R. (1991a). Submitted for publication.

Yarbough, P. O., Tam, A. W., Krawczynski, K., Fry, K. E., McCaustland, K. A., Miller, A., Fernandez, J., Huang, C.-C., Bradley, D. W., and Reyes, G. R. (1991b). *In* "VII International Symposium on Viral Hepatitis and Liver Disease" (F. B. Hollinger, S. M. Lemon, and H. S. Margolis, eds.). Williams & Wilkins, Baltimore, Maryland. In press.

ADVANCES IN VIRUS RESEARCH, VOL. 40

THE 5'-UNTRANSLATED REGION OF PICORNAVIRAL GENOMES

Vadim I. Agol

Institute of Poliomyelitis and Viral Encephalitides
U.S.S.R. Academy of Medical Sciences
Moscow Region 142782
and Moscow State University
Moscow 119899, U.S.S.R.

I. Introduction

Picornaviruses are small naked icosahedral viruses with a single-stranded RNA genome of positive polarity (Rueckert, 1985; Koch and

Koch, 1985). According to current taxonomy (Cooper *et al.*, 1978; Matthews, 1982), the family includes four genera: *Enterovirus* (polioviruses, coxsackieviruses, echoviruses, and other enteroviruses), *Rhinovirus, Cardiovirus* [encephalomyocarditis virus (EMCV), mengovirus, Theiler's murine encephalomyelitis virus (TMEV)], and *Aphthovirus* [foot-and-mouth disease viruses (FMDV)]. There are also some as yet unclassified picornaviruses [e.g., hepatitis A virus (HAV)], which should certainly be assessed as a separate genus.

Studies on the molecular biology of picornaviruses might be divided into two periods: those before and after the first sequencing of the poliovirus genome. This milestone event was accomplished in the laboratories of E. Wimmer (Kitamura *et al.*, 1981) and D. Baltimore (Racaniello and Baltimore, 1981). The knowledge of the primary structure of poliovirus RNA not only solved (or helped to solve) many outstanding problems, but also confronted researchers with new puzzles. The 5'-untranslated region (5-UTR) of the viral genome was one of such unexpected problems. This segment proved to be immensely long: about 750 nucleotides, or ~10% of the genome length. There were also other unusual features (e.g., multiple AUG triplets preceding the single open reading frame (ORF) that encodes the viral polyprotein). A question arose: "What useful purpose could this giant noncoding genomic segment serve?"

In this chapter I attempt to answer this question. As we shall see, the picornaviral 5-UTRs are not only involved in such essential events as the synthesis of viral proteins and RNAs (which could be expected to some extent, although some of the underlying mechanisms appeared to be quite a surprise), but also may determine diverse biological phenotypes, from the plaque size or thermosensitivity of reproduction to attenuation of neurovirulence. Furthermore, a close inspection of the 5-UTR structure unravels certain hidden facets of the evolution of the picornaviral genome. Finally, the conclusions drawn from the experiments with the picornaviral 5-UTRs provide important clues for understanding the functional capabilities of the eukaryotic ribosomes.

II. PRIMARY STRUCTURE

A. *Size and Some Gross Features*

As mentioned in Section I, the initial sequencing of the poliovirus RNA revealed the presence of a long 5-UTR, about 750 nucleotides in length (Kitamura *et al.*, 1981; Racaniello and Baltimore, 1981). Subsequent analyses showed that the 5-UTR size was remarkably constant

TABLE I

SOME FEATURES OF PICORNAVIRAL 5-UTRs[a,b]

Virus	Size (nucleotides)	Poly(C)	Number of AUGs	Initiator AUG context
Enterovirus				
Polioviruses (1–3)	742–747	–	6–8	A(C,U)(A,U)AUGG
CAV (9 and 21)	711–743	–	6–8	(A,C)AAAUGG
CBV (1, 3, and 4)	740–743	–	6–7	AAAAUGG
SVDV	742	–	10–11	AAAAUGG
Enterovirus 70	726	–	11	AUAAUGG
BEV	819	–	7	ACAAUGG
Rhinovirus				
HRV (1B, 2, 14, and 89)	610–628	–	11–13	A(U,C)CAUGG
Cardiovirus				
EMCV and mengovirus	758–842	+	5–10	AA(G,U)AUGG
TMEV (BeAn, DA, and GDVII)	1064–1068	–	8–10	A(A,C)UAUGG
Aphthovirus				
FMDV (A_{10} and O_1K)	1167–1194	+	8–11	(A,C)(A,U)(C,U)AUGA
Unclassified				
HAV	726–734	–	10	AUAAUGG

[a] References: Polioviruses: Kitamura *et al.* (1981), Racaniello and Baltimore (1981), Nomoto *et al.* (1982), Stanway *et al.* (1983, 1984a), Toyoda *et al.* (1984), Cann *et al.* (1984), Hughes *et al.* (1986), La Monica *et al.* (1986), Pevear *et al.* (1990), CAV: Hughes *et al.* (1989), K. H. Chang *et al.* (1989); CBV: Tracy *et al.* (1985), Iizuka *et al.* (1987), Jenkins *et al.* (1987), Lindberg *et al.* (1987); SVDV: Inoue *et al.* (1989), Seechurn *et al.* (1990); enterovirus 70: N. Takeda (unpublished observations, communicated by A. C. Palmenberg); BEV: Earle *et al.* (1988); HRV: Stanway *et al.* (1984b), Callahan *et al.* (1985), Skern *et al.* (1985), Duechler *et al.* (1987), Hughes *et al.* (1988); EMCV and mengovirus: Vartapetian *et al.* (1983), Palmenberg *et al.* (1984), A. C. Palmenberg (personal communication), Cohen *et al.* (1988), Bae *et al.* (1989, 1990); TMEV: Pevear *et al.* (1987, 1988), Ohara *et al.* (1988); FMDV: Forss *et al.* (1984), Newton *et al.* (1985), Robertson *et al.* (1985), Clarke *et al.* (1987); HAV: Najarian *et al.* (1985), Cohen *et al.* (1987a,b), Paul *et al.* (1987), Jansen *et al.* (1988), Ross *et al.* (1989).

[b] Abbreviations: CAV, coxsackie A viruses; CBV, coxsackie B viruses; SVDV, swine vesicular disease virus; BEV, bovine enterovirus; HRV, human rhinovirus; EMCV, encephalomyocarditis virus; TMEV, Theiler's murine encephalomyelitis virus; FMDV, foot-and-mouth disease virus; HAV, hepatitis A virus.

in different polio- and coxsackie B virus strains, varying from 740 to 747 residues (Table I). There is much greater diversity, however, among other picornaviruses. Thus, rhinovirus 5-UTRs are 610–628 nucleotides long, whereas the 5'-noncoding segments in the genomes of some cardioviruses and especially of aphthoviruses are much longer, in some cases nearly twice as long (up to about 1200 nucleotides) (Table I).

Partly, the greater 5-UTR size in the latter group of viruses is due to the presence of a poly(C) tract. Such a homopolymeric tract was originally discovered in the EMCV RNA (Porter et al., 1974), and a similar segment was found soon afterward in the genomes of some other cardioviruses and FMDV (Brown et al., 1974). The length of the tract varies among populations of the given virus, usually from about 100 to 250 nucleotides in both aphtho- and cardiovirus RNAs (Brown et al., 1974; Harris and Brown, 1977; Black et al., 1979; Costa Giomi et al., 1984), but values as high as 600 nucleotides have also been reported (Brown, 1979).

Originally, poly(C) was erroneously suggested to lie at the 3' end of the cardio- and aphthovirus genomes [supposedly replacing the absent poly(A)]. The actual (i.e., the nearly 5'-terminal) location of poly(C) in EMCV RNA was established in our laboratory (Chumakov and Agol, 1976), followed by similar data on RNAs of mengovirus (Perez-Bercoff and Gander, 1977) and FMDV (Harris and Brown, 1976). The distance between the RNA 5' end and the poly(C) tract is about 150 and 360–370 nucleotides in EMCV and FMDV RNAs, respectively (Rowlands et al., 1978; Chumakov et al., 1979; Vartapetian et al., 1983; Newton et al., 1985). The 3'-terminal portion of the poly(C) tract may contain intermittent inclusions of other nucleotides, mostly uridylic acid residues (Svitkin et al., 1983). Remarkably, unlike conventional cardioviruses (e.g., EMCV or mengovirus), TMEV RNA lacks a poly(C) tract (Pevear et al., 1987; Ohara et al., 1988).

Another contribution to the length polymorphism of the picornaviral 5-UTRs is provided by insertion/deletions of different kinds (discussed in Section X).

An important feature of the picornaviral RNAs, distinguishing them from the overwhelming majority of mRNA species operating in eukaryotic cells, is the absence of a 5'-terminal cap structure. Instead, the virion RNA contains a covalently linked low-molecular-mass virus-specific polypeptide, VPg (Lee et al., 1976, 1977; Flanegan et al., 1977; Sangar et al., 1977; Drygin et al., 1979; reviewed by Wimmer, 1982; Vartapetian and Bogdanov, 1987); the bond between nucleic acid and protein moieties is represented by O^4-(5'-uridylyl)tyrosine (Ambros and Baltimore, 1978; Rothberg et al., 1978; Vartapetian et al., 1980). In contrast, the majority of cytoplasmic nonencapsidated (both polyribosome-associated and "free") positive strands of picornavirus-specific RNA species from the infected cells, which actually serve as templates for the viral protein synthesis, terminate with pUp (Hewlett et al., 1976; Nomoto et al., 1976; Fernandez-Muñoz and Darnell, 1976;

Fernandez-Muñoz and Lavi, 1977; Grubman and Bachrach, 1979). The removal of VPg from a proportion of the viral RNA molecules to generate pUp-ended mRNA species is accomplished by host cell enzymatic activity (Ambros et al., 1978). This reaction readily occurs during the cell-free translation of the VPg-terminated RNA isolated from the virions (Dorner et al., 1981).

B. AUGs and Open Reading Frames

As mentioned in Section I, a characteristic feature of the picornaviral 5-UTRs is the presence of multiple cryptic AUGs preceding the polyprotein reading frame (Table I). According to the "leaky scanning" hypothesis proposed by Kozak (1986c, 1989a) (see also Section IV,A), the utilization of upstream AUGs as translational initiation signals depends heavily on the surrounding nucleotide sequence and, in particular, is facilitated by a "favorable" context, for example, an A or G residue at position −3 and a G residue at position 4, the first nucleotide in the AUG being at position 1. The majority of the upstream AUGs in picornaviral 5-UTRs actually lie in an "unfavorable" environment.

Nevertheless, it could easily be seen that most of these viruses possess at least one such AUG in quite a favorable context. Are they functional? Are the corresponding upstream reading frames (URFs) utilized for the synthesis of any polypeptides? The most likely answers are "no," even though a polypeptide encoded in a URF was reported to accumulate on in vitro translation of FMDV RNA (Forss et al., 1984). Indeed, so far there is no evidence that any of the picornaviral URFs are translated, within the virus-infected cell, into biologically relevant polypeptides. Moreover, no phenotypic changes, at least in vitro, were observed on mutational inactivation (by an A-to-U transversion) of all but one (the last; position 588) internal AUG in the poliovirus 5-UTR (Pelletier et al., 1988a). Nor did such phenotypic changes result from the disruption of short URFs due to insertion of tetranucleotides (Trono et al., 1988a).

Why then does the position of some AUGs (and URFs) in the picornaviral 5-UTRs appear to be conserved, as originally noted for the three poliovirus serotypes by Toyoda et al. (1984) and as could be deduced from an inspection of certain other sequences (cf. Hughes et al., 1989)? One can guess that the URFs are vestiges of the coding sequences from which the 5-UTRs evolved (cf. Section X), and that the triplets had been conserved not because of their amino acid-coding potential, but rather because of the constraints imposed on the primary and secondary structures of the relevant segments of the viral

RNAs as components of the cis-acting element(s) involved in the interaction with translation initiation factors and/or ribosomes (see Section IV,B).

C. Conservation and Divergence

A remarkable conservation of the primary structure of the poliovirus 5-UTR was recognized just after the genomes of the three existing poliovirus serotypes had been sequenced (Toyoda *et al.*, 1984). Further accumulation of the sequencing data revealed that 5-UTRs of other enteroviruses as well as rhinoviruses share with polioviruses many identical (or nearly identical) elements, and all of them could therefore be easily aligned (cf. Rivera *et al.*, 1988; Pilipenko *et al.*, 1989a). For cases in which the RNA primary structure is not yet available, hybridization studies revealed a close relationship among the 5-UTRs of numerous enteroviruses (coxsackie A and B, echo-, and polioviruses), except echovirus 22 and enterovirus 71 (Auvinen *et al.*, 1989; Bruce *et al.*, 1989; cf. Hyypiä *et al.*, 1989). Interestingly, the 5-UTR primary structure of coxsackievirus A9 is much more closely related to that of coxsackie B viruses (84–86% similarity) than to the nucleotide sequence of another coxsackie A virus, A21 (70% similarity) (K. H. Chang *et al.*, 1989).

A significant similarity among 5-UTRs of cardio- and aphthoviruses was also detected (Pilipenko *et al.*, 1989b), although no obvious relationship between the nucleotide sequences of the 5-UTRs of entero- and rhinoviruses, on the one hand, and those of cardio- and aphthoviruses, on the other, could be demonstrated. No noticeable similarity of the hepatitis A virus 5-UTR primary structure to the structure of this region in other picornaviral genomes was reported.

Despite the high degree of intragroup conservation, a closer comparison has revealed several examples of gross rearrangements in the individual 5-UTRs. These are discussed in Section X.

III. SECONDARY AND TERTIARY STRUCTURES

It could *a priori* be expected that the 5-UTRs would contain elements involved in the replication and translation of the picornaviral genome. These putative elements could hardly be anticipated to function solely as linear entities.

Just after the very 5'-terminal nucleotide sequences of the picornaviral genomes had become available, appropriate secondary structure models were proposed. The first such structural element recognized (and experimentally supported) in the poliovirus 5-UTR was a relatively stable (approximately -20 kcal/mol) 10-bp hairpin located 9 nucleotides from the 5' end (Larsen *et al.*, 1981) (or, avoiding a subterminal bulge, a 9-bp hairpin 10 nucleotides from the end). This feature was found to be characteristic of the genomes of all entero- and rhinoviruses investigated thus far (Rivera *et al.*, 1988; K. H. Chang *et al.*, 1989). There is evidence that the hairpin is involved in a physiologically significant function (Racaniello and Meriam, 1986) (Section VIII). Secondary structure models for the very 5'-proximal segments of the 5-UTRs were also proposed for the EMCV (Vartapetian *et al.*, 1983), TMEV (Pevear *et al.*, 1988), FMDV (Harris, 1979, 1980; Newton *et al.*, 1985; Clarke *et al.*, 1987), and HAV (Paul *et al.*, 1987) genomes. Although experimental support is available only for the EMCV model (Vartapetian *et al.*, 1983), the similarity of several elements in the EMCV, TMEV, and FMDV structures strongly suggests the existence of a consensus folding (Pevear *et al.*, 1988; Pilipenko *et al.*, 1990). These elements are likely to participate in the replication of the viral genomes (see Section VIII).

So far as the internal portions of the picornaviral 5-UTRs are concerned, there have been attempts to fold appropriate segments of individual poliovirus (Evans *et al.*, 1985), FMDV (Clarke *et al.*, 1987), and HAV (Cohen *et al.*, 1987a) RNAs by using computer programs. The models thus obtained proved to be at variance with subsequent more direct data (at least in the two former cases for which such subsequent data are available). A powerful approach to RNA secondary structure modeling (especially if this structure is expected to be of functional significance) consists of deduction of a consensus folding for numerous related RNA species by taking into account not only fully conserved nucleotide sequences, but also, and in particular, compensating mutations in pairs of noncontiguous bases (cf. Fox and Woese, 1975; Noller and Woese, 1981; James *et al.*, 1988). When this approach was applied to the picornaviral 5-UTRs, no general consensus was found; however, highly conserved secondary structure models could be derived for entero- and rhinoviruses, on the one hand (Rivera *et al.*, 1988; Blinov *et al.*, 1988; Pilipenko *et al.*, 1989a), and cardio- and aphthoviruses, on the other (Pilipenko *et al.*, 1988b, 1990). The models for entero- and rhinovirus 5-UTRs proposed by Rivera *et al.*, (1988) and Pilipenko *et al.* (1989a; see also Blinov *et al.*, 1988), while sharing many secondary structure elements, differed nevertheless from one another in some significant aspects (significant because they involved certain physiologically important regions). The models proposed by Pilipenko *et al.* (1989a,b) for entero- and rhinovirus as well as cardio- and aphthovirus RNAs were supported experimentally by testing the susceptibility to chemical modifications and to a hydrolytic attack by single-strand- and double-strand-specific nucleases; a similar folding was also deduced for poliovirus 5-UTRs by Skinner *et al.* (1989). Figures 1 and 2 present the consensus models for the two groups of picornavirus 5-UTRs.

It is perhaps appropriate to note that the internal poly(C) tract does not appear to be involved in extensive base-pairing with other portions of either EMCV (Goodchild *et al.*, 1975) or FMDV (Mellor *et al.*, 1985) RNAs.

One should note that support for the above structures came, apart from evolutionary considerations, from experiments performed with RNA molecules in salt solutions. To what extent, however, do they reflect the RNA folding inside the infected cell? It is difficult, if at all possible to obtain a complete answer to this question. Nevertheless, genetic evidence strongly favors the *in vivo* reality of at least some elements of the proposed secondary structures. Thus, the model (Fig. 1) predicts pairing between nucleotides 480 and 525 in the poliovirus type 1 RNA. In all but one sequenced entero- and rhinovirus genome the corresponding nucleotides are expected to form an A–U pair (Pilipenko *et al.*, 1989a). The only known exception is the Sabin type 1 strain, having G_{480} instead of A; obviously, the potential to form a base pair between nucleotides 480 and 525 must in this case be diminished. In the gut of those vaccinated, however, the Sabin 1 genome appears to be unstable, and mutations in one of the two nucleotides under discussion are readily selected for. Either G_{480} is mutated back to A or U_{525} is replaced by C; the potential to form a base pair between nucleotides 480 and 525 is fully restored in either case (Muzychenko *et al.*, 1991; see also Agol, 1990). Similar evidence for the biological relevance of the same secondary structure element in the poliovirus genome can be derived from the experiments described by Kuge and Nomoto (1987) and Skinner *et al.* (1989).

The flat structures just discussed should obviously be regarded as merely a first approximation to the actual spatial organization of the picornaviral 5-UTRs. Unfortunately, no reliable tools are presently available for unraveling the tertiary structure of long RNA segments. Perhaps some useful information in this respect can be obtained by fine mapping of psoralen cross-links [e.g., exploiting the recently developed technique (Ericson and Wollenzien, 1988)], but, to our knowledge, only very long-range (i.e., visible under the electron microscope) cross-links in poliovirus RNA have been investigated thus far (Currey *et al.*, 1986).

An independent way of approaching the tertiary structure may consist of a search for evolutionary conserved potentials to form pseudoknotlike structures (cf. James *et al.*, 1988; Haselman *et al.*, 1989;

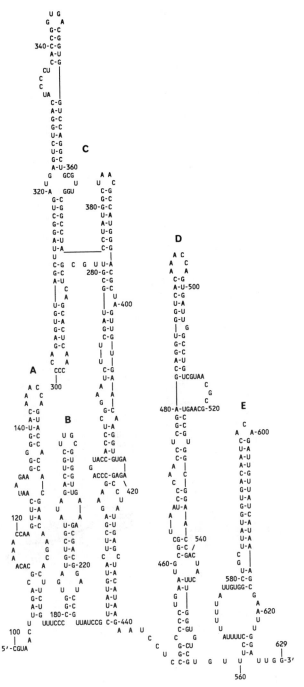

FIG. 1. A secondary structure model for the internal region of the 5-UTR of poliovirus type 1 (Pilipenko *et al.*, 1989a, 1991). This is a consensus model for all entero- and rhinoviruses.

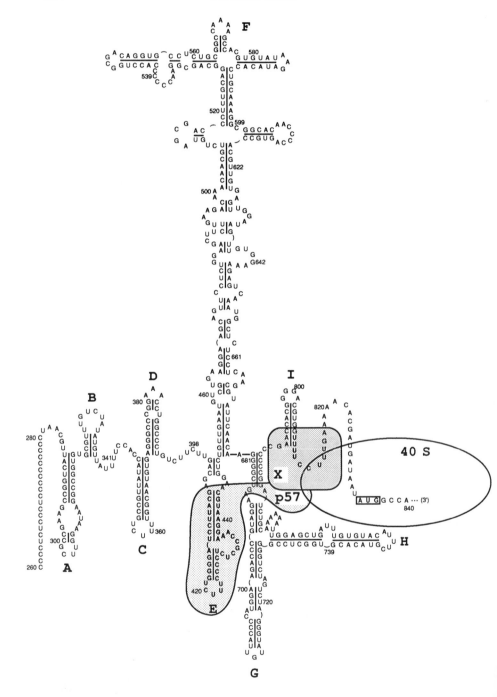

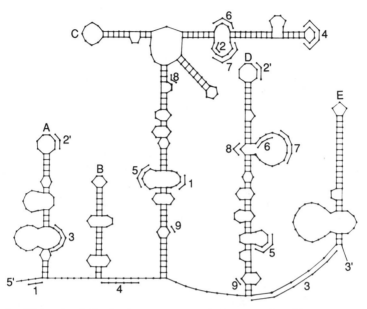

FIG. 3. Elements of the tertiary structure of picornaviral 5-UTRs. Pairs of nucleotide stretches potentially able to form conserved tertiary bonds in the entero- and rhinovirus 5-UTRs are assigned identical numbers (Pilipenko *et al.*, 1991).

Woese and Gutell, 1989). In fact, several conserved sites potentially capable of ensuring such interdomain, or other long-range, interactions in entero- and rhinovirus 5-UTRs can be envisioned (Pilipenko *et al.*, 1991) (Fig. 3). In certain cases the real possibility of such interactions *in vivo* could be supported indirectly by the observation that they tend to be maintained by coupled mutations of noncontiguous nucleotides (Muzychenko *et al.*, 1991; see also Agol, 1990). Thus, a theoretically possible pseudoknot may involve nucleotide 398 in one secondary structure domain and nucleotide 481 in another domain of the 5-UTR of poliovirus type 2. In the Sabin type 2 vaccine strain these two nucleotides might be expected to form a U_{398}–A_{481} base pair. However, among mutant Sabin 2 strains isolated from vaccine-associated cases of paralytic poliomyelitis, one can often find genomes with the two coincidentally changed bases, C_{398} and G_{481}; the potential to

←───

FIG. 2. A consensus secondary structure model for the EMCV 5-UTR downstream of the poly(C) tract. This is essentially the model predicted by Pilipenko *et al.* (1989b) with the very terminal structures added by Jang and Wimmer (1990). Putative binding sites for polypeptides p57 and X as well as for the 40 S ribosome subunit are shown (see text for details). (From Jang and Wimmer, 1990.)

form a tertiary base pair was evidently maintained. In FMDV 5-UTRs the potential to form pseudoknots has also been identified, and similar structures can be generated in the genomes of three FMDV strains (Clarke *et al.*, 1987). The problem with these pseudoknots, however, is that they are not compatible with the predicted secondary structure of the appropriate RNA segment.

Although no direct experimental evidence is available, in light of the above considerations, the structure of picornaviral 5-UTRs is likely to be more or less condensed and maintained by specific tertiary interactions between secondary structure domains. In a sense it may be reminiscent of globular proteins. If such a viewpoint is valid, it should have obvious functional implications, and thus elucidation of the spatial (i.e., three-dimensional) organization of picornaviral 5-UTRs should be of paramount significance.

IV. cis-Acting Translational Control Element(s)

It was originally suggested by Jacobson and Baltimore (1968) that mRNA species, operating in eukaryotic cells, are functionally monocistronic, by virtue of having only a single translation initiation site. Although we are now aware of numerous exceptions to this rule, it could be accepted as an extremely useful first approximation. This biological regularity was very aptly explained by the scanning hypothesis of translation initiation proposed by Kozak (1978, 1981). The hypothesis states that a Met–tRNA$_i^{Met}$-bearing 40 S ribosomal subunit aided by specific initiation factors recognizes the capped 5′ end of an mRNA molecule and moves along it until the first AUG is encountered. Here, a 60 S subunit is recruited to form a complete ribosome, and the polypeptide chain synthesis is initiated. Accordingly, two major types of translational cis-acting elements were first considered: the cap structure and the initiator AUG.

The adaptation of the hypothesis to the newly accumulating data has led to its modification, which takes into account that the first (i.e., 5′-proximal) AUG might not necessarily be the initiator codon (Kozak, 1986c, 1989a). The modified version suggests that the ribosome scanning might be "leaky" (i.e., able to skip an AUG). Whether or not the AUG is recognized as the initiator codon largely depends on its "context" (i.e., the surrounding sequences), the optimal context being (CC)$_G^A$CCAUGG (Kozak, 1984a, 1986a) [a more extended version of the context consensus is (GCC)GCC$_G^A$CCAUGG (see Kozak, 1987a, 1987c, 1989a)]. Upstream AUGs, depending on the context, may to some extent interfere with the "correct" initiation due to "false starts," distracting ribosomes from their proper destination (Kozak, 1984b; Liu *et al.*, 1984; Hunt, 1985), but such a situation, in eukaryotic mRNAs, was considered to be "breaking the rule" (Kozak, 1989a). Moreover, there is a possibility that, on completing the translation of an ORF, the ribosome (or its smaller subunit) does not dissociate from the template, but rather continues scanning and reinitiates a new polypeptide at the next AUG; in such cases the relative efficiencies of the usage of the upstream and downstream initiation codons depends not only on their contexts, but also on other peculiarities of the given template (e.g., whether the respective ORFs are overlapping or not, how far they are located from each other and from the 5′ end, and some other conditions) (Kozak, 1984b, 1986c, 1987b, 1989b; Johansen *et al.*, 1984; Peabody and Berg, 1986; Peabody *et al.*, 1986; van Duijn *et al.*, 1988; Dabrowski and Alwine, 1988; Williams and Lamb, 1989). Thus, the upstream AUGs (and the reading frames they open) may affect the activity of the genuine initiator codons in different ways.

While testing the scanning hypothesis, one more kind of regulatory cis element was recognized in the 5-UTRs, namely, stable secondary structures. Such structures may hamper either the initial interaction of the 40 S ribosomal subunit with the mRNA (if the structured RNA element is located near the template 5′ end) (Kozak, 1989c) or the scanning process itself (Pelletier and Sonenberg, 1985; Kozak, 1986b, 1989c; Sedman *et al.*, 1990). On the other hand, unstructured regions of the template, even if they precede a stable hairpin, may facilitate initiation (cf. Kozak, 1988). Finally, a eukaryotic analog of the Shine–Dalgarno element (i.e., an oligopyrimidine sequence, just upstream of the initiator AUG, which is complementary to a segment of the 18 S ribosomal RNA) was also proposed to promote proper ribosome binding to the template (Hagenbüche *et al.*, 1978). Its significance, however, is far from certain (cf. Laz *et al.*, 1987). Thus, several, but not too many, general cis-acting signals are known to control the efficiency of translation initiation on a eukaryotic template.

As mentioned in Section II,A, picornaviral RNAs lack a cap structure. Although the idea that the terminal protein VPg might be a functional substitute for the cap was considered soon after the discovery of this protein (cf. Perez-Bercoff and Gander, 1978), it has found no experimental support and has been abandoned. Moreover, the picornaviral RNA species, serving as translational templates, appear to be not only uncapped, but also lacking VPg (see Section II,A). In fact, the realization that picornaviral templates lack a 5′-terminal cap structure was one of the first, if not the first, hint that these templates may use some noncanonical mode of translation initiation.[1]

[1]It has been reported recently that capping of the picornaviral 5-UTR may result in severe inhibition, rather than stimulation, of the template activity of an appropriate mRNA construct transfected into HeLa cells (Macejak *et al.*, 1990).

Multiple AUGs that precede the genuine initiator codon in picornavirus RNAs (Section II,B) do not appear to be directly involved in polypeptide synthesis (see also Section VI). Thus, they cannot be regarded as codons, and are unlikely to constitute, as such, any translational control signals.

What do we know about other possible cis-acting translational control elements?

A. Sequences Surrounding the Initiator Codon

When the sequences surrounding the genuine initiator codon opening the polyprotein reading frames in the picornaviral genomes are inspected, one finds that quite a favorable context (AXXAUGG) exists here in the overwhelming majority of the picornaviral genera (Table I). The exceptions to this rule are of obvious interest. First, the sequence that opens the HAV polyprotein ORF has an A residue at position 4. One may wonder whether this has anything to do with the notoriously weak template activity of the HAV RNA *in vitro* and/or the very slow reproduction of this virus within the infected cell. Second, an A at position 4 may be accompanied by a pyrimidine at position −3 in the segment surrounding the first AUG of the polyprotein-coding sequence of several FMDV serotypes (see also Robertson *et al.*, 1985; Sanger *et al.*, 1987). This may reflect the fact that only a portion of ribosomes actually initiate translation at the first AUG of the FMDV polyprotein reading frame; the others, on the other hand, appear to ignore this triplet and continue scanning until they encounter the next AUG (which, by the way, is in the frame with the previous one and is located in a favorable context). Accordingly, two sites of translation initiation could be demonstrated to operate on the FMDV genome, and the choice between these site appears to depend, at least in part, on the context of the respective AUGs (Sangar *et al.*, 1987). Third, a pyrimidine (C) occupies position −3 (and a "good" G occupies position 4) in the vicinity of the coxsackievirus A21 initiator triplet; by the way, this RNA contains at least two more upstream AUGs having a context of similar "strength" (Hughes *et al.*, 1989). The physiological significance of this "abnormality" is not immediately apparent, because the template activity of the appropriate RNA remains to be evaluated.

An oligopyrimidine tract preceding the polyprotein reading frame is present in RNAs of different picornaviruses. In aphtho- and cardiovirus 5-UTRs such a tract is situated close to the initiator codon (Beck *et al.*, 1983; Forss *et al.*, 1984; Sangar *et al.*, 1987), and there is some evidence that it is actually involved in the initiation of translation, inasmuch as its mutational alteration (in EMCV 5-UTR) might result in inefficient protein synthesis (Jang and Wimmer, 1990). The situation in entero- and rhinovirus genomes is somewhat different: Here, a conserved oligopyrimidine sequence could be found in the 560–570 region (i.e., well apart from the genuine initiator triplet). Nevertheless, there is a possibility that this oligopyrimidine tract may be involved in the interaction with ribosomes.

B. Internal Positive Control Element

1. In Poliovirus

It was against the above background that we first observed that mutations in the middle of the poliovirus 5-UTR (at positions 472–480) strongly affected the efficiency of cell-free translation initiation on poliovirus RNA [the experiments were carried out with the genomes of neurovirulent type 1 and type 3 poliovirus strains, their attenuated (Sabin) derivatives, and a neurovirulent revertant of the latter, using two in vitro translation systems, extracts of Krebs-2 cells and rabbit reticulocyte lysates] (Svitkin et al., 1985). Since position 472 does not coincide with any internal AUGs and a mutation diminishing the efficiency of initiation should decrease, rather than increase, the stability of the appropriate secondary structure element [corresponding evidence was published later (Blinov et al., 1988; Pilipenko et al., 1989a)], we suggested "that a sequence around position 472 is involved, directly or indirectly, in interactions with ribosomes or initiation factors, and that the strength of this interaction may be altered by mutations at position 472" (in the poliovirus type 3 RNA) (Svitkin et al., 1985). It should be noted that at that time the possibility, though unlikely, could not be rigorously ruled out that mutations other than in the region 472–480 were actually responsible for the alterations of the template activity of RNAs from attenuated poliovirus strains. However, unambiguous proof of the validity of our original assignment was provided recently by investigations of appropriately engineered templates (Svitkin et al., 1990).

The suggestion that a cis-acting translational control element of a novel type exists in the middle of the poliovirus 5-UTR was supported by subsequent studies by different laboratories. First, different kinds of deletions and insertions within an extended internal segment of the poliovirus 5-UTR were shown to affect translation initiation in vitro (Pelletier et al., 1988b,c; Trono et al., 1988a; Bienkowska-Szewczyk and Ehrenfeld, 1988) and in vivo (Pelletier et al., 1988c; Trono et al., 1988b) or to change phenotypic properties (e.g., the plaque size or the efficiency of reproduction at an elevated temperature) (Kuge and Nomoto, 1987; Trono et al., 1988a; Dildine and Semler, 1989), most likely due to translation deficiencies. These studies are discussed in greater detail below.

Second, and most importantly, it was directly demonstrated that an internal portion of the poliovirus 5-UTR is involved in the cap-independent internal ribosome binding. The elegant idea of Pelletier and Sonenberg (1988) was to insert poliovirus 5-UTR into a chimeric mRNA template between two cistrons encoding readily detectable polypeptides. The efficient expression of the second cistron under conditions in which the first one was made "silent" would mean that the intercistronic poliovirus 5-UTR does secure the internal binding of ribosomes.

Specifically, several sets of plasmids were constructed. All of these sets included, among others, two monocistronic plasmids, corresponding to each of the two relevant genes, and two bicistronic plasmids, with or without intercistronically placed poliovirus 5-UTR. In one such set the first and second cistrons were represented by the thymidine kinase (TK) gene of herpes simplex virus type 1 and the bacterial chloramphenicol acetyltransferase (CAT) gene, respectively (Pelletier and Sonenberg, 1988). On transfection of tissue culture cells with a specific plasmid, relevant 5'-capped mono- or bicistronic mRNA species should be synthesized, and their translation productions, if formed, can readily be identified.

A key experiment was performed in poliovirus-infected cells (Pelletier and Sonenberg, 1988). It is well known (see Section V,B) that the canonical cap-dependent initiation is severely impaired in such cells. When poliovirus-infected cells were transfected with the above series of plasmids, the only active plasmid was that encoding the bicistronic template with the intercistronically placed poliovirus 5-UTR; the only product directed by this template was the CAT protein encoded in the second cistron (although in mock-infected cells the same plasmid directed predominantly the accumulation of the TK protein). Thus, the inactivation of the cap-dependent mechanism of translation initiation by poliovirus infection rendered the first cistron inactive, and the expression of the second cistron could hardly be attributed to the ribosomes coming from the template 5' end. Rather, this expression should be due to the turning-on of a novel mechanism, most likely involving internal ribosome binding mediated by the intercistronic poliovirus-derived 5-UTR. The selective translation of the second cistron of the bicistronic template could be demonstrated even in the mock-infected cells, provided that the transfected cells were incubated under hypertonic conditions [it has long been known that translation initiation on poliovirus RNA, as opposed to cellular mRNAs, is much more resistant to hypertonic salt concentrations (Nuss *et al.*, 1975)]. This experiment should be interpreted in the same way as the previous one.

Translation of bicistronic mRNAs *in vitro* lent additional support to the notion on the ability of poliovirus 5-UTR to ensure internal initiation of translation. The first cistron in the bicistronic templates used was represented either by the *TK* gene again (Pelletier and Sonenberg, 1988) or by the σS gene of reovirus type 2 (Pelletier and Sonenberg, 1989), whereas the *CAT* gene served as the second cistron in both types of constructs. In reticulocyte lysates the second cistron of the bicistronic template could be expressed, provided it had been juxtaposed to the poliovirus 5-UTR. Moreover, it continued to be expressed even when the translation of the first cistron had been abrogated by the insertion of a stable hairpin structure in its own 5-UTR (Pelletier and Sonenberg, 1988). Clear-cut results were also obtained in extracts from poliovirus-infected HeLa cells, in which the bicistronic template with the intercistronic poliovirus 5-UTR directed the synthesis of only the second gene (*CAT*) product (Pelletier and Sonenberg, 1989). The results obtained in the same study with extracts from uninfected HeLa cells were also fully compatible with the theory on the ability of poliovirus 5-UTR to direct internal ribosome binding.

Now that this theory is supported with reasonable certainty, we can attempt to deduce, from the available data, the approximate borders of the translational cis-acting element involved. To this end, it would seem logical to assume that the sequences, whose removal or mutational alterations do not appear to affect significantly the function of the element, are located outside its body. Such an approach, simple as it is, is not devoid of pitfalls. The nucleotide sequence of the element, while forming a complex three-dimensional entity, is expected to be (and actually is), in a sense, discontinuous; some of its internal portions may be (and perhaps are) of no serious functional significance. Therefore, the finding of a silent portion of the 5-UTR does not necessarily mean that it lies beyond the borders of the cis element. There is another ambiguity. The element could be considered to be composed of the essential core, whose absence results in the complete invalidation of the internal initiation mechanism and dispensable, yet functionally significant, surrounding sequences. In practice it is not always easy to judge whether a particular genomic alteration totally inactivates the element's function, affects it significantly, or leaves it unchanged. Taking into account these limitations, let us consider the experimental data accumulated thus far. First, we shall attempt to define the location of the element as a whole; then the possible functional significance of its individual components is discussed.

What can be said about the element's 3' (downstream) border? It would be not too surprising if the positive translational control element were located close to the initiator codon at position 743. But this is not the case. Surprisingly, Kuge and Nomoto (1987) discovered that some large deletions within the downstream portion of the poliovirus 5-UTR are fully compatible with the virus viability. Specifically, the removal of nucleotides 630–726, 622–726, or 600–726 from the poliovirus type 1 (Sabin) genome does not alter such *in vitro* phenotypic properties of the virus as the plaque size or the time course of virus growth and development of cytopathic changes (see also Trono *et al.*, 1988a). More extended deletions, 570–726 or 564–726, also produce viable progeny, but the mutant viruses exhibit a small-plaque (sp) phenotype due to their retarded growth rate (see also Iizuka *et al.*, 1989). Assuming that phenotypic changes in this series of mutants are largely due to the altered function of the cis-acting translation control element (and we believe that this is a likely assumption), it could be concluded that the downstream border of this element should map between positions 564 and 600. The border may perhaps lie even farther to the 3' end since the 5-UTR region affected by the deletions includes repeated sequences, and therefore reformation of nearly normal structural features may be feasible in some of the deletion mutants (Pilipenko *et al.*, 1990) (see also Section X).

Another approach to the boundaries' mapping, which does not require viability of the deleted or otherwise modified genomes, consists of evaluation in their *in vitro* translational template activity. A difficulty inherent in the interpretation of the results thus obtained is due to the dependence of the outcome of the experiments on the particular cell-free system used and, sometimes, on the incubation conditions as well. An even more serious problem concerns the need of unambiguous differentiation between the internal and 5'-terminal initiations of translation, because some deletions may concurrently inactivate the former and active the latter mode.

To circumvent the latter problem, some *in vitro* translation experiments were performed in extracts from poliovirus-infected HeLa cells in which the cap-dependent initiation mode was invalidated (see Section V,B); both monocistronic (Pelletier *et al.*, 1988b) and bicistronic (Pelletier and Sonenberg, 1988) templates with a reporter gene preceded by a segment of the poliovirus type 2-derived 5-UTR were used. In agreement with the aforementioned data of Kuge and Nomoto (1987), the removal of nucleotides 632–732 did not affect the template activity, whereas the template lacking nucleotides 462–732 (or 382–732) was completely devoid, under the experimental conditions used, of the capacity to direct the synthesis of polypeptides encoded in the downstream cistron. These data placed the 3' border of the cis element between positions 462 and 632.

So far as the 5' border of the element is concerned, the deletion of the first 32 or 79 nucleotides from the 5-UTR was without any significant effect on the cap-independent internal ribosome binding in experiments with bicistronic templates and extracts from poliovirus-infected HeLa cells (Pelletier et al., 1988b; Pelletier and Sonenberg, 1988). Dispensability, for the cap-independent internal initiation of translation, of the very proximal segment of the poliovirus 5-UTR follows also from results obtained in experimental systems for which unambiguous conclusions could not be drawn so readily, such as translation of templates with partially deleted poliovirus 5-UTR in extracts from uninfected HeLa cells (Pelletier et al., 1988b; Pelletier and Sonenberg, 1988) and Krebs-2 cells (Pestova et al., 1989) as well as in reticulocyte lysates supplemented with initiation factors from HeLa cells (Bienkowska-Szewczyk and Ehrenfeld, 1988). On the other hand, the removal of 139 or 319 nucleotides caused either quite small (30–40%) or marked (3- to 5-fold), depending on the construct, inhibition of the downstream gene expression (Pelletier et al., 1988b; Pelletier and Sonenberg, 1988).

To sum up, we can conclude that the cis-acting element ensuring the cap-independent ribosome binding is contained within a segment with coordinates of approximately 140 to 600–630. If we accept that the cis-acting element under discussion is hardly functioning as merely a linear nucleotide sequence, then the secondary structure domains A–E (Fig. 1) should be regarded as likely components of this element.

An intriguing question concerns the functional significance of individual domains or even smaller structural entities of the whole cis-acting element. Although there is no complete answer to this question, some tentative and partial judgments could perhaps be made on the basis of the experimental data collected.

As already mentioned, the removal of about 140 5'-terminal nucleotides somewhat decreased the template activity of the poliovirus 5-UTR (Pelletier et al., 1988b; Pelletier and Sonenberg, 1988). This observation may be interpreted to mean that domain A is involved in the translation initiation. Furthermore, Jackson et al. (1990; R. J. Jackson, personal communication) showed that nucleotides approximately between positions 70 and 160 are required in order that the translation of poliovirus RNA in reticulocyte lysates be stimulated by the initiation factors from HeLa cells (the initiation factor requirements for picornaviral RNA translation are considered in Section V,B). Nevertheless, the appropriate structure does not appear to be essential for the internal ribosome binding.

Also nonessential is domain B, although it is most likely involved in translation control. Short insertions/deletions at position 220 (note that different authors designate the positions of the same insertions slightly differently) led to sp, ts, or even lethal phenotypes (depending on the nature of the insert), and these alterations can be traced to the impairment of the cap-independent internal initiation of translation (Kuge and Nomoto, 1987; Trono et al., 1988a,b; Dildine and Semler, 1989). Significantly, large-plaque revertants of a small-plaque mutant with an insertion at position 220 were demonstrated to acquire two second-site mutations, one, invariably, at position 186 and the other, affecting nucleotides, at either 480 or 525 (Kuge and Nomoto, 1987). The first of these second-site mutations (at position 186) affected a nucleotide which is expected to interact with the region surrounding mutated position 220 (Pilipenko et al., 1991),[2] suggesting the involvement of a secondary structure element. In line with such reasoning, a ts + pseudorevertant of a mutant with a 4-nucleotide deletion at position 220 was shown to acquire an additional 41-nucleotide deletion (altogether, positions 184 to 228 were lost) (Dildine and Semler, 1989), again suggesting that a (structured) RNA segment around positions 180–220 must be somehow involved in the translational control.

One more, albeit indirect, piece of evidence for the participation of domain B in a significant physiological function came from experiments with engineered recombinants between poliovirus and coxsackievirus genomes (Semler et al., 1986). In one such recombinant a segment of poliovirus type 1 RNA with coordinates 220–627 was replaced by the homologous segment derived from coxsackievirus B3 RNA. The recombinant was viable, but exhibited a strong ts phenotype (although both its parents were ts +). It seems likely that the defect was primarily due to some translational impairment. Moreover, this impairment resulted most probably from structural alterations at the 5′-proximal junction between the polio- and coxsackievirus RNAs (i.e., near position 220). Indeed, a ts + pseudorevertant of the recombinant acquired a deletion just downstream of this junction, and a related poliovirus–coxsackievirus recombinant, but with a more 5′-proximally located junction (at position 66), that is, having a "normal" sequence around base 220, exhibited a ts + phenotype (Johnson and Semler, 1988).

Although important, domain B is not essential. Its dispensability is unambiguously demonstrated by the aforementioned fact that a 45-bp deletion encompassing nucleotides 184–228 is compatible not merely

[2]So far as the 480 and 525 mutations in the revertant genomes are concerned, some yet undefined long-rang nucleotide interactions cannot definitely be ruled out. However, these remote second-site mutations may have a simpler explanation. They could independently enhance the translation efficiency by increasing the stability of an important secondary structure element (see Section III), compensating, thereby, the impaired template activity of the original mutant with an insertion at position 220.

with the viability of a virus, but also with its wild-type (wt) phenotype (Dildine and Semler, 1989).

There is ample evidence that domain C (approximate coordinates 236–443) is involved in promoting the cap-independent internal initiation of translation. Thus, oligonucleotide insertions at position 267, 322, or 388 resulted in either sp and ts mutant phenotype or unviable genome, in both cases the deficiency being due to a translational impairment (Kuge and Nomoto, 1987; Trono et al., 1988a,b). Importantly, the locations of physiologically significant mutations appear to affect diverse subdomains of the large domain (cf. Fig. 1), suggesting multiple intra- and/or interdomain interactions. A particular example of such an interdomain bonding (between nucleotides 398 and 481 in the poliovirus type 2 RNA) has already been discussed in Section III.

Somewhat more confusing is the question of whether domain C is essential for the cap-independent translation initiation or whether it performs merely an auxiliary role. On the one hand, the lethal phenotype of some insertion mutants mentioned above [e.g., at position 322 (Trono et al., 1988a)] did testify to the indispensability of the relevant structure. On the other hand, however, the complete removal of this domain did not seem to abrogate the cap-independent initiation of translation, at least in certain cell-free systems (Bienkowska-Szewczyk and Ehrenfeld, 1988; Pestova et al., 1989), although it should be noted that no rigorous proof was provided in these studies that the initiation on the domain C-lacking templates was actually accomplished by a 5′ terminus-independent (i.e., internal) mode. Perhaps the strongest evidence that the integrity of the entire domain C is not essential for the internal initiation was provided by the observation that a construct retaining only nucleotides 320–631 from poliovirus 5-UTR (and hence no intact domain C) still proved to be capable of ensuring cap-independent internal ribosome binding (Pelletier et al., 1988b; Pelletier and Sonenberg, 1988). Thus, the apparent conflict of the data concerning the essentiality of domain C perhaps reflects not an entirely uncommon situation in which certain damages to a biological structure are lethal, while its complete removal is not. Another possible reason for the discrepancy is that two sets of data came from the in vivo and in vitro experiments, respectively.

An interesting locus within the cis element corresponds to a highly conserved 7-nucleotide "linker" between domains C and D (cf. Fig. 1; positions 444–450 in the poliovirus type 3 RNA). Insertion of an oligonucleotide into this linker rendered the mutant virus unviable (Trono et al., 1988a). This could suggest that the linker may interact with either a trans-acting factor or a distant nucleotide sequence.

The fourth domain (D; positions 451–559 in poliovirus type 3 RNA) appears to perform a key role; its partial or complete destruction proved to be incompatible with the ability of the cis element to direct internal ribosome binding both *in vivo* (Kuge and Nomoto, 1987; Trono *et al.*, 1988b; see also Dildine and Semler, 1989) and *in vitro* (Pelletier *et al.*, 1988b; Trono *et al.*, 1988a; Pestova *et al.*, 1989). Even point mutations within this region could modulate the efficiency of translation initiation (Svitkin *et al.*, 1985, 1988). As discussed in Section III, there is good evidence that maintenance of the secondary structure is important for the function of this domain, sometimes perhaps even more important than maintenance of the primary structure. It should be added, however, that Jackson *et al.* (1990) have reported that the deletion, from the poliovirus 5-UTR inserted between two reporter genes, of the 5′ end-adjacent nucleotides up to position 539 did not abolish the expression of the second cistron in reticulocyte lysates. Such a deletion should certainly destroy domain D. This contradiction may, again, reflect the basic differences between the requirements for the internal initiation *in vivo* and *in vitro*.

There is a highly conserved 21-nucleotide stretch which encompasses a portion of the 3′ branch of domain D as well as a linker separating domains D and E (coordinate 543–563). This stretch may perform an essential function, as first suggested by Kuge and Nomoto (1987) on the basis of the assessment of the viability of poliovirus deletion mutants. This theory was further supported by experiments with modified poliovirus (Iizuka *et al.*, 1989) and coxsackievirus B1 (Iizuka *et al.*, 1990; A. Nomoto and N. Iizuka, personal communication) 5-UTRs. Moreover, these experiments showed directly that mutations within this stretch did affect the translation efficiency. Curiously, the sequence with coordinates 539–563 in the poliovirus type 1 RNA exhibits a striking complementarity to sequence 1301–1320 in human 28 S rRNA (Iizuka *et al.*, 1989); the significance of this observation remains entirely obscure.

As far as domain E itself (approximate coordinates 582–620) is concerned, its possible significance has already been briefly discussed when we considered the 3′ border of the entire cis-acting element. It may be added that mutations within this domain, particularly those that affected a conserved AUG located in its 5′ branch, diminished the translation efficiency (Pelletier *et al.*, 1988a; N. Sonenberg, personal communication). Moreover, Bienkowska-Szewczyk and Ehrenfeld (1988) inferred from their *in vitro* experiments with truncated templates that the corresponding sequence would be essential for the internal initiation translation. Nevertheless, the fact that the removal of nucleotides 564–726 (i.e., deletion of the entire domain E) from poliovirus RNA or the corresponding segment from coxsackievirus B1

RNA did not abrogate the virus viability (but only retarded its growth) (Kuge and Nomoto, 1987; Iizuka et al., 1989, 1990) indicates, perhaps, unambiguously that this domain is not essential for the virus-specific translation. Some additional information about the significance of its structural details is presented in Section V,C, and speculations about its specific role are considered in Section X.

2. In Encephalomyocarditis Virus

Simultaneously with the just described work on the poliovirus 5-UTR, similar, but independent, studies were carried out on the EMCV 5-UTR. Though seemingly identical, the relevant experiments on polio and cardiovirus 5-UTRs should not be regarded as a mere repetition of one another. It should be borne in mind that the structures of the corresponding 5-UTRs appear to be dramatically different (see Figs. 1 and 2). Just as the mechanisms of the host cell protein synthesis in poliovirus- and cardiovirus-infected cells are essentially not the same (reviewed by Ehrenfeld, 1984; Sonenberg, 1987), the modes of translation initiation might also be different. But this is not the case.

The conclusion that could be drawn from the experiments involving both template truncation (Shih et al., 1987) and construction of bi- or even tricistronic templates with segments of the EMCV 5-UTR inserted between two reporter genes (Jang et al., 1988, 1989) was very similar to that just discussed with regard to poliovirus: Inside the cardiovirus 5-UTR there is a cis-acting element controlling the initiation of the viral polyprotein synthesis; this element appears to be responsible for the cap-independent internal binding of ribosomes. This conclusion appears all the more substantiated after experiments with the engineered templates were performed in poliovirus-infected cells (Jang et al., 1989) known to fail to initiate polypeptide synthesis in a cap-dependent mode (the cardiovirus infection itself does not bring about any apparent alterations to the cap-dependent translation machinery) (reviewed by Ehrenfeld, 1984; see also Mosenkis et al., 1985).

What have we learned about the borders of the cis-acting element involved? The 5' border should certainly be placed downstream of the poly(C) tract, since the removal of this tract, together with the adjoining upstream sequence (so-called short, or S, segment), does not significantly alter the translation efficiency of the EMCV RNA to any significant degree (Chumakov et al., 1979; Shih et al., 1987). Moreover, a 5'-truncated segment of the EMCV 5-UTR containing about 480 (of 833) 5-UTR nucleotides confers a high template activity to the engineered templates (Parks et al., 1986; Kräusslich et al., 1987; see also Elroy-Stein et al., 1989). More detailed studies on EMCV RNA templates that were annealed with complementary oligodeoxyribonucleotides or were 5′-truncated (Shih *et al.*, 1987) as well as on constructs containing partially deleted versions of EMCV 5-UTR (Jang *et al.*, 1988; Jang and Wimmer, 1990), while yielding somewhat deviating results, nevertheless permitted placement of the 5′ border of the essential core of the cis element somewhere downstream of position 402, that is, approximately 430 nucleotides before the initiator AUG (position 834). In any case deletions up to nucleotide 421 completely abolished the translational activity of appropriate constructs (Jang and Wimmer, 1990). Thus, the 5′ border of the control element coincides with a conserved hairpin–loop structure shown in Fig. 2. There is convincing experimental evidence that the secondary structure of this locus is no less important than its primary structure (Jang and Wimmer, 1990). It should be noted that nonessential, though significant, portions of the control element could be revealed in a more upstream segment of the EMCV 5-UTR (e.g., around positions 373–392) (Jang and Wimmer, 1990).

Exploiting the same technique of deletion of the EMCV 5-UTR placed between cistrons of a bicistronic mRNA construct, Jang and Wimmer (1990) demonstrated that the 3′ border of the translational cis element was located around position 810, that is, very close to the initiator codon that opens the polyprotein reading frame. Interesting, this border coincides with a conserved oligopyrimidine region.

Thus, cardioviruses have in common with enteroviruses an extended cis-acting control element within their 5-UTRs which ensures internal binding of ribosomes, but the primary and secondary structures as well as the locations of these elements relative to the true initiator codon are quite different in the two groups of picornaviruses.

3. In Other Picornaviruses

The nature of the cis-acting elements responsible for the ribosome binding to internal segments of the 5-UTRs of other picornaviruses has, to our knowledge, not been thoroughly investigated. Experiments with 5′-truncated 5-UTRs of rhinovirus 14 pointed to an essential role for a sequence encompassed by positions 546–621 (AlSaadi *et al.*, 1989), or perhaps to domain E and possibly a portion of domain D (Fig. 1). Moreover, a striking conservation of the primary and secondary structures also of more upstream domains (see Section III) is, in our opinion, a strong argument for the notion that entero- and rhinovirus 5-UTRs should share their major functional properties with those of poliovirus. By the same token, the TMEV and aphthovirus 5-UTRs should work similarly to that of EMCV. But there are no analogies with the HAV genome, and this system therefore needs special scrutiny.

C. A Negative Control Element

As discussed in the preceding section, the 5-UTRs of picornaviral genomes *can* and *do* promote the initiation of polyprotein synthesis through the internal cap-independent binding of ribosomes to a cis-acting control element. But what about the canonical (i.e., 5' end-dependent) initiation of translation on these templates? Is it possible, in principle, and, if not, does it mean that a distinct negative cis-acting control element is an integral part of the picornaviral 5-UTRs?

One may argue that the ability to promote translation initiation by a cap-dependent mechanism would make no sense for picornaviral templates primarily because they simply lack a cap structure. However, it could not be ruled out in theory that portion of the VPg-devoid viral RNA molecules could be capped in the infected cell. Another theoretical argument against the possible involvement of the cap-dependent initiation system in the polyprotein synthesis is the inactivation of this system on poliovirus infection. Again, counterarguments can be put forward. First, the putative cap-dependent initiation could be imagined to occur at the very early stages of infection, prior to the inactivation of the cap-dependent factors. Second, there are examples of enterovirus infection in which no appreciable shut-off of the cellular protein synthesis could be registered due to viral mutations (Cooper, 1977; Bernstein *et al.*, 1985). Third, the inactivation of the cap-dependent machinery does not appear to take place during cardiovirus infection at all (Mosenkis *et al.*, 1985).

Therefore, the questions posed in the beginning of this section should be approached experimentally. As already mentioned, the transfection of cultured cells with capped poliovirus RNA yielded less (rather than more) viral proteins compared with the uncapped species (Macejak *et al.*, 1990). The capping of transcripts containing the entire poliovirus 5-UTR did not appreciably affect, as compared with the uncapped templates, the efficiency of the expression of the adjacent reporter gene in extracts from either mock- or poliovirus-infected HeLa cells (Pelletier *et al.*, 1988b). These facts should mean that the cap-dependent initiation of translation, if it occurs at all, could be responsible for only a negligible part of the overall protein synthesis directed by the intact poliovirus RNA. On deletions of specific segments of the 5-UTR, however, the cap dependence of translation increased markedly; it was especially pronounced with extensively truncated templates (e.g., those lacking the first 631 nucleotides); such truncated RNAs, however, could not serve as active templates in extracts from poliovirus-infected cells (Pelletier *et al.*, 1988b), in which the cap-dependent translation initiation system is inactivated (see Section V,B). These observations may be interpreted by postulating the existence of a negative cis-acting control element or, in other words, an obstacle to ribosome scanning from the template 5' end to the initiator codon.

This negative element could conveniently be studied in reticulocyte lysates, where, as shown by several groups (Shih *et al.*, 1978; Brown and Ehrenfeld, 1979; Dorner *et al.*, 1984; Svitkin *et al.*, 1985; Phillips and Emmert, 1986), the efficiency of polyprotein synthesis on poliovirus template is rather low due to the deficiency in an important trans factor. The diminished ability of this cell-free system to promote cap-independent internal initiation of translation was an obvious advantage for studying the efficiency of the cap-dependent 5' terminus-dependent event.

An internally deleted template, having only the first 70 nucleotides of the poliovirus 5-UTR, was quite competent in promoting the synthesis of a reporter protein in reticulocyte lysates, whereas the presence of nucleotides corresponding to positions 70–381 was clearly inhibitory (Pelletier *et al.*, 1988c). These data are in full agreement with a related study by Howell *et al.* (1990), who joined portions of the poliovirus 5-UTR to a reporter gene (the *NS* gene of the influenza virus) and investigated the ability of the constructed templates to promote translation initiation at specific AUGs. The results showed that the template with the very 5'-terminal 66 nucleotides of the poliovirus 5-UTR serving as a leader was quite active, whereas the first 390 nucleotides or so presented a real obstacle for ribosome scanning.

Remarkably, the inhibitory cis-acting element resides in the region of poliovirus 5-UTR (nucleotides 70–380) that precedes the positive element, and this appears to make sense. The negative element may well represent a tool to prevent unnecessary ribosome scanning along certain important portions of the 5-UTR. Such scanning might result in some adverse effects, for example, in the trapping of ribosomes or in the unwinding, and thus inactivation, of the positive control element.

On the other hand, similar experiments with mRNA constructs containing portions of the 5-UTR of human rhinovirus 14 suggested that the appropriate negative element was located more downstream, namely, within, or close to, positions 491–546 (AlSaadi *et al.*, 1989). It remains to be seen whether or not this difference reflects important peculiarities of the mechanisms of expression of entero- and rhinovirus genomes, respectively.

Although not studied in detail, the efficient translation of 5'-truncated RNAs of EMCV (Parks *et al.*, 1986) and FMDV (Clarke and Sangar, 1988) suggests that negative translational control elements are commonly present in the picornaviral genomes. Accordingly, in order to construct templates, on the basis of the picornaviral 5-UTRs, with a high capacity for the cap-dependent (or simply 5' terminus dependent) initiation of translation, the negative cis-acting elements are removed (cf. Parks et al., 1986; Nicklin et al., 1987; Clarke and Sangar, 1988).

The molecular basis of negative element activity has yet to be defined. It could be proposed that the secondary structure of the appropriate segment of the 5-UTR is involved (Pelletier et al., 1988c), but its actual contribution is unknown. In any case one should take into consideration that, in order to be inhibitory, the 5-UTR secondary structure elements should either be located very close to the 5' end or exhibit a very high stability, with a ΔG better than -30 kcal/mol (Kozak, 1989c).

V. TRANSLATIONAL trans-ACTING FACTORS INVOLVED IN INTERACTION WITH THE cis-ELEMENT

A. General Initiation Factor Requirements for the Translation of Eukaryotic Cellular mRNAs

A brief overview of the trans factors participating in the initiation of polypeptide synthesis in mammalian cells in general seems to be appropriate prior to a discussion on the peculiarities of the initiation factors' involvement in the picornaviral RNA translation. This overview is, of necessity, sketchy and oversimplified. A deeper analysis of the diverse aspects of the eukaryotic initiation mechanism as well as appropriate references can be found in other reviews (Moldave, 1985; Proud, 1986; Sonenberg, 1988; Rhoads, 1988; Sonenberg and Pelletier, 1989).

There are three key participants in the initiation process: the ribosome, an mRNA template, and the aminoacylated (methionyl) initiator tRNA species, and each of these must be converted, with the aid of initiation factors, into an active form in order to perform its specific role (Fig. 4).

The primary participant in the initiation process is not the whole (80 S) ribosome, but rather its smaller (40 S) subunit. The 80 S ribosomes are normally in equilibrium with their 40 S and 60 S subunits. To prevent association of the subunits, the smaller one should bind two initiation factors: a huge multicomponent factor eIF-3 and factor eIF-4C (the 60 S subunit may perhaps also be temporarily blocked by binding the initiator factor eIF-6). The second participant, the aminoacylated initiator tRNA (Met-tRNA$_i$), should first form a ternary

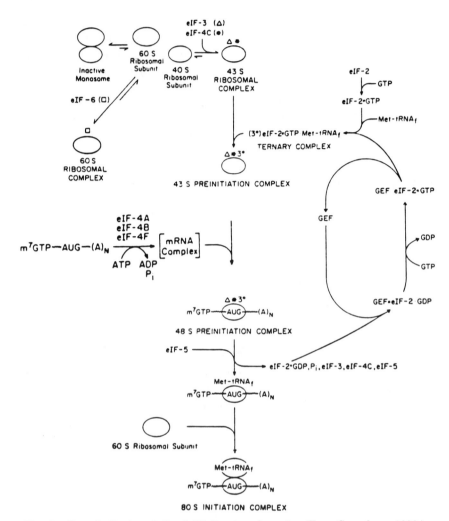

FIG. 4.   Steps in the translation initiation in eukaryotes. (From Sonenberg, 1990.)

complex with factor eIF-2 and GTP. These two components, the ribosome subunit and the ternary complex, combine with one another to form a so-called 43 S initiation complex. Concurrently, the third participant, an mRNA species, which is generally capped at its 5' end, interacts with the cap-recognizing factor, eIF-4E, which can be found in the cell both in a free form and in a complex with two other factors, eIF-4A and p220 (the complex of eIF-4E, eIF-4A, and p220 is called eIF-4F). It is believed that it is eIF-4F that actually interacts with mRNA in an ATP-dependent reaction. The next step is the interaction of the Met-tRNA$_i$-charged 40 S ribosome subunit (the 43 S initiation complex) with the mRNA template associated with eIF-4F; this reaction also requires ATP and free eIF-4A as well as one more initiation factor, eIF-4B. Since eIF-4F as well as eIF-4A alone, and especially together with eIF-4B, exhibit RNA helicase properties (Ray et al., 1985; Lawson et al., 1989; Rozen et al., 1990, quoted by Sonenberg, 1990), it was proposed that their function consists in unwinding mRNA secondary structure elements to permit ribosome "landing," which is presumed to be RNA sequence unspecific and requiring merely a "melted" segment of the template (Sonenberg, 1988; Sonenberg and Pelletier, 1989). The ribosome–template–Met-tRNA$_i$–factors complex is called a 48 S initiation complex.

At this stage a poorly understood process, the ATP-dependent movement ("scanning") of the ribosome subunit along the template ensues (Kozak, 1978, 1989a); it is not known whether the moving force for this process is provided by the ribosome itself or by initiation factors (e.g., the eIF-4A/eIF-4B helicase). The scanning continues until an AUG in a favorable context is encountered. Here, the larger ribosomal subunit (60 S) is, with high probability, recruited to form the 80 S initiation complex in a reaction which requires GTP and one more initiation factor, eIF-5, and is accompanied by the release of an eIF-2–GDP complex, as well as eIF-3, eIF-4C, eIF-5, and probably some other initiation factors. This is the end of the initiation step and the beginning of the elongation step.

Before leaving this topic, however, we should briefly mention the problem of initiation factor recycling. In fact, this problem concerns primarily factor eIF-2, because it is released from the 80 S initiation complex in a form stably bound to GDP and is therefore unable to enter a new initiation cycle. eIF-2–GDP could be converted into a reusable species through an interaction with factor eIF-2B (known also as the guanine nucleotide exchange factor, GEF) and GTP, whereby the nucleotide moiety in eIF-2–GTP is eventually regenerated. A powerful and widespread tool to regulate translation initiation consists of simply inhibiting the eIF-2 recycling by phosphorylation of its α subunit, which can be accomplished by a variety of protein kinases. The phosphorylated form of eIF-2 is, in turn, a potent inhibitor of the eIF-2B activity. The complex and physiologically important controls involving the eIF-2 system have recently been the subject of several reviews (Kaempfer, 1984; London et al., 1987; Gupta et al., 1987; Hershey, 1989; Sarre, 1989).

B. Initiation Factor Requirements for the Translation of Picornaviral RNAs

To comprehend the mode of initiation on picornaviral RNAs we should know, among other things, whether the initiation factor requirements for the efficient translation of these templates, on the one hand, and of the common cellular mRNA species, on the other, are identical. In fact, they could not be expected to be the same in at least one respect. Due to the lack of a cap structure, picornaviral RNAs could hardly be expected to need the cap-recognizing factors eIF-4E and eIF-4F. They do not. Moreover, the p220 subunit of eIF-4F is degraded on enterovirus (Etchison *et al.*, 1982; Lee *et al.*, 1985; Buckley and Ehrenfeld, 1987), rhinovirus (Etchison and Fout, 1985), and aphthovirus (Lloyd *et al.*, 1988; Devaney *et al.*, 1988) [but not cardiovirus (Mosenkis *et al.*, 1985; Lloyd *et al.*, 1988)] infections, rendering the host cell protein-synthesizing machinery unable to efficiently initiate translation of capped mRNAs, while retaining, if not increasing, the capacity to utilize uncapped picornaviral templates (Kaufmann *et al.*, 1976; Helentjaris and Ehrenfeld, 1978; Rose *et al.*, 1978; reviewed by Ehrenfeld, 1984; Kozak, 1986d; Sonenberg, 1987). The diminished ability of cell extracts from poliovirus-infected cells to translate capped mRNA species can be restored on the addition of eIF-4F preparations (Tahara *et al.*, 1981; Grifo *et al.*, 1983; Etchison *et al.*, 1984; Edery *et al.*, 1984), supporting the theory that it is the deficiency in this factor that is responsible, at least partially, for the virus-induced shut-off of the host protein synthesis. Importantly, no evidence for poliovirus infection-induced alterations of other constituents of eIF-4F, either eIF-4A or eIF-4E, nor of another participant of the RNA–ribosome interaction, eIF-4B, was found (Helentjaris *et al.*, 1979; Duncan *et al.*, 1983; Etchison *et al.*, 1984; Lee *et al.*, 1985; Buckley and Ehrenfeld, 1986). Since poliovirus mutants unable to induce efficient inactivation of eIF-4F, being viable, grow relatively poorly (Bernstein *et al.*, 1985), it seems likely that the inhibition of the cap-dependent initiation machinery is beneficial for the virus because it eliminates competition with other templates.

The apparently diminished requirement for ATP during EMCV RNA (as opposed to capped mRNA) translation (Jackson, 1982, 1989) seems to be relevant to our discussion. This observation may suggest that relatively little work would be done by the ATP-consuming enzymes (e.g., eIF-4A/eIF-4B helicase or whatever entity is involved in the ribosome scanning) during the translation initiation on the cardiovirus template. On the other hand, a high requirement for eIF-4A was reported for the initiation of poliovirus polyprotein (Daniels-McQueen *et al.*, 1983). If there actually is a difference in the eIF-4A requirement for the poliovirus and EMCV RNA translation, it is tempting to relate it to a longer distance that a 40 S ribosome has perhaps to cover (i.e., scan) between the entry site and the initiator AUG on the former template (see Section VI). [In considering the validity of these speculations, one should take into account, however, that Blair *et al.* (1977) reported quite the opposite results; according to their data, eIF-4A markedly stimulated the *in vitro* translation of EMCV, mengovirus, and FMDV RNAs, but not that of poliovirus.]

Inasmuch as p220 exhibits some affinity for eIF-3 (and formerly was even considered to be a component of the latter), one should not be particularly surprised if eIF-3 function were somehow modified in poliovirus-infected cells (cf. Kozak, 1986d). However, the existing data do not appear to substantiate such a possibility (Etchison *et al.*, 1984), nor is anything known about alterations in specific requirements for this factor during picornaviral RNA translation.

The next part of our discussion concerns a specific role, if any, of eIF-2 in the initiation of picornaviral polyprotein synthesis. The problem has several facets. One was set forth by R. Kaempfer and associates, who proposed that eIF-2, in addition to its canonical function in guiding Met-tRNA$_i$ to ribosomes, promotes ribosome binding to certain templates, particularly to cardiovirus (mengovirus) RNA, and probably takes part in the selection of proper AUGs as well (reviewed by Kaempfer, 1984). These proposals were based on several considerations. (1) eIF-2 was claimed to bind, with high affinity, to the neighborhood of the initiator AUG, for example, in the satellite tobacco necrosis virus RNA (which is a naturally uncapped template) (Kaempfer *et al.*, 1981) and to mengovirus RNA (Perez-Bercoff and Kaempfer, 1982); the loci of specific eIF-2 and ribosome binding appeared to overlap on these templates. (2) The addition of exogenous eIF-2 was reported to affect the competition between mRNAs with different template activities, such as α- and β-globin mRNAs or globin mRNA and mengovirus RNA, in a manner that could be interpreted to suggest that these RNAs competed for eIF-2 (Di Segni *et al.*, 1979; Rosen *et al.*, 1982). It should be noted that these ideas generally have been met with a great deal of skepticism, but we would consider it unwise to disregard them, especially in light of some recent observations. Thus, a mutational alteration in the eIF-2 β subunit led to a remarkable change in yeast's ability to choose the start codon; the mutant could now begin translation of a specific template from a UUG introduced in place of the genuine AUG (Donahue *et al.*, 1988). The ability of eIF-2 to affect the choice of initiator AUGs by the translational apparatus of rabbit reticulocytes has also been reported (Dasso *et al.*, 1990).

Another aspect of the eIF-2 system, as related to the picornavirus RNA translation, concerns regulation of its activity by phosphorylation and dephosphorylation. There is evidence that enhanced phosphorylation of the α subunit of eIF-2 accompanies poliovirus infection

(Black *et al.*, 1989; O'Neill and Racaniello, 1989), the enhancement being due to the self-phosphorylation and concomitant activation of a protein kinase induced by the accumulating poliovirus-specific double-stranded RNA (the same double-stranded RNA-stimulated kinase as that activated by interferon). Interestingly, the overall activity of this kinase in the infected cells was increased despite extensive degradation of the enzyme due to an unknown poliovirus-induced process (Black *et al.*, 1989) (one may wonder whether a poliovirus-specific protease is involved in this degradation). A similar combination of the activation and degradation of the kinase takes place during the EMCV infection too (Hovanessian *et al.*, 1987). It seems likely that the phosphorylation-triggered partial inactivation of eIF-2 in the infected cells is related to the general inhibition of protein synthesis characteristic of the final steps of the reproduction cycle (O'Neill and Racaniello, 1989).

Other reports describe, on the other hand, accumulation of an inhibitor of the dsRNA-stimulated kinase in poliovirus-infected cells (Ransone and Dasgupta, 1987, 1988) and activation of a kinase, which is also able to phosphorylate the α subunit of eIF-2, but is distinct from the double-stranded RNA-activated enzyme, during mengovirus infection (Pani *et al.*, 1986).

To sum up, we might state that the only relatively well-characterized alteration of the translation initiation factors in picornavirus-infected cells concerns the p220 subunit of eIF-4F. Inactivation of this factor provides an advantage for the virus-specific protein synthesis and furnishes additional evidence for its cap independence. Modulations of the activity of some other initiation factors, first of all eIF-2, and their possible physiological significance deserve further attention. There is no clear evidence that translation initiation on picornaviral templates exhibits an especially high specific requirement for certain initiation factors. It should be stressed, however, that thus far we have been dealing only with the generally recognized initiation factors, which have more or less characterized function(s). The possible involvement of novel factors in the translation initiation on picornaviral templates is discussed in the next section.

An additional general remark seems to be appropriate here. It is a common approach to study the initiation factor requirements for translation of picornaviral mRNAs by using an artificial host system such as HeLa cells, or even such nonhost cells as rabbit reticulocytes. Evidence is accumulating, however, that properties of the translational machinery may vary in cells of different origin or differentiation status, and that such variation may markedly affect expression of the picornaviral genomes. Thus, translation of poliovirus RNA in human blood cells (López-Guerrero *et al.*, 1989) and neural cells (La Monica and Racaniello, 1989; Agol *et al.*, 1989) appears to be relatively restricted. A likely reason for these disparities may consist of the different availability of initiation factors (see also Section IX). The cell dependence of picornavirus translation appears to be a newly emerging and exciting area of research.

C. *Search for Translational trans Factors Interacting with the 5′- Untranslated Region*

As a matter of fact, any cis-acting element can work only if it interacts with something else. In principle, this "something else" may be represented by another genomic cis element or a variety of molecules of different kinds, but the first thing that comes into mind on considering the activity of an mRNA translational cis element is naturally a protein "factor." Several approaches have been pursued to identify novel factor(s) recognizing the translational cis-acting control element(s) in the middle of the picornaviral 5-UTR. Perhaps the first attempt of this kind was undertaken in our laboratory when we looked for a factor able to "sense" mutations modulating the cis element activity (Svitkin *et al.*, 1988). Our study was based on two lines of previous observations.

First, Dorner *et al.* (1984) showed that, in rabbit reticulocyte lysates, poliovirus RNA was translated predominantly from "aberrant" sites located within the polyprotein coding region, although "correct" initiation at the beginning of this region also occurred to a lesser extent. As demonstrated first by Brown and Ehrenfeld (1979) and confirmed in other laboratories (Dorner *et al.*, 1984; Svitkin *et al.*, 1985; Phillips and Emmert, 1986), crude preparations of translation initiation factors from nucleated (e.g., HeLa or Krebs-2) cells could "normalize" the pattern of products directed by poliovirus RNA in reticulocyte lysates. Thus, an "initiation correcting factor" (ICF) should be present in the active fraction from nucleated cells, whereas reticulocyte lysates appear to be ICF deficient. Second, the correct initiation was shown to be especially weak when RNA templates from attenuated poliovirus strains (known to bear point mutations within the 472–480 region of the cis-acting control element; see Section IV,B) were translated in reticulocyte lysates (Svitkin *et al.*, 1985). We wondered, therefore, whether the poor template activity of the RNAs with mutations in the 472–480 region could be explained by their diminished responsiveness to ICF. In other words, we asked whether ICF could "recognize" mutations within the cis-acting control element. The answer was "yes": the initiation of viral polyprotein synthesis was stimulated by partially purified ICF preparations from Krebs-2 cells to a markedly greater extent when RNAs from virulent, compared with attenuated, poliovirus strains were used as templates. We inferred that ICF does interact with a region of the viral 5-UTR that encompasses mutations in the 472–480 region of poliovirus 5-UTR.

The exact nature of ICF has not been established. During several purification steps it copurified with an activity of initiation factor eIF-2, although pure preparations of eIF-2 exerted a negligible ICF effect. Our data are consistent with the assumption that ICF corresponds to a complex between translation initiation factors eIF-2 and eIF-2B (Svitkin et al., 1988), but additional experiments are needed to unequivocally prove this point.

Two other (often used together) approaches to the identification of the trans factors are so-called gel retardation (mobility shift) experiments and ultraviolet (UV) cross-linking. While the meaning of the latter is more or less self-evident, the former consists of investigating the ability of different protein fractions to form complexes with RNA molecules, thereby diminishing the mobility of the latter on nondenaturing gel electrophoresis. In both cases the problem is to find a balance between two opposite requirements. On the one hand, the RNA fragment should be as short as possible, because this gives a relatively greater mobility shift on binding a protein, and, in UV cross-linking experiments, the sequence nonspecific binding will be minimized. On the other hand, the shortening of the fragment increases the chance that it will lose its original secondary and tertiary structures, and hence certain potential RNA–protein interactions would not be realized. This conflict of requirements is a real problem, especially since the length of the RNA fragment is often determined by the availability of convenient sites for restriction nucleases on appropriate plasmids, and since the secondary and especially tertiary structure of the RNA in question is rarely known.

Nothwithstanding the above reservation, experiments of this sort did demonstrate the existence of specific host cell proteins with an affinity for specific segments of poliovirus RNA. It was shown in a pioneering study by Meerovitch et al. (1989) that the electrophoretic mobility of a segment of poliovirus type 2 (Lansing) RNA with coordinates 559–624 (i.e., encompassing the entire domain E; see Fig. 1) was diminished on a reaction with a HeLa cell extract; this retardation was not due to a nonspecific RNA–protein interaction, since it could not be prevented by a huge excess of unrelated RNA. Two proteins, with M_r values of 52K and 100K (p52 and p100, respectively), have been identified as covalently linked to the 559–624 RNA segment after UV irradiation, but, on the basis of several controls, only the former was considered to be specifically bound. Purified preparations of p52 stimulated *in vitro* translation of poliovirus RNA, but p52 could not be identified with any known translation initiation factors (Sonenberg and Meerovitch, 1990). Interestingly, reticulocyte lysates, notorious for their poor ability to translate poliovirus RNA (see above), appeared to be deficient with respect to p52 (Meerovitch *et al.*, 1989). Meerovitch *et al.* considered the possibility that p52 may be responsible for the ICF activity described by Svitkin *et al.* (1988).

Importantly, the A→U substitution at position 588, that is, in AUG_7 [this mutation was shown to lower the cell-free translational template activity of the viral RNA (Pelletier *et al.*, 1988a) and could be expected to destabilize the secondary structure of domain E] decreased the apparent affinity of the RNA segment to p52 (Meerovitch *et al.*, 1989). Further studies (Sonenberg and Meerovitch, 1990; N. Sonenberg, personal communication) showed that similar effects (i.e., a decrease in both template activity and p52 binding) also accompanied mutations of the two other nucleotides of AUG_7; importantly, introduction of the compensating mutations in the 3' branch of domain E did not restore the ability to bind p52. Therefore, p52 appeared to recognize the primary, rather than secondary, structure. This notion was strongly supported by the observation that a shorter RNA segment, entirely lacking the 3' branch of domain E and consequently devoid of the original double-stranded stem, did retain p52-binding capacity (N. Sonenberg, personal communication). It should perhaps be noted that AUG_7 lies just downstream of a conserved stretch of pyrimidine residues, and these two elements may conceivably form a common cis-acting protein-recognizing signal. However important this signal may be, it does not appear to be essential, since, as discussed in Section IV,B, deletions of the corresponding regions from the poliovirus (Kuge and Nomoto, 1987; Iizuka *et al.*, 1989) or coxsackie B1 virus (Iizuka *et al.*, 1990) 5-UTR resulted in viable progeny, although exhibiting an sp phenotype.

However, another cis-acting element, a highly conserved 21-nucleotide sequence shared by entero- and rhinoviruses, lying immediately upstream of the oligopyrimidine stretch and occupying portions of the 3' branch of domain D and the linker between domains D and E (coordinates 543–563 in the poliovirus type 1 RNA), may well be essential (Kuge and Nomoto, 1987; Iizuka *et al.*, 1989, 1990) (see Section IV,B). This element appears to possess the capacity to specifically bind a 57-kDa cellular protein (A. Nomoto, personal communication). The nature of this protein and its relationship, if any, to p52, described in the preceding paragraph, remain unknown. It may be noted that, using a longer poliovirus RNA segment (coordinates 320–629), del Angel *et al.* (1989) also detected specific binding of a cellular protein complex to a region with coordinates 510–629, that is, spanning both the cis-acting elements studies by Sonenberg's and Nomoto's groups; the presence of the α subunit of the translation initiation factor eIF-2 in this complex has been revealed.

Still another region of the enterovirus 5-UTRs, having a potential to bind specific cellular proteins, lies near the 5' end of the cis-acting element and perhaps involves domain A or B (Fig. 1) or both. A 46-nucleotide segment with coordinates 178–224 specifically interacted with a 50K protein, which was distinct from p52 described by Meerovitch *et al.* (1989) as well as from translation initiation factors eIF-2, eIF-3, eIF-4A, and elongation factor EF-1α (Najita and Sarnow, 1990). The interaction seemed to involve the loop of domain B and was accompanied by an apparent partial melting of its stem. The RNA–protein complex was stable enough to suggest that it was held by a covalent bond, and there was evidence that the bond involved an SH group of the protein and a U_{202} residue of the loop (Najita and Sarnow, 1990). Although such a unique feature might suggest an important characteristic of the interaction, the relevant cis-acting element is, nevertheless, dispensable for the virus' viability (Dildine and Semler, 1989) (see Section IV,B).

Finally, a slightly more upstream segment (coordinates 97–182) was also shown to bind a distinct complex of cellular proteins, which seemed to include the α subunit of eIF-2 (del Angel *et al.*, 1989).

Thus, several, though not yet fully characterized, cellular proteins exhibit specific affinities for at least four seemingly separate sites within the enterovirus 5-UTR cis-acting translational control element. Although the functional significance of these numerous protein–RNA interactions remains obscure, it is tempting to speculate that at least some of them are directly involved in the internal entry of ribosomes (see also Section VI).

Two recent papers describe attempts to identify trans factors specifically interacting with the cis element of the EMCV 5-UTR (Borovjagin *et al.*, 1990; Jang and Wimmer, 1990). A cellular polypeptide of about 57–58K (p57) was shown to bind an EMCV RNA segment containing an imperfect stem–loop element with coordinates 401–550. The secondary structure of this element appeared to be more important than the nucleotide sequence for its protein-binding capacity, as judged by the effect of mutations either destroying or restoring the double-stranded stem (Jang and Wimmer, 1990). The changes in this capacity were in parallel with the changes in the template activity of the mutated RNA species in reticulocyte lysates, suggesting that p57 does participate in the internal ribosome binding.

VI. Current Views on the Mechanisms of Translation Initiation on Picornaviral RNAs

Thus, the existence of a cis-acting element in the middle of picornaviral 5-UTRs playing an essential part in the initiation of polyprotein synthesis is firmly established, its boundaries are tentatively defined, and some of the putative trans-acting factors involved are being characterized. Nevertheless, the specific functions of this element are far from being understood. Two major, and not necessarily mutually exclusive, ideas have been put forward to explain the mechanism of initiation in this unusual system.

According to the scenario proposed by Sonenberg and Pelletier (1989) for the poliovirus protein initiation (Fig. 5), some unidentified factor [e.g., host polypeptide p52 (Meerovitch et al., 1989)] specifically binds to the cis-acting element inside the viral 5-UTR, and this binding somehow attracts a helicase complex composed of translation ini-

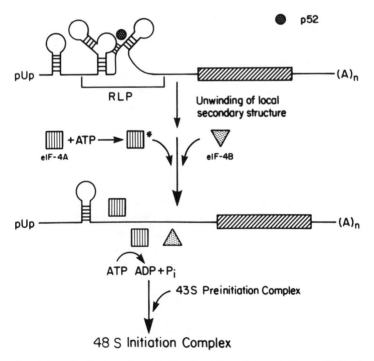

Fig. 5. A hypothetical mechanism for the internal translation initiation on the picornaviral RNAs based on the unwinding of the "ribosome landing pad" (RLP) by the helicase activity of initiation factors eIF-4A and eIF-4B. (From Sonenberg, 1990.)

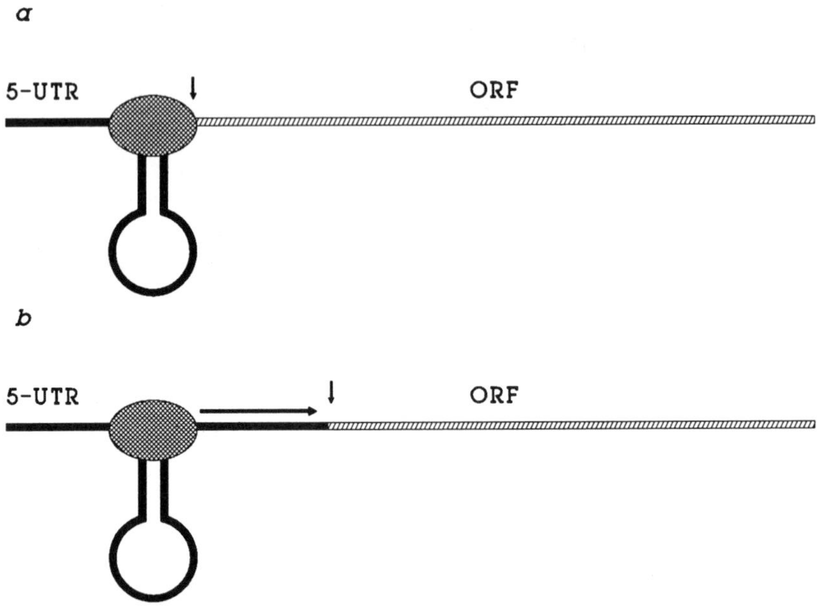

a

5-UTR   ↓   ORF

b

5-UTR   ↓   ORF

FIG. 6.   Two hypothetical modes of the translation initiation on the picornaviral RNAs. (a) Direct ribosome binding to both the initiation site and cis-acting upstream regions brought into proximity by secondary and tertiary interactions. Such a mechanism is probably realized in cardio-, aphtho-, and rhinoviruses. (b) Internal ribosome binding to the cis-acting control region and subsequent ribosome scanning (horizontal arrow) toward the initiation site (vertical arrow). This mode is believed to be characteristic of poliovirus and other enteroviruses. (Modified from Howell *et al.*, 1990.)

tiation factors eIF-4A and eIF-4B. In an ATP-dependent reaction the bidirectional helicase (Rozen *et al.*, 1990, quoted by Sonenberg, 1990) unwinds a segment of the otherwise extensively structured cis element, providing, thereby, an unspecific (just single-stranded) ground for the 40S ribosomal subunit "landing." Being landed, the subunit has to obey Kozak's rules and performs a canonical walk along the 5-UTR until it encounters the correct initiator AUG that opens the polyprotein reading frame.

The other point of view, most clearly and explicitly expressed by Jackson and associates (Howell *et al.*, 1990), assumes that the cis element under discussion is involved in ribosome binding in a more direct way, serving as a framework for putting the bound ribosome (or its 40 S subunit) precisely at a key place, in some cases perhaps very closely to the initiation codon (Fig. 6). This view appears to be directly applicable to the initiation of translation on the EMCV template. Howell *et al.* (1990) compared the probabilities of initiation at the genuine AUG (position 834), on the one hand, and at any of the three closely adjacent upstream AUGs (positions 826, 752, and 743, respectively), on the other (actually, a set of artificial templates containing a portion of EMCV 5-UTR fused to a reporter protein coding sequence was translated in reticulocyte lysates). The main conclusion emerging from these elegant experiments was that, given the presence of an intact upstream cis-acting control element, the initiation occurred almost exclusively at the correct AUG, with a striking disregard of even an AUG located less than 10 nucleotides upstream. This and the two more upstream ignored AUGs were by no means "crippled" ones, since they could ensure fairly efficient initiation, provided a significant portion of the preceding sequence was deleted. The invalidation of the cis-acting control elements (both the positive and negative ones) rendered the mode of initiation fully conforming to Kozak's rules; the most 5'-proximal AUG was not preferentially used, and the initiation was markedly stimulated on template capping.

Somewhat different results were obtained with the poliovirus RNA initiation system. It is true that different lines of evidence strongly argue for the internal entry of ribosomes and against the very possibility that ribosomes can reach the correct initiation codon (at position 743) by scanning from the 5' end of the viral RNA (see Section IV). Nevertheless, scanning along a limited segment of poliovirus RNA adjacent to the initiator AUG seems to be substantiated by the following observations: (1) Insertion of an AUG-containing sequence downstream of position 702 resulted in the lowering of the growth potential of the mutant poliovirus (the appearance of sp phenotype) due most likely to the impairment of the viral protein synthesis; large-plaque revertants of this mutant invariably lacked the newly acquired AUG triplet (Kuge et al., 1989a,b); (2) insertion of a foreign reporter gene together with its own initiator AUG at position 630 of the poliovirus 5-UTR produced a template capable of directing in vitro the synthesis of the reporter protein in a manner indistinguishable from that characteristic of the synthesis of genuine poliovirus proteins (Howell et al., 1990); and (3) insertion of a stable secondary structure element (thought to be an "obstacle" for ribosome scanning) at position 631 of poliovirus type 2-derived 5-UTR resulted in abrogation of the expression of the following ORF (Pelletier and Sonenberg, 1988).

Thus, the cis-acting element in the poliovirus 5-UTR appears to guide ribosomes (or, perhaps more exactly, their 40 S subunits) to a specific locus over 100 nucleotides upstream of the initiator triplet, the distance they must cover by "ordinary" scanning. It seems that the point from which such scanning does start can be specified more precisely. As discussed in Section X, the poliovirus genome could well have originated from a "rhinoviruslike" ancestor by the acquisition of an additional segment between the cis-acting translational control element and the polyprotein reading frame. In the rhinovirus type 2 genome the initiator AUG is located at position 611 (see Table I) just in the conserved secondary structure element E (Fig. 1). We propose that, in the case of entero- and rhinovirus RNAs, the ribosomes are fixed precisely near the AUG they "expect" to find at a proper place within this conserved element. If they do find it (as in rhinovirus RNA), they begin to translate; but if they do not (as in enteroviruses), they have nothing to do but to start searching for an appropriate AUG by scanning.

The now available data do not enable us to define accurately the 5-UTR locus from which the bound ribosome begins its search for an AUG triplet in other picornaviral RNAs. But in EMCV (strain R) RNA it is likely to lie between positions 834 (the genuine start codon) and 826 (the preceding AUG) (Howell et al., 1990). In entero- and rhinovirus genomes the locus is expected to be located within a conserved stem–loop structure (domain E in Fig. 1), perhaps between nucleotides 611 (rhinovirus type 2 numbering; the initiator codon) and 576 (the preceding AUG).

The above considerations prompt the following specification concerning the boundaries (primarily, the 3′ boundary) of the cis-acting translational control element: One should distinguish between the intrinsic ability of this element to promote cap-independent internal ribosome binding to a template, on the one hand, and to mediate the precise positioning of the bound ribosome with respect to the initiator AUG, on the other. Experiments aimed at a more precise definition of the appropriate boundaries should take this specification into account.

One point, already briefly discussed above, deserves, perhaps, more detailed discussion. If the ribosomes were able to accurately place themselves at the correct locus of the template, how would they do this? We have argued that the picornaviral cis-acting element responsible for the internal ribosome binding should have a quasiglobular conformation due to multiple tertiary interdomain interactions (Section III). In order to secure the precise orientation of a 40 S ribosome subunit respective to the template, more than one contact point appears to be needed. We propose that such multipoint ribosome–RNA interaction is a common, and important, feature of the mechanism of translation initiation on the picornaviral RNA (cf. Jang and Wimmer, 1990). Within the framework of such a hypothesis, the abundance of both the protein-recognizing segments in the 5-UTR (Section IV) and protein factors involved (Section V) becomes easily understandable, as does the poor predictability of the outcome of mutational alterations in the cis-acting control element.

VII. "PICORNAVIRUS-LIKE" PROPERTIES OF THE 5'-UNTRANSLATED REGIONS IN OTHER VIRAL AND CELLULAR mRNAs

Thus, we have seen that the peculiar structure of picornaviral 5-UTRs serves a very specific function in the translation of the viral genome. However, there are few, if any, biochemical mechanisms exploited by only a single group of viruses. Moreover, the history of molecular biology teaches us that cellular counterparts could eventually be found for the overwhelming majority of mechanisms for the storage, replication, and expression of the genetic information that had been first discovered in viral systems and considered to be unique at that time. In this section I consider, above all, instances pertaining to the occurrence, among different viral and cellular templates, of such fundamental "picornavirus-like" properties as the capacity for internal ribosome binding, cap independence, and the possession of a negative cis-acting control element. In addition, I touch on the peculiarities of some mRNA species which point to other functional potentials of the long 5-UTRs with multiple AUGs. For convenience I discuss different picornavirus-like traits more or less separately; actually, they may join one another in various combinations.

A. Internal Ribosome Binding

Here, I consider only instances of what suggested to be true internal initiation, not leaky scanning or reinitiation. In this regard the case of the viral polymerase of a hepadnavirus, duck hepatitis B virus (DHBV), should perhaps be mentioned first. The enzyme is translated from a functionally bicistronic 3.3-kb template, whose 5'-proximal and -distal cistrons encode, in two different ORFs, the core protein and polymerase, respectively. The cistrons overlap one another, and the 305-nucleotides overlap region contains two AUGs in the polymerase ORF. Several lines of evidence ruled out the possibility that the synthesis of DHBV polymerase depended on the translation of the preceding cistron; in particular, the utilization of a frame-shifting mechanism was demonstrated to be highly unlikely (Schlicht et al., 1989; L.-J. Chang et al., 1989). The internal initiation at the first AUG of the polymerase ORF was suggested by these authors to be the favored, and, in fact, the only feasible, explanation. Similar data were obtained with another hepadnavirus, human hepatitis B virus (Jean-Jean et al., 1989).

Remarkably, the same mechanism of internal ribosome binding may explain the synthesis of the polymerase encoded in a bicistronic template generated by still another retroid virus, cauliflower mosaic virus (Penswick *et al.*, 1988; Schultze *et al.*, 1990).

It was proposed that the synthesis of a polypeptide of a negative-strand RNA virus, vesicular stomatitis virus (VSV), is also initiated in a similar way (Herman, 1986). This small protein, called 7K, is encoded in the same ORF as phosphoprotein P (also known as polypeptide NS) and corresponds to the carboxy-terminal portion of the latter. Hybridization of the P protein mRNA with appropriate oligodeoxyribonucleotides (hybrid-arrest experiments) permitted blocking of the accumulation of P without affecting the appearance of 7K. Moreover, the synthesis of 7K, unlike that of P, was reported to be insensitive to cap analogs (Herman, 1986, 1987). The simplest, but perhaps not the only, way to explain both these observations is to postulate that the synthesis of 7K is initiated by internally bound ribosomes.

A peculiar case of internal initiation of translation was suggested to explain the synthesis of the so-called X protein of another negative-strand RNA virus, paramyxovirus Sendai virus (Curran and Kolakofsky, 1988). Similar to the previous example, this protein corresponds to a carboxy-terminal portion of a much larger viral polypeptide, protein P, and the former does not seem to be a proteolytic product of the latter. Accumulation of X, in *in vitro* hybrid-arrest translation experiments, also could be demonstrated under conditions excluding synthesis of P, consistent with an internal mode of initiation. The peculiarity of this initiation and its apparent distinction from the VSV case, however, is its sensitivity to cap analogs, suggesting the involvement of the template 5' terminus (Curran and Kolakofsky, 1988).

Two interesting explanations of this apparent discrepancy, internal initiation and cap dependence, were offered by these authors. According to one explanation, eIF-4F is needed to stimulate the eIF-4A/eIF-4B-promoted unwinding of an RNA region (cf. Abramson *et al.*, 1988) to prepare the internal ribosome binding locus near the X polypeptide initiation site; the novel suggestion is that eIF-4F should reach this region only by initial binding to the 5'-terminal cap structure followed by the movement along the template (i.e., by a process similar to the ribosome scanning). The other explanation implies that ribosomes do begin their scanning along the template from the 5' end, but, with some probability, they could skip (i.e., "jump over") an RNA segment; this hypothesis is supported by a case in which a similar ability of prokaryotic ribosomes has been well documented (Huang *et al.*, 1988). In considering these hypotheses, one should take into account, however, that the inhibition by cap analogs could be observed even with a cap-lacking template [e.g., satellite tobacco necrosis virus RNA (Smith and Clark, 1979)], indicating that the competition with a genuine cap is not the exclusive mode of the inhibitory effect of cap analogs.

The above [and some other (cf. Herman, 1989)] examples from the molecular biology of negative-strand RNA viruses are very interesting, but the conclusions on the internal initiation of translation, being derived largely from experiments of the hybrid-arrest type, need, perhaps, more rigorous proof (cf. critical comments by Kozak, 1989a). As in the picornavirus system, such proof may consist of the identification of cis-acting elements able to confer the property of internal ribosome binding to a heterologous gene of a bicistronic template.

There are several other strong candidates for the internal initiation of translation. Some of these are dealt with in the next section.

B. Cap Independence

There is a variety of positive-strand RNA viral genomes lacking a 5'-terminal cap structure and having a covalently linked protein instead; these include RNAs of calici-, como-, nepo-, poty-, and some other plant viruses. Although a detailed mechanism(s) of the translation initiation on such template is of obvious interest, little is actually known in this regard. Nevertheless, it seems appropriate to emphasize that the absence of a cap would not necessarily imply that internal ribosome binding is involved. For example, the satellite tobacco necrosis virus RNA, a naturally uncapped (and VPg-less) template (Wimmer et al., 1968), appears to exploit the canonical set of translation initiation factors, eIF-4F included (Browning et al., 1988); its translation could be inhibited by a cap analog (Smith and Clark, 1979), in contradistinction to the picornavirus RNAs (cf. Svitkin et al., 1986).

Perhaps more intriguing is the situation in which a capped mRNA exhibits the ability to be efficiently translated in cells with inactivated cap-dependent machinery, namely, in poliovirus-infected cells (the physical and functional invalidation of the cap-dependent initiation factor eIF-4F in such cells is discussed in Section V,B). Two clear examples of such templates are late adenovirus mRNAs and a glucose-dependent heat-shock protein mRNA species.

The late adenovirus-specific mRNA species contain an ~200-bp 5-UTR, usually designated as tripartite leader [because it is composed of transcripts of three exons (see Horwitz, 1985)]. The leader has no upstream AUGs and contains a cis-acting control element ensuring translational activation late, but not early, in infection (Logan and

Shenk, 1984). This element permitted efficient translation of appropriate adenovirus mRNA species in poliovirus-infected cells (Castrillo and Carrasco, 1987; Dolph et al., 1988; Jang et al., 1989), suggesting that its function is independent of the cap-binding protein factors. The proposed explanation for this property is based on the determination of the leader's secondary structure; since it contains two apparently unpaired regions, it could simply bind factors eIF-4A and eIF-4B in a cap-independent manner, and the helicase activity of these factors would prepare a single-stranded landing pad of a sufficient length to secure internal binding of ribosomes (Zhang et al., 1989). This explanation, which is in line with the popular view that the major function of the cap-dependent machinery is to promote melting of the mRNA secondary structure (Sonenberg et al., 1982; Sonenberg, 1988), may, however, be incomplete, because it fails to take into account that the tripartite leader appears to activate translation during late, but not early, stages of the adenovirus reproduction cycle. The involvement of trans-acting factors whose concentration or activity would be changed on infection seems, therefore, quite likely. In any case the possibility that the initiation of translation of late adenovirus-specific proteins is accomplished through internal ribosome binding deserves further attention.

Another exciting system of translational control is operating during heat shock (for reviews see Lindquist, 1987; Edery et al., 1987). At supraoptimal temperatures (as well as in response to some other stresses) severe inhibition of translation of the majority of cellular templates usually occurs. Partly, but not solely, this appears to be due to the inactivation of eIF-4F (Panniers et al., 1985; Duncan et al., 1987). Against this inhibited background, however, a class of transcripts encoding so-called heat-shock proteins is synthesized and translated with enhanced efficiencies. At least several mRNA species encoding heat-shock proteins [e.g., heat-shock protein 70 and glucose-regulated protein 78 (GRP78; also known as immunoglobulin heavy-chain binding protein)] are preferentially (compared to other cellular mRNAs) translated in poliovirus-infected cells (Muñoz et al., 1984; Sarnow, 1989), once more demonstrating their independence of the eIF-4F activity.

The peculiar translational properties of the heat-shock protein templates are controlled by their 5-UTRs (McGarry and Lindquist, 1985; Lindquist, 1987), which, by the way, are relatively long (200–250 bp) (cf., e.g., Ting and Lee, 1988). Macejak et al. (1990), using a novel assay system (transient transfection of tissue culture cells with a specific mRNA species), showed that GRP78 5-UTR is able to confer, to a heterologous gene, translatability, and even enhanced translatability, in the poliovirus-infected cells. Moreover, the bicistronic templates, having GRP78 5-UTR as an intercistronic element, express the second cistron under conditions prohibiting expression of the first one (P. Sarnow, personal communication). The latter experiment establishes unambiguously that these elements are able to ensure cap-independent internal ribosome binding.

One more example of the cap-independent internal ribosome binding, though not yet published in detail, concerns the translation initiation on the mRNA encoding the Antennapedia protein of *Drosophila melanogaster* involved in the differentiation of the fruit fly. This mRNA has a remarkably long (i.e., >1500-nucleotide) 5-UTR with 15 upstream AUGs (Stroeher *et al.*, 1986). When segments of this 5-UTR were introduced between cistrons of an engineered bicistronic template. independent expression of the second cistron could be demonstrated, testifying unequivocally to the internal initiation of translation (Oh *et al.*, 1990; P. Sarnow, personal communication). The positive cis-acting element responsible for such initiation appeared to adjoin closely to the initiator AUG.

It is perhaps appropriate to note that the 5-UTRs of several oncogenes are very similar, with respect to their length and sometimes to the number of upstream AUGs, to that of the Antennapedia; the mRNA encoding human c-*abl* (Bernards *et al.*, 1987) and c-*sis* (Ratner *et al.*, 1987) oncogenes could be taken as examples. mRNAs of cellular proteins other than oncogenes, for example, that of 3-hydroxy-3-methylglutaryl-coenzyme A reductase (Reynolds *et al.*, 1984), the rate-limiting enzyme of cholesterol biosynthesis, also has an apparently similar organization. It could be expected that all these templates exploit a common mechanism for translation initiation.

Thus, mRNAs for certain important cellular proteins, or even classes of such proteins, are likely to share with picornaviral and other viral RNAs the propensity for cap-independent internal ribosome binding. Physiological, and possibly evolutionary, significance of this fact has yet to be elucidated.

C. Negative cis-Acting Control Elements

Two kinds of negative translational control elements within eukaryotic 5-UTRs are often considered—secondary structures and upstream AUGs—which may function separately or together.

A remarkable example of what appears to be the regulatory element of the first kind is provided by mRNA encoding a cellular oncogene, c-*myc*. Like many other oncogene templates (cf. Kozak, 1987c), c-*myc* mRNAs of different origin possess a long (i.e., several hundred bases) 5-UTR; at least in some species, these GC-rich 5-UTRs contain no AUGs (Watt *et al.*, 1983; Bernard *et al.*, 1983; Saito *et al.*, 1983). Although we are aware of no direct determinations of the secondary structure of the *myc* mRNA 5-UTR, computer modeling predicts the existence of multiple base-paired elements both between 5-UTR and coding sequences (Saito *et al.*, 1983) and within 5-UTR itself (Parkin *et al.*, 1988a). The *myc* 5-UTR contains a negative translational control element, as judged by a significant increase in the *myc* mRNA template activity on its 5′ truncation (Darveau *et al.*, 1985), and it was proposed that the activation of Myc protein synthesis (and concomitant tumorigenesis) on the translocation of a portion of the *myc* gene into a different genetic background (e.g., as in Burkitt's lymphomas) may result, at least in part, from the removal of this element (Saito *et al.*, 1983).

Remarkably, like the poliovirus negative cis element, the *myc* 5-UTR element exerts its translation inhibiting activity in reticulocyte lysates or *Xenopus* oocytes quite well, being nearly inactive in cultured mammalian cells or extracts therefrom (Parkin *et al.*, 1988a). The reason(s) for these differences is unknown, but it may be speculated that reticulocyte lysates are deficient in either the RNA duplex unwinding (if the negative effect of the cis element is due to an RNA secondary structure) or an unidentified translation initiation factor, which should interact with the cis element to override an obstacle of some other kind (Parkin *et al.*, 1988a); in any case a striking resemblance to the peculiarities of poliovirus RNA translation in reticulocyte lysates is obvious.

It can well be imagined that the negative element takes part in the physiological control of the *myc* gene expression as well. If this is true, then a mechanism for overcoming the element's inhibitory effect should exist. Evidence strongly supporting this theory was reported by Lazarus *et al.* (1988), who observed that, unlike *Xenopus* oocytes, *Xenopus* eggs or embryos could quite efficiently support translation of the *myc* mRNA.

A functionally similar negative translational control element has recently been discovered in the 5-UTR of another human oncogene, *BCR/ABL*. This GC-rich element has the potential to form a stable secondary structure, and its ability to control the efficiency of translation initiation is cell type dependent (Muller and Witte, 1989).

Inhibitory effects of secondary structure elements within 5-UTRs have also been reported for coronavirus (Soe *et al.*, 1987) and human immunodeficiency virus type 1 (Parkin *et al.*, 1988b), porcine proopiomelanocortin (Chevrier *et al.*, 1988), and bovine liver mitochondrial aldehyde dehydrogenase mRNAs (Guan and Weiner, 1989) as well as for other templates.

The involvement of another cis-acting element, upstream AUGs, in the control of translatability of numerous mRNA species also was often suggested, though more rarely documented. A few examples are viral mRNAs of retroviruses (Hackett et al., 1986; Petersen et al., 1989; Hensel et al., 1989), caulimoviruses (Baughman and Howell, 1988; Fütterer et al., 1988, 1989), papovaviruses (Khalili et al., 1987; Sedman et al., 1989), parvoviruses (Ozawa et al., 1988), and herpesviruses (Geballe and Mocarski, 1988); the list also includes templates encoding cellular proteins, such as a variety of oncogenes (cf. Marth et al., 1988; Rao et al., 1988), ornithine decarboxylase (Kahana and Nathans, 1985; Fitzgerald and Flanagan, 1989), and a nuclear protein, PET111, controlling mitochondrial translation in *Saccharomyces cerevisiae* (Strick and Fox, 1987). Detailed analysis of the nature of these controls, however, has yet to be performed. Moreover, a contribution of elements other than the upstream AUGs (e.g., stable hairpins) can be no means be excluded in some of these cases.

D. Some Specific Potentials of Long 5'-Untranslated Regions with Multiple AUGs

The examples that follow do not necessarily differ fundamentally from at least some of those already discussed; they are, however, studied in enough detail to reveal certain specific potentials of the URFs.

The GCN4 protein of *S. cerevisiae* (GCN stands for general control nonderepressible) plays a key role in the regulation of amino acid metabolism, being a transcriptional activator of many unlinked genes involved in amino acid biosynthesis. Its own expression, normally repressed, is tremendously activated on amino acid starvation. GCN4 protein synthesis is controlled at the translational level by cis-acting signals in the 5-UTR of its mRNA as well as by several trans-acting protein factors (reviewed by Hinnebusch and Mueller, 1987; Hinnebusch, 1988).

The 5-UTR of GCN4 mRNA is about 600 nucleotides long (perhaps there are several closely spaced transcription start sites) and it contains four AUGs followed by very short (i.e., only two- or three-codon) URFs (Thireos et al., 1984; Hinnebusch, 1984). These URFs, together with adjoining sequences, constitute a cis-acting control element; their deletion relieves translational repression of GCN4 protein synthesis under nonstarvation conditions (Thireos et al., 1984; Hinnebusch, 1984), and an ~240-bp segment of the GCN4 mRNA containing these URFs confers the GCN4-specific controls to a heterologous gene (Mueller et al., 1987). Negative effects of upstream AUGs in eukaryotic mRNAs are not a surprise (see Section IV,C); in the case of GCN4 template, however, the role of such AUGs is far from trivial.

Thus, different URFs appear to have very distinct functional properties. URF4 and URF3 (numbering is from the 5' end) are the strongest
with respect to repression activity; when all other URFs were inactivated, the modified template with a single URF4 or URF3 exhibited,
under nonstarvation conditions, nearly as low a translational activity
as an mRNA with intact GCN4 5-UTR; the negative effect of the single
URF1 was about 30 times lower (Mueller and Hinnebusch, 1986).

This differential activity could only partly be explained by the distance of the respective URF from the GCN4 start codon; nor was there
any appreciable difference in the intrinsic initiation efficiencies of the
first and fourth AUGs; unexpectedly, a functional difference appeared
to be traced to events occurring during, or soon after, the termination
of translation of short URFs (Mueller et al., 1988; Miller and Hinnebusch, 1989). These authors propose that a sequence to the 3' side of
the URF4 promotes ribosome conversion into a form unable to reinitiate at the downstream, bona fide, GCN4 initiation codon. On the other
hand, on completing translation of the first URF, ribosomes retain, if
not increase, the ability for further scanning of the template and for
the efficient reinitiation at the next AUG. The molecular basis of
these striking differences in the functions of the 5'-proximal and distal GCN4 URFs remains obscure (cf Williams et al., 1988).

What is known, however, is that these differences appear to be directly related to the physiological function of the cis-acting control
element. While URF4 (and perhaps also URF3) is likely to ensure
repression of GCN4 protein synthesis under nonstarvation conditions,
the URF1 behaves as a positive control element and is primarily involved in the translational activation (derepression) on amino acid
deficiencies (Mueller and Hinnebusch, 1986).

This system can work properly, however, only when aided by several
trans-acting factors (Hinnebusch, 1985; Harashima and Hinnebusch,
1986). These factors include, among others, products of negative
(GCD1 gene) as well as positive (GCN2 and GCN3) regulators. Under
good nutrition conditions the GCD1 gene is active and its product
somehow ensures that ribosomes, on completing their working cycles
(initiation, reading, and termination), first at the upstream and then
at the downstream URFs, severely diminish their ability to reinitiate
at the GCN4 ORF [this notion primarily stems from the observation
that GCD1 mutants efficiently synthesize the GCN4 protein constitutively (see Hinnebusch, 1985)]. Amino acid deficiency, however,
stimulates certain GCN genes [in particular, GCN2, which contains a
protein kinase domain (Roussou et al., 1988) and a domain closely
related to histidyl-tRNA synthetase (Wek et al., 1989), as well as
GCN3 (Hannig and Hinnebusch, 1988)], and this stimulation in turn

results in abrogation of the activities of the negative elements (the GCD1 protein and/or the downstream URFs), accompanied by a dramatic increase in the reinitiating capacity of the ribosomes and hence in the GCN4 mRNA translatability (Tzamarias and Thireos, 1988). Hypothetical models explaining the complex interaction of the diverse cis- and trans-acting elements involved, possibly including modulation of the eIF-2 activity, have recently been put forward (Tzamarias *et al.*, 1989; Hinnebusch, 1990).

Still another kind of translational regulation based on the interaction between a 5-UTR cis-acting element and a trans-acting protein factor is exemplified by the ferritin system. Ferritin, a ubiquitous intracellular iron storage protein, is composed of multiple copies of two subunits, light (L) and heavy (H). Ferritin synthesis is controlled posttranscriptionally, with iron deficiency resulting in the decreased translatability of both L and H ferritin mRNA species (reviewed by Theil, 1987). The cis-acting iron-responsive control element was traced to relatively long (i.e., around 200-bp) 5-UTRs of these mRNAs, particularly to their highly conserved stem−loop structure just below 30 bp in length (Hentze *et al.*, 1987; Aziz and Munro, 1987; Rouault *et al.*, 1988). This structure is specifically recognized by a protein repressor, and the affinity between the cis- and trans-acting elements increases on iron deficiency (Leibold and Munro, 1988; Walden *et al.*, 1988, 1989; Rouault *et al.*, 1988, 1989; P. H. Brown *et al.*, 1989) due to the reversible reduction of a sulfhydryl group of the repressor (Haile *et al.*, 1989).

The two regulatory systems just described seem to bear no obvious resemblance to the known function of the appropriate control element within the picornaviral 5-UTRs. Nevertheless, these systems tell us what additional functional potentials are inherent in long 5-UTRs. It would be tempting to speculate that at least some of these potentials may be exploited during the picornaviral reproduction cycle as well, for example, for alterations of the relative efficiencies of the initiations at different sites of the template; it should be noted, however, that there is as yet no experimental evidence to support such a theory.

VIII. STRUCTURE OF 5'-UNTRANSLATED REGION AND RNA REPLICATION

The second essential function I am going to deal with is the replication of the viral genome. One should note that, although the structure of 5-UTR itself could, in principle, affect the replication process, it is the 3', rather than 5', end of the RNA molecule that is expected to contain essential cis-acting signals responsible for the specific template recognition by an RNA-dependent RNA polymerase or whatever other entity is involved. However, since the 3' end of the complementary, or negative, RNA strand serving as a template for the synthesis of the viral, or positive, strand is mirrored in the 5-UTR structure, it is convenient to say that a given 5-UTR mutation affects the synthesis of viral RNA, although the 5-UTR complement may actually be involved.

One may wonder whether the recognition signals are shared by the positive and negative RNA templates of a given virus. If this is so, some structural similarity between the 3' termini of the opposite strands, or, in other words, some complementarity between the 5' and 3' termini of the same strand should exist. The 5' and 3' termini of picornaviral RNAs are, however, obviously unrelated, being represented by a heteropolymeric sequence and a poly(A) tract, respectively. This may indicate that the signals responsible for the positive and negative template recognition are substantially different. On the other hand, the cis-acting signals, especially those located in the positive RNA 3' end, could well be subterminal. Indeed, there is some complementarity between the 5' end- and poly(A)-adjacent sequences of poliovirus RNA molecules, but its functional significance remains unknown.

It should be noted that the functional analysis of the 5-UTR elements involved in replication is far from complete. The removal of over 100 5'-terminal nucleotides resulted in the loss of viability (Racaniello and Baltimore, 1981), most likely because of an RNA replication defect. Deeper insights came from more subtle interventions in the 5-UTRs structure. Racaniello and Meriam (1986), while constructing different poliovirus cDNA clones, isolated a mutant of the type 1 virus lacking nucleotide 10. In the original genome C_{10} is expected to interact with G_{34}, forming the lowest base pair in a 9-bp stem of a hairpin (Larsen et al., 1981), so this stem should be truncated in the mutant RNA. The C_{10}-lacking mutant exhibited a ts and sp phenotype, was deficient in the synthesis of both positive and negative RNA strands at a nonpermissive temperature, and demonstrated somewhat retarded virus-specific protein synthesis at any temperature (Rancaniello and Meriam, 1986). Most likely, the primary lesion in the mutant reproduction was in the viral RNA replication. Several ts [+] pseudorevertants have been isolated from the mutant virus stock, and they invariably retained the original mutation (loss of C_{10}), but all acquired a second-site mutation, $G_{34} \rightarrow U$. Notably, this second-site mutation could be envisioned to restore the original length of the hairpin stem, if the formation of a G_9-U_{34} base pair were postulated (Racaniello and Meriam, 1986). Other independent evidence also appears to suggest an important functional role for the secondary structure, not merely the nucleotide sequence (R. Andino, personal communication).

There is no ready answer as to the possible mechanisms of the adverse effect of the C_{10} loss as well as of the beneficial effect of the second-site mutation at position 34 on the viral genome replication. These mutations perhaps impair and restore, respectively, an important cis-acting replication signal. The nature of the compensating mutation suggests that this signal may be related to a secondary structure element. If this suggestion is correct, we would think that the signal is located in the positive, rather than negative, strand, since no restoration of the stem could be expected in the negative strand (the relevant positions are occupied by C and A residues) (Racaniello and Meriam, 1986). How a secondary structure element in the vicinity of the 5' end of the positive RNA strand can affect the replication process remains entirely unclear. Alternatively, the effect of mutations may be attributed to alterations in the primary, rather than secondary, structure of the RNA templates (be they of positive or negative polarity or even double-stranded). In this case, also, no obvious explanation for the relationship between mutations at positions 10 and 34 could be proposed.

The involvement in the genome replication of another, more downstream, secondary structure element was also documented. There is a very convenient *KpnI* restriction endonuclease site near position 70 of the poliovirus cDNA. This site corresponds to the second (from the 5' end), imperfect, hairpin of the viral RNA, encompassing nucleotides 51–78 (cf. Rivera *et al.*, 1988). There is strong evidence that the secondary, rather than primary, structure of this element is important for efficient genome replication (R. Andino, personal communication). A 4-bp deletion (expected to nearly destroy the element) resulted in the loss of viral viability (Kuge and Nomoto, 1987), whereas a 4-bp insertion at the same site (leading merely to an enlargement of the loop) produced a mutant with ts and sp phenotypes as well as a severe lesion in viral RNA synthesis (Trono *et al.*, 1988a). Viruses with an incomplete reversion to the wt phenotype (i.e., partially restoring the ability to synthesize RNA and produce progeny at 37°C) could readily be selected from the mutant population; remarkably, independent clones of these "revertants" retained the original mutation (i.e., the insertion at position 70) (Andino *et al.*, 1990). Surprisingly, the suppressing mutations in the pseudorevertant genomes mapped to the sequence encoding polypeptide 3C, a viral protease. Although the suppressors appeared to be active only in cis, an analysis of the phenotypic expression of several site-directed mutations in the *3C* gene led Andino *et al.* to the conclusion that the polypeptide itself, rather than

RNA, was responsible for the suppression. If so, then one could postulate that the 3C polypeptide interacts with a hairpin in the poliovirus 5-UTR and that this interaction is involved in the viral genome replication (Andino et al., 1990).

The molecular basis for this involvement is again obscure. Among likely possibilities, the protease bound to the positive RNA 5' terminus or negative RNA 3' terminus may promote the liberation of VPg from its precursor, or the protease moiety of an uncleaved RNA polymerase precursor promotes proper fixation of the latter on the RNA template.

Thus, there is strong evidence that about 80 5'-terminal nucleotides are involved in poliovirus RNA synthesis, but a detailed understanding of the underlying mechanisms is still lacking. Taking into account that both aforementioned hairpins are strongly conserved among entero- and rhinovirus genomes (Rivera et al., 1988), we may assume that these mechanisms, whatever their nature, are also conserved in these two picornavirus genera.

We are aware of no experimental data directly demonstrating the involvement of 5'-terminal structures of cardio- and aphthovirus RNAs in the replication reactions, but remarkable conservation of certain secondary structure elements in these RNAs (see Section X) is highly suggestive of such an involvement.

IX. PHENOTYPIC EXPRESSION OF MUTATIONS

Since picornaviral 5-UTRs take part in at least two essential steps of viral reproduction—translation and replication—one could expect that diverse mutational alterations of this segment would be phenotypically expressed as inhibiting viral growth. The retarded or otherwise impaired virus growth is generally reflected in sp, ts, or cs (cold-sensitive) phenotypes. The identification of such mutations could tell us about the *in vivo* significance of this or that RNA structural element. Especially instructive are cases in which revertants to the wt phenotype were selected and characterized. Numerous examples of such types of analysis were mentioned in the preceding sections.

Here, I intend to discuss picornaviral 5-UTR mutations that alter the virus interaction with specific target cells or organisms. As we shall see, this problem is directly related to certain facets of viral pathogenesis and cell differentiation.

Involvement of the poliovirus 5-UTR in determining the level of viral neurovirulence could first be suspected when the primary structure differences in this segment of the genomes of attenuated (Sabin) strains and their neurovirulent counterparts belonging to either type 1

(Nomoto *et al.*, 1982) or type 3 (Stanway *et al.*, 1984a) had been detected. This suspicion was greatly reinforced by the discovery that reversion to neurovirulence of the Sabin type 3 vaccine strain was often accompanied by a back mutation in position 472 (Cann *et al.*, 1984; Evans *et al.*, 1985); thus, the neurovirulent parent and revertants have a C residue at this position, whereas the attenuated strain has a U residue there. Further experiments with engineered genomes composed of segment derived from attenuated and neurovirulent strains have established unambiguously that the attenuated phenotype of the Sabin type 1 (Kawamura *et al.*, 1989), type 2 (Moss *et al.*, 1989), and type 3 (Westrop *et al.*, 1989) strains is partly due to peculiarities of their 5-UTRs, particularly the segment encompassing nucleotides 472–481 (reviewed by Almond, 1987; Nomoto and Wimmer, 1987; Racaniello, 1988; Agol, 1988).[3] Furthermore, different arbitrary alterations of this segment resulted in significant changes in the level of viral neurovirulence (Skinner *et al.*, 1989).

Several lines of evidence suggest that phenotypic expression of the attenuating mutations located within the 472–481 region of the 5-UTR is tissue specific.[4] Thus, two engineered strains differing from one another only in position 472 of their type 3-derived 5-UTR were shown to produce one log different harvests in human neuroblastoma cells (but not in HeLa cells), the attenuated strain being less proficient (La Monica and Racaniello, 1989; see also Agol *et al.*, 1989). But the neural tissue does not appear to be the only one that could "sense" attenuating mutations. Since the reversion $U_{472} \rightarrow C$ in the Sabin 3 genome is readily selected for in the gut of those vaccinated (Minor and Dunn, 1988), but not in conventional tissue culture cells, it seems likely that the intestinal virus-sensitive cells, whatever their nature, also efficiently discriminate against the U_{472}-containing poliovirus type 3 genomes. There appears to be a similar, gut tissue-specific, discrimination against G in position 480 and against A in position 481 of poliovirus type 1 and type 2 genomes, respectively (Minor and Dunn, 1988; Muzychenko *et al.*, 1991; Pollard *et al.*, 1989).

Other evidence for the host dependence of the expression of mutations in the 472–481 region of the poliovirus RNA consists of the fact

[3]These data refer to the Sabin vaccine strains and they do not mean that *any* attenuated poliovirus should obligatorily have appropriate mutations within its 5-UTR; thus, the 472–481 region of the Koprowski attenuated type 2 poliovirus strain W-2 is identical to that of the virulent Lansing strain (Pevear *et al.*, 1990).

[4]Other 5-UTR mutations (e.g., deletions of nucleotides 564–726 from the genomes of both virulent and attenuated poliovirus type 1 strains) may also lead to a decrease in the level of neurovirulence, but appropriate mutants appear to exhibit a diminished reproductive capacity in any host cells (Iizuka *et al.*, 1989).

that the appropriate mutants exhibited a ts phenotype in certain tissue cultures, being ts $^+$ in others (P. D. Minor, personal communication).

Although it was originally suggested that the attenuating point mutations within the poliovirus 5-UTR could bring about dramatic changes in the secondary structure of the relevant segment of the poliovirus genome (Evans et al., 1985), it was later found that, instead, they generally resulted in certain destabilization, not gross rearrangement, of the structure characteristic of the neurovirulent counterpart (Pilipenko et al., 1989a; Skinner et al., 1989; Muzychenko et al., 1991).

As discussed in Sections IV and V, the most likely reason for the attenuating character of these mutations is the impaired ability of the mutated RNA to initiate polyprotein synthesis (Svitkin et al., 1985). A natural question then arises as to why this translation defect should be tissue specific? Although the exact answer is unknown, a possible clue is provided by the observation that translation initiation factor eIF-2 seems to be involved, directly or otherwise, in the interaction with the cis-acting control element (Svitkin et al., 1988; del Angel et al., 1989; see also Section V). Since the eIF-2 system is known to be subject to diverse controls (London et al., 1987; Hershey, 1989), it could well be imagined that its activity is dependent on the status of cell differentiation. Therefore, the apparently extraordinary sensitivity of neural cells to attenuating mutations in the 5-UTR control element seems, in general, to be understandable and, in fact, has been predicted (Svitkin et al., 1985, 1988; Agol, 1988). Moreover, it could be suggested that the ability of even wild-type poliovirus strains to cause a paralytic disease in only a minute proportion of infected nonimmune humans (Melnick, 1985) could partly be due to physiological or pathological alterations of the translation machinery in their central nervous systems (Agol, 1988, 1990).

It should be emphasized, however, that direct involvement of the eIF-2 system is likely, but not rigorously proved, and by no means is the only possible way to explain tissue specificity of the expression of attenuating mutations located within the poliovirus 5-UTR. Evidence is accumulating that other newly emerging and less characterized initiation factors, such as p52 (Meerovitch et al., 1989) or p100 (T. V. Pestova, unpublished observations), could be involved in the recognition of the poliovirus 5-UTR cis-acting signals, on the one hand, and may exhibit a certain degree of cell specificity, on the other (see Section V,C).

Whatever the nature of the tissue specificity of the attenuating mutations in the poliovirus genome, the very existence of such specificity and its apparent relation to the 5-UTR structure raises the possibility of creating completely nonvirulent polioviruses by fusing the poliovirus coding sequence with a (portion of) 5-UTR borrowed from a nonneurotropic picornavirus. Such polioviruses would prove to be very promising for live vaccines.

As to the aphtho- and cardioviruses, there is ample evidence that the poly(C) tract is somehow involved in their pathogenicity. The first hint at the possibility of such an involvement was reported by Harris and Brown (1977), who observed that selection, by passages in a tissue culture, of an attenuated FMDV strain from the virulent parent was accompanied by a marked decrease in the poly(C) length (from 170 to 100 nucleotides). Although a later study by another group failed to reveal a simple correlation between FMDV pathogenicity and poly(C) length (Costa Giomi et al., 1984), this group did detect an apparent contribution of the homopolymeric tract length to the reproductive capacity of the virus in the organism (Costa Giomi et al., 1988). Indeed, these authors noted that strains having a longer poly(C) tract had been selected for on chronic infection of steers, though at late stages shortening of the originally long poly(C) was also observed. Mixed infection experiments with the two FMDV isolates differing from each other with respect to the poly(C) length demonstrated that the strain with a longer poly(C) had a selective growth advantage (Costa Giomi et al., 1988).

More clear-cut evidence for the involvement of poly(C) in virulence was obtained with EMCV. Nearly complete removal of the poly(C) tract (e.g., leaving only eight C residues), while not impairing its reproductive capacity in tissue culture cells (Duke and Palmenberg, 1989), rendered the virus strongly attenuated (Duke et al., 1990). Such a poly(C)⁻ attenuated virus turned out to be a very efficient live vaccine able to prevent superinfection of mice with lethal doses of wild-type EMCV. The molecular mechanism underlying the attenuated phenotype of the poly(C)⁻ EMCV mutant has yet to be established.

Possible involvement of 5-UTR sequences other than the poly(C) tract in the control of mice neurovirulence of another cardiovirus, TMEV (which has no such tract), follows from experiments reported by Calenoff et al. (1990). These authors constructed a set of infectious recombinant cDNA clones composed of segments derived from the genomes of attenuated and neurovirulent strains. Although the precise mapping of the appropriate determinants has yet to be done, some indirect evidence allowed these authors to suggest that a region downstream of position 790 might be especially important.

A comparison of nucleotide sequences of several EMCV variants with different biological properties permitted the suggestion that the ability to induce interferon in vitro may be related to a U insertion at position 765, which could potentially add a base pair to a secondary structure element (Bae et al., 1989).

Several groups (Cohen et al., 1987a,b; Jansen et al., 1988; Ross et al., 1989) noted that HAV strains differing in the levels of attenuation to monkeys or of adaptation to growth in tissue culture also exhibited several 5-UTR alterations [which may or may not be reflected in modifications of the RNA secondary structure (cf. Cohen et al., 1987a,b)]. An actual contribution of the first 347 nucleotides to the latter property has been documented (Cohen et al., 1989).

It is obvious that biological relevance should not necessarily be attributed to every 5-UTR mutation found in a picornavirus with altered behavior. There is a 1- to 3-nucleotide difference in the poly(C) length among the related genomes of diabetogenic and nondiabetogenic EMCV variants, respectively (Cohen et al., 1988; Bae et al., 1989, 1990). Nothing suggests, however, that this minor deviation has any physiological significance; rather, the biological difference should perhaps be attributed to an amino acid change in the VP1 site possibly involved in the interaction with β cells (Bae et al., 1990). One should note that there is no general theory able to predict the biological outcome of this or that 5-UTR mutation.

X. SOME EVOLUTIONARY CONSIDERATIONS

The data presently available do not allow us a serious discussion on the origin of picornaviral 5-UTRs. Nevertheless, an analysis of 5-UTR structure variations among different picornavirus representatives could provide insights into how some specific segments of this region could evolve. Several clues emerged from the discovery of gross rearrangements (duplications and insertions/deletions) within certain 5-UTRs (Pilipenko et al., 1990).

We already know that the essential cis-acting translational control element lies far upstream (i.e., > 100 nucleotides) from the initiator AUG in the poliovirus RNA. How did this element find itself at such a remote, and unique, position? Our hypothesis is based on the existence of directly repeated sequences located in the segment preceding, and partly intruding into, the polyprotein coding region (Pilipenko et al., 1990) (Fig. 7). Actually, there are two such duplications. The repeating unit of the first duplication is over 100 nucleotides long, and its two copies (S' + L' + C' and S + L + C) occupy, in poliovirus type 1 Mahoney strain, positions 533–645 and 670–772, respectively (the initiator AUG starts at residue 743 and defines the upstream border of the C region). The segment separating these repeating units (positions

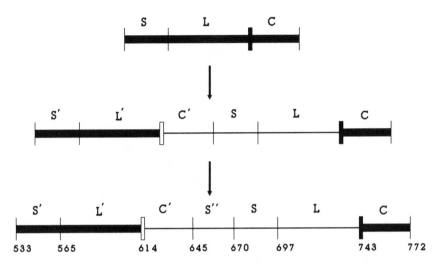

F<small>IG</small>. 7. A hypothetical reconstruction of the origin of the enterovirus genome. S and L, Short and long segments of the repeating unit; C, the region encoding an amino-terminal portion of the viral polyprotein; solid horizontal bars, the most conserved and functionally most important segments; solid vertical bars, actual initiator AUGs of the polyprotein reading frame; open vertical bars, mutated or otherwise inactivated initiation codons. The numbers correspond to the nucleotide positions in the poliovirus type 1 RNA. For other explanations, see text. (From Pilipenko *et al.*, 1990.)

646–669) is, in turn, a unit (S") of another tandem duplication; the other, downstream, unit (S) of the latter corresponds to a 5'-terminal portion of the distal unit of the larger, just mentioned, repeating element. Importantly, the rhinovirus 5-UTR has only a single copy of the sequence corresponding to the poliovirus S + L + C element, even though in other respects the polio- and rhinovirus 5-UTRs are strikingly similar. On the basis of these observations, we proposed the following partial reconstruction of the poliovirus genome evolution (Fig. 7).

A predecessor of poliovirus had a rhinoviruslike genome, that is, the translational cis element (S + L) adjoined the beginning of the polyprotein reading frame (C), and it was present as a single-copy sequence. Then, a duplication of this element occurred, accompanied by the appearance of an additional functional initiation codon which, according to the Kozak's theory, could interfere with normal initiation. The adverse effect of such a development could be abolished by inactivation of the upstream AUG by a mutation (in fact, AAG is located here in the present-day poliovirus RNA). In addition the unwanted reading frame could be destroyed in other ways as well, for example, by an additional shorter duplication (e.g., by the appearance

of S''). The most important parts of this newly evolved structure corre-
sponded to the 5' segment of the upstream repeating unit (S' + L'),
where the cis element is located, and to the 3' segment of the downstream unit (C), harboring the beginning of the polyprotein coding-
sequence. Just these two parts of the polio genome turned out to be
much more conserved as compared to the "insert" (C' + S + L).

As discovered by Kuge and Nomoto (1987), this insert could be deleted without inflicting any apparent damage to the viral phenotype.
This should not be a great surprise, since, due to the duplication, such
deletions might render nearly complete restoration of the secondary
structure of the essential control element (Pilipenko et al., 1990). Despite the fact that this insert portion of the poliovirus genome is certainly not essential, it is present (in a more or less diverged form) in
the RNAs of all enteroviruses investigated so far. This suggests that it
is hardly devoid of any biological significance; perhaps it could participate in fine translational controls.

The above considerations allow us to suggest that the primary interaction of ribosomes with rhinovirus RNA should be in a sense more
similar to their interaction with EMCV RNA than with seemingly
more related poliovirus RNA: In the two former cases the ribosome is
expected to be positioned just adjacent to the initiator triplet and ready
to start translation, whereas in the latter case it should scan a distance
between the landing site and the initiator triplet (cf. Section VI).

Some other rearrangements were found closer to the end of the
picornaviral 5-UTRs, and are, therefore, more likely to affect replication, rather than translation, of the viral genome. Thus, a direct tandem repeat >100 nucleoties in length (positions 7–115 and 116–230)
could be revealed in the genome of bovine enterovirus (BEV) (Pilipenko et al., 1990). The repeating units can fold into nearly identical
secondary structure elements that, in turn, are very similar to a single-copy element found by Rivera et al. (1988) to be conserved among
different entero- and rhinovirus genomes; as already discussed in Section VIII, mutations in this conserved sequence affect replication of
poliovirus RNA (Racaniello and Meriam, 1986; Andino et al., 1990). It
seems likely that at least the upstream repeating unit of the BEV 5-
UTR could be specifically recognized by the viral genome replication
machinery. The biological relevance, if any, of the downstream unit
has yet to be established.

An interesting rearrangement was also detected on a comparison of
the primary and secondary structures of the poly(C)-preceding segments in EMCV and FMDV 5-UTRs. These segments could form
stem–loop structures (Vartapetian et al., 1983; Newton et al., 1985)
with several conserved, among EMCV and FMDV RNAs, secondary

structure elements and an approximately 250-nucleotide insertion into the loop in the latter case (Pilipenko *et al.*, 1990). Despite this large insertion into the terminal structure, apparently involved in the genome replication, the essential recognition elements appear to be conserved in the FMDV RNA due to long-range interactions. The origin of this insertion is unknown, although a weak similarity to a region of FMDV RNA downstream from poly(C) could be revealed.

It may be noted that in the genome of another cardiovirus, TMEV, there is a long segment that appears to replace the poly(C) tract of such cardioviruses as EMCV or mengovirus (Pevear *et al.*, 1987). Although the origin of this replacement remains a mystery, it may be related to pathogenic or other host-related properties of the virus.

Thus, a variety of duplications and large insertions/deletions could be discerned in the picornaviral 5-UTRs. In addition, relatively short repeating elements 3' from FMDV poly(C) were described previously (Clarke *et al.*, 1987). Moreover, it was speculated that the entire poliovirus genome had originated through the multiplication of short genetic elements (Gorbalenya *et al.*, 1986).

The mechanism of generation of repeats and other rearrangements is unknown, but it is most likely related to template switches postulated to occur during the replication of picornaviral RNA (Romanova *et al.*, 1986; Kirkegaard and Baltimore, 1986; Kuge *et al.*, 1986). The generation of adjacent (i.e., tandem) direct repeating elements requires a single "jump" of the nascent chain, perhaps in association with the RNA-dependent RNA polymerase, from one template to another (or from one locus of a template to another locus of the same template) (cf. Romanova *et al.*, 1986), whereas for the appearance of noncontiguous repeating elements (or the insertion of "foreign" sequences) at least two such jumps are needed. The duplication-generating jumps would obviously be facilitated if the template contained short direct repeats flanking the sequence to be duplicated.

There is another intriguing evolutionary problem. Coding sequences for picornaviral nonstructural proteins are remarkably like those of several plant virus families, especially como- and nepoviruses, suggesting their close interrelatedness (Argos *et al.*, 1984; Franssen *et al.*, 1984; Greif *et al.*, 1988). However, the 5-UTRs of true picornaviruses and picornavirus-like plant viruses, as already mentioned, hardly share any significant structural features, except for 5'-terminal VPg (instead of a cap structure). It could be speculated that the picornaviral 5-UTRs are either a relatively late acquisition, at least in comparison with the age of their coding sequences, or that this structural and functional genomic unit had, for whatever reasons, been lost at birth or during subsequent evolution of the plant viral genomes. One may wonder whether the above distinction reflects certain intrinsic differences in the mammalian and plant translational machineries.

XI. Concluding Remarks

On reviewing the wealth of existing knowledge related to picornaviral 5-UTRs, one might perhaps be left with somewhat mixed feelings. On the one hand, the efforts of numerous research groups appear to be fairly rewarded, and the field, which was discovered only a decade ago, can undoubtedly be regarded as one of the most advanced areas of molecular virology. It gives insights not only into the mechanisms of viral reproduction and behavior, but also into the working abilities of the eukaryotic translational machinery in general. On the other hand, the understanding, in molecular terms, of the activities of the translational and replicational cis-acting elements within the picornaviral 5-UTRs is yet to come. Among the questions most urgently awaiting adequate answers, I would list the following.

1. What is the spatial organization of the 5-UTRs, and of their cis-acting regulatory elements in particular?

2. What is the nature of the protein trans factors interacting with the relevant cis elements of the 5-UTRs, and how do these two partners recognize one another?

3. What are the physiological outcomes of the interactions between the trans factors and the cis elements?

4. Why is the expression of some 5-UTR mutations tissue dependent?

5. What is the origin of the 5-UTRs?

Obviously, most of these questions could be, and actually have been, put forward even before any systematic investigation of the picornaviral 5-UTRs was initiated. Not less evident, however, is the fact that adequate answers to these questions are now within much easier reach.

Acknowledgments

I am deeply indebted to my colleagues who participated in the work done at the Institute of Poliomyelitis and Viral Encephalitides as well as at Moscow State University. I thank numerous investigators who generously provided me with their prepublication data. I am also grateful to Vera Ginevskaya, Anatoly Gmyl, and Natalya Bodanovich for their help with the manuscript preparation.

REFERENCES

Abramson, R. D., Dever, T. E., and Merrick, W. C. (1988). Biochemical evidence supporting a mechanism for cap-independent and internal initiation of eukaryotic mRNA. *J.
Biol. Chem.* **263**, 6016–6019.

Agol, V. I. (1988). Genetic determinants of neurovirulence and attenuation of poliovirus.
Mol. Genet. Mikrobiol. Virusol. **1**, 3–9.

Agol, V. I. (1990). Current approaches to the problem of poliovirus attenuation. In "New
Aspects of Positive-Strand RNA Viruses" (M. A. Brinton and F. X. Heinz, eds.), pp.
311–318. Am. Soc. Microbiol., Washington, D.C.

Agol, V. I., Drozdov, S. G., Ivannikova, T. A., Kolesnikova, M. S., Korolev, M. B., and
Tolskaya, E. A. (1989). Restricted growth of attenuated poliovirus strains in cultured
cells of a human neuroblastoma. *J. Virol.* **63**, 4034–4038.

Almond, J. W. (1987). The attenuation of poliovirus neurovirulence. *Annu. Rev. Microbiol.* **41**, 153–180.

AlSaadi, S., Hassard, S., and Stanway, G. (1989). Sequences in the 5' non-coding region
of human rhinovirus 14 RNA that affect *in vitro* translation. *J. Gen. Virol.* **70**, 2799–
2804.

Ambros, V., and Baltimore, D. (1978). Protein is linked to the 5' end of poliovirus RNA
by a phosphodiester linkage to tyrosine. *J. Biol. Chem.* **253**, 5263–5266.

Ambros, V., Pettersson, R. F., and Baltimore, D. (1978). An enzymatic activity in uninfected cells that cleaves the linkage between poliovirion RNA and the 5' terminal
protein. *Cell* **15**, 1439–1446.

Andino, R., Rieckhof, G. E., Trono, D., and Baltimore, D. (1990). Substitutions in the
protease (3C^{pro}) gene of poliovirus can suppress a mutation in the 5' noncoding region. *J. Virol.* **64**, 607–612.

Argos, P., Kamer, G., Nicklin, M. J. H., and Wimmer, E. (1984). Similarity in gene
organization and homology between proteins of animal picornaviruses and a plant
comovirus suggest common ancestry of these virus families. *Nucleic Acids Res.* **12**,
7251–7267.

Auvinen, P., Stanway, G., and Hyypiä, T. (1989). Genetic diversity of enterovirus sub-
groups. *Arch. Virol.* **104**, 175–186.

Aziz, N., and Munro, H. N. (1987). Iron regulates ferritin mRNA translation through a
segment of its 5' untranslated region. *Proc. Natl. Acad. Sci. U.S.A.* **84**, 8478–8482.

Bae, Y. S., Eun, H. M., and Yoon, J. W. (1989). Genomic differences between the diabetogenic and nondiabetogenic variants of encephalomyocarditis virus. *Virology*
170, 282–287.

Bae, Y. S., Eun, H. M., Pon, R. T., Giron, D., and Yoon, J. W. (1990). Two amino acids,
Phe 16 and Ala 776, on the polyprotein are most likely to be responsible for the
diabetogenicity of encephalomyocarditis virus. *J. Gen. Virol.* **71**, 639–645.

Baughman, G., and Howell, S. H. (1988). Cauliflower mosaic virus 35 S RNA leader
region inhibits translation of downstream genes. *Virology* **167**, 125–135.

Beck, E., Forss, S., Strebel, K., Cattaneo, R., and Feil, G. (1983). Structure of the FMDV
translation initiation site and of the structural proteins. *Nucleic Acids Res.* **11**, 7873–
7885.

Bernard, O., Cory, S., Gerondakis, S., Webb, E., and Adams, J. M. (1983). Sequence of
the murine and human cellular *myc* oncogenes and two modes of *myc* transcription
resulting from chromosome translocation in B lymphoid tumours. *EMBO J.* **2**, 2375–
2383.

Bernards, A., Rubin, C. M., Westbrook, C. A., Paskind, M., and Baltimore, D. (1987). The first intron in the human c-*abl* gene is at least 200 kilobases long and is a target for translocations in chronic myelogenous leukemia. *Mol. Cell. Biol.* **7**, 3231–3236.

Bernstein, H. D., Sonenberg, N., and Baltimore, D. (1985). Poliovirus mutant that does not selectively inhibit host cell protein synthesis. *Mol. Cell. Biol.* **5**, 2913–2923.

Bienkowska-Szewczyk, K., and Ehrenfeld, E. (1988). An internal 5'-noncoding region required for translation of poliovirus RNA *in vitro*. *J. Virol.* **52**, 3068–3072.

Black, D. N., Stephenson, P., Rowlands, D. J., and Brown, F. (1979). Sequence and location of the poly C tract in aphtho- and cardiovirus RNA. *Nucleic Acids Res.* **6**, 2381–2390.

Black, T. L., Safer, B., Hovanessian, A., and Katze, M. G. (1989). The cellular 68,000-M_r protein kinase is highly autophosphorylated and activated yet significantly degraded during poliovirus infection: Implications for translational regulation. *J. Virol.* **63**, 2244–2251.

Blair, G. E., Dahl, H. H. M., Truelsen, E., and Lelong, J. C. (1977). Functional identity of a mouse ascites and a rabbit reticulocyte initiation factor required for natural mRNA translation. *Nature (London)* **265**, 651–653.

Blinov, V. M., Pilipenko, E. V., Romanova, L. I., Sinyakov, A. N., Maslova, S. V., and Agol, V. I. (1988). A comparison of the secondary structures of the 5'-untranslated segment of neurovirulent and attenuated poliovirus strains. *Dokl. Akad. Nauk SSSR* **298**, 1004–1006.

Borovjagin, A. V., Evstafieva, A. G., Ugarova, T. Y., and Shatsky, I. N. (1990). A factor that specifically binds to the 5'-untranslated region of encephalomyocarditis virus RNA. *FEBS Lett.* **261**, 237–240.

Brown, F. (1979). Structure–function relationships in the picornaviruses. *In* "The Molecular Biology of Picornaviruses" (R. Perez-Bercoff, ed.), pp. 49–72. Plenum, New York.

Brown, B. A., and Ehrenfeld, E. (1979). Translation of poliovirus RNA *in vitro*: Changes in cleavage pattern and initiation sites by ribosomal salt wash. *Virology* **97**, 396–405.

Brown, E. A., Jansen, R. W., and Lemon, S. M. (1989). Characterization of a simian hepatitis A virus (HAV): Antigenic and genetic comparison with human HAV. *J. Virol.* **63**, 4932–4937.

Brown, F., Newman, J., Stott, J., Porter, A., Frisby, D., Newton, C., Carey, N., and Fellner, P. (1974). Poly(C) in animal viral RNAs. *Nature (London)* **251**, 342–344.

Brown, P. H., Daniels-McQueen, S., Walden, W. E., Patino, M. M., Gaffield, L., Bielser, D., and Thach, R. E. (1989). Requirements for the translation repression of ferritin transcripts in wheat germ extracts by a 90-kDa protein from rabbit liver. *J. Biol. Chem.* **264**, 13383–13386.

Browning, K. S., Fletcher, L., and Ravel, J. M. (1988). Evidence that the requirements for ATP and wheat germ initiation factors 4A and 4F are affected by a region of satellite tobacco necrosis virus RNA that is 3' to the ribosomal binding site. *J. Biol. Chem.* **263**, 8380–8383.

Bruce, C., Al-Nakib, W., Forsyth, M., Stanway, G., and Almond, J. W. (1989). Detection of enterovirus using cDNA and synthetic oligonucleotide probes. *J. Virol. Methods* **25**, 233–240.

Buckley, B., and Ehrenfeld, E. (1986). Two-dimensional gel analyses of the 24-kDa cap binding protein from poliovirus-infected and uninfected HeLa cells. *Virology* **152**, 497–501.

Buckley, B., and Ehrenfeld, E. (1987). The cap-binding protein complex in uninfected and poliovirus-infected HeLa cells. *J. Biol. Chem.* **262**, 13599–13606.

Calenoff, M. A., Faaberg, K. S., and Lipton, H. L. (1990). Genomic regions of neurovirulence and attenuation in Theiler murine encephalomyelitis virus. *Proc. Natl. Acad. Sci. U.S.A.* **87**, 978–982.

Callahan, P. L., Mizutani, S., and Colonno, R. J. (1985). Molecular cloning and complete sequence determination of RNA genome of human rhinovirus type 14. *Proc. Natl. Acad. Sci. U.S.A.* **82**, 732–736.

Cann, A. J., Stanway, G., Hughes, P. J., Minor, P. D., Evans, D. M. A., Schild, G. C., and Almond, J. W. (1984). Reversion to neurovirulence of the live-attenuated Sabin type 3 oral poliovirus vaccine. *Nucleic Acids Res.* **12**, 7787–7792.

Castrillo, J. L., and Carrasco, L. (1987). Adenovirus late protein synthesis is resistant to the inhibition of translation induced by poliovirus. *J. Biol. Chem.* **262**, 7328–7334.

Chang, K. H., Auvinen, P., Hyypiä, T., and Stanway, G. (1989). The nucleotide sequence of coxsackievirus A9; implications for receptor binding and enterovirus classification. *J. Gen. Virol.* **70**, 3269–3280.

Chang, L.-J., Pryciak, P., Ganem, D., and Varmus, H. E. (1989). Biosynthesis of the reverse transcriptase of hepatitis B viruses involves *de novo* translational initiation not ribosomal frameshifting. *Nature (London)* **337**, 364–368.

Chevrier, D., Vezina, C., Bastille, J., Linard, C., Sonenberg, N., and Boileau, G. (1988). Higher order structures of the 5'-proximal region decrease the efficiency of translation of the porcine pro-opiomelanocortin mRNA. *J. Biol. Chem.* **263**, 902–910.

Chumakov, K. M., and Agol, V. I. (1976). Poly(C) sequence is located near the 5' end of encephalomyocarditis virus RNA. *Biochem. Biophys. Res. Commun.* **71**, 551–557.

Chumakov, K. M., Chichkova, N. V., and Agol, V. I. (1979). 5'-Terminal sequence of encephalomyocarditis virus RNA: Localization of the poly(C) tract and its role in translation. *Dokl. Akad. Nauk SSSR* **246**, 994–996.

Clarke, B. E., and Sangar, D. V. (1988). Processing and assembly of foot-and-mouth disease virus proteins using subgenomic RNA. *J. Gen. Virol.* **69**, 2313–2325.

Clarke, B. E., Brown, A. L., Currey, K. M., Newton, S. E., Rowlands, D. J., and Carroll, A. R. (1987). Potential secondary and tertiary structure in the genomic RNA of foot and mouth disease virus. *Nucleic Acids Res.* **15**, 7067–7079.

Cohen, J. I., Rosenblum, B., Ticehurst, J. R., Daemer, R. J., Feinstone, S. M., and Purcell, R. H. (1987a). Complete nucleotide sequence of an attenuated hepatitis A virus: Comparison with wild-type virus. *Proc. Natl. Acad. Sci. U.S.A.* **84**, 2497–2501.

Cohen, J. I., Ticehurst, J. R., Purcell, R. H., Buckler-White, A., and Baroudy, B. M. (1987b). Complete nucleotide sequence of wild-type hepatitis A virus: Comparison with different strains of hepatitis A virus and other picornaviruses. *J. Virol.* **61**, 50–59.

Cohen, J. I., Rosenblum, B., Feinstone, S. M., Ticehurst, J., and Purcell, R. H. (1989). Attenuation and cell culture adaptation of hepatitis A virus (HAV): A genetic analysis with HAV cDNA. *J. Virol.* **63**, 5364–5370.

Cohen, S. H., Naviaux, R. K., Vanden Brink, K. M., and Jordan, G. W. (1988). Comparison of the nucleotide sequences of diabetogenic and nondiabetogenic encephalomyocarditis virus. *Virology* **166**, 603–607.

Cooper, P. D. (1977). Genetics of picornaviruses. *Compr. Virol.* **9**, 133–207.

Cooper, P. D., Agol, V. I., Bachrach, H. L., Brown, F., Ghendon, Y., Gibbs, A. J., Gillespie, J. H., Lonberg-Holm, K., Mandel, B., Melnick, J. L., Mohanty, S. B., Povey, R. C., Rueckert, R. R., Schaffer, F. L., and Tyrrell, D. A. J. (1978). Picornaviridae: Second report. *Intervirology* **10**, 165–180.

Costa Giomi, M. P., Bergman, I. E., Scodeller, E. A., Auge de Mello, P., Gomez, I., and La Torre, J. L. (1984). Heterogeneity of the polyribocytidylic acid tract in aphthovirus: Biochemical and biological studies of viruses carrying polyribocytidylic acid tracts of different lengths. *J. Virol.* **51**, 799–805.

Costa Giomi, M. P., Gomes, I., Tiraboschi, B., Auge de Mello, P., Bergman, I. E., Scodeller, E. A., and La Torre, J. L. (1988). Heterogeneity of the polyribocytidylic acid tract in aphthovirus: Changes in the size of the poly(C) of viruses recovered from persistently infected cattle. *Virology* **162**, 58–64.

Curran, J., and Kolakofsky, D. (1988). Scanning independent ribosomal initiation of the Sendai virus X protein. *EMBO J.* **7**, 2869–2874.

Currey, K. M., Peterlin, B. M., and Maizel, J. V., Jr. (1986). Secondary structure of poliovirus RNA: Correlation of computer-predicted with electron microscopically observed structures. *Virology* **148**, 33–46.

Dabrowski, C., and Alwine, J. C. (1988). Translational control of synthesis of simian virus 40 late proteins from polycistronic 19S late mRNA. *J. Virol.* **62**, 3182–3192.

Daniels-McQueen, S., Detjen, B. M., Grifo, J. A., Merrick, W. C., and Thach, R. E. (1983). Unusual requirements for optimum translation of polio viral RNA *in vitro*. *J. Biol. Chem.* **258**, 7195–7199.

Darveau, A., Pelletier, J., and Sonenberg, N. (1985). Differential efficiencies of *in vitro* translation of mouse c-*myc* transcripts differing in the 5′ untranslated region. *Proc. Natl. Acad. Sci. U.S.A.* **82**, 2315–2319.

Dasso, M. C., Milburn, S. C., Hershey, J. W. B., and Jackson, R. J. (1990). Selection of the 5′-proximal translation initiation site is influenced by mRNA and eIF-2 concentrations. *Eur. J. Biochem.* **187**, 361–371.

del Angel, R. M., Papavassiliou, A. G., Fernández-Tomas, C., Silverstein, S. J., and Racaniello, V. R. (1989). Cell proteins bind to multiple sites within the 5′-untranslated region of poliovirus RNA. *Proc. Natl. Acad. Sci. U.S.A.* **86**, 8299–8303.

Devaney, M. A., Vakharia, V. N., Lloyd, R. E., Ehrenfeld, E., and Grubman, M. J. (1988). Leader protein of foot-and-mouth disease virus is required for cleavage of the p220 component of the cap-binding protein complex. *J. Virol.* **62**, 4407–4409.

Dildine, S. L., and Semler, B. L. (1989). The deletion of 41 proximal nucleotides reverts a poliovirus mutant containing a temperature-sensitive lesion in the 5′ noncoding region of genomic RNA. *J. Virol.* **63**, 847–862.

Di Segni, G. D., Rosen, H., and Kaempfer, R. (1979). Competition between α- and β-globin messenger ribonucleic acids for eucaryotic initiation factor 2. *Biochemistry* **13**, 2847–2854.

Dolph, P. J., Racaniello, V., Villamarin, A., Palladino, F., and Schneider, R. J. (1988). The adenovirus tripartite leader may eliminate the requirement for cap-binding protein complex during translation initiation. *J. Virol.* **62**, 2059–2066.

Donahue, T. F., Cigan, A. M., Pabich, E. K., and Valavicius, B. C. (1988). Mutations at a Zn(II) finger motif in the yeast eIF-2B gene alter ribosomal start-site selection during the scanning process. *Cell* **54**, 621–632.

Dorner, A. J., Rothberg, P. G., and Wimmer, E. (1981). The fate of VPg during *in vitro* translation of poliovirus RNA. *FEBS Lett.* **132**, 219–223.

Dorner, A. J., Semler, B. L., Jackson, R. J., Hanecak, R., Duprey, E., and Wimmer, E. (1984). In vitro translation of poliovirus RNA: Utilization of internal initiation sites in reticulocyte lysate. *J. Virol.* **50**, 507–514.

Drygin, Y. F., Vartapetian, A. B., and Chumakov, K. M. (1979). The covalent bond between RNA and protein in encephalomyocarditis virus. *Mol. Biol. (Moscow)* **13**, 777–789.

Duechler, M., Skern, T., Sommergruber, W., Neubauer, C., Gruendler, P., Fogy, I., Blaas, D., and Kuechler, E. (1987). Evolutionary relationships within the human rhinovirus genus: Comparison of serotypes 89, 2, and 14. *Proc. Natl. Acad. Sci. U.S.A.* **84**, 2605–2609.

Duke, G. M., and Palmenberg, A. C. (1989). Cloning and synthesis of infectious cardiovirus RNAs containing short, discrete poly(C) tracts. *J. Virol.* **63**, 1822–1826.

Duke, G. M., Osorio, J. E., and Palmenberg, A. C. (1990). Attenuation of Mengo virus through genetic engineering of the 5' noncoding poly(C) tract. *Nature (London)* **343,** 474–476.

Duncan, R., Etchison, D., and Hershey, J. W. B. (1983). Protein synthesis eukaryotic initiation factors 4A and 4B are not altered by poliovirus infection of HeLa cells. *J. Biol. Chem.* **258,** 7236–7239.

Duncan, R., Milburn, S. C., and Hershey, J. W. B. (1987). Regulated phosphorylation and low abundance of HeLa cell initiation factor eIF-4F suggest a role in translational control. Heat shock effects on eIF-4F. *J. Biol. Chem.* **262,** 380–388.

Earl, J. A. P., Skuce, R. A., Fleming, C. S., Hoey, E. M., and Martin, S. J. (1988). The complete nucleotide sequence of a bovine enterovirus. *J. Gen. Virol.* **69,** 253–263.

Edery, I., Lee, K. A. W., and Sonenberg, N. (1984). Functional characterization of eucaryotic mRNA cap binding protein complex: Effects on translation of capped and naturally uncapped RNAs. *Biochemistry* **23,** 2456–2462.

Edery, I., Pelletier, J., and Sonenberg, N. (1987). Role of eukaryotic messenger RNA cap-binding protein in regulation of translation. *In* "Translational Regulation of Gene Expression" (J. Ilan, ed.), pp. 335–366. Plenum, New York.

Ehrenfeld, E. (1984). Picornavirus inhibition of host cell protein synthesis. *Compr. Virol.* **19,** 177–221.

Elroy-Stein, O., Fuerst, T. R., and Moss, B. (1989). Cap-independent translation of mRNA conferred by encephalomyocarditis virus 5' sequence improves the performance of the vaccinia virus/bacteriophage T7 hybrid expression system. *Proc. Natl. Acad. Sci. U.S.A.* **86,** 6126–6130.

Ericson, G., and Wollenzien, P. (1988). Use of reverse transcription to determine the exact locations of psoralen photochemical crosslinks in RNA. *Anal. Biochem.* **174,** 215–223.

Etchison, D., and Fout, S. (1985). Human rhinovirus 14 infection of HeLa cells results in the proteolytic cleavage of the p220 cap-binding complex subunit and inactivates globin mRNA translation *in vitro*. *J. Virol.* **54,** 634–638.

Etchison, D., Milburn, S. C., Edery, I., Sonenberg, N., and Hershey, J. W. B. (1982). Inhibition of HeLa cell protein synthesis following poliovirus infection correlates with the proteolysis of a 220,000-dalton polypeptide associated with eucaryotic initiation factor 3 and a cap binding protein complex. *J. Biol. Chem.* **257,** 14806–14810.

Etchison, D., Hansen, J., Ehrenfeld, E., Edery, I., Sonenberg, N., Milburn, S., and Hershey, J. W. B. (1984). Demonstration *in vitro* that eucaryotic initiation factor 3 is active but that a cap-binding protein complex is inactive in poliovirus-infected HeLa cells. *J. Virol.* **51,** 832–837.

Evans, D. M. A., Dunn, G., Minor, P. D., Schild, G. C., Cann, A. J., Stanway, G., Almond, J. W., Currey, K., and Maizel, J. V., Jr. (1985). Increased neurovirulence associated with a single nucleotide change in a noncoding region of the Sabin type 3 poliovaccine genome. *Nature (London)* **314,** 548–550.

Fernandez-Muñoz, R., and Darnell, J. E. (1976). Structural difference between the 5' termini of viral and cellular mRNA in the poliovirus-infected cells: Possible basis for the inhibition of host protein synthesis. *J. Virol.* **18,** 719–726.

Fernandez-Muñoz, R., and Lavi, U. (1977). 5' Termini of poliovirus RNA: Difference between virion and nonencapsidated 35S RNA. *J. Virol.* **21,** 820–824.

Fitzgerald, M. C., and Flanagan, M. A. (1989). Characterization and sequence analysis of the human ornithine decarboxylase gene. *DNA* **8,** 623–634.

Flanegan, J. B., Pettersson, R. F., Ambros, V., Hewlett, M. J., and Baltimore, D. (1977). Covalent linkage of a protein to a defined nucleotide sequence at the 5'-terminus of virion and replicative intermediate RNAs of poliovirus. *Proc. Natl. Acad. Sci. U.S.A.* **74,** 961–965.

Forss, S., Strebel, K., Beck, E., and Schaller, H. (1984). Nucleotide sequence and genome organization or foot-and-mouth disease virus. *Nucleic Acids Res.* **12**, 6587–6601.

Fox, G. E., and Woese, C. R. (1975). 5S RNA secondary structure. *Nature (London)* **256**, 505–507.

Franssen, H., Leunissen, J., Goldbach, R., Lomonossoff, G., and Zimmern, D. (1984). Homologous sequences in non-structural proteins from cowpea mosaic virus and picornaviruses. *EMBO J.* **3**, 855–861.

Fütterer, J., Gordon, K., Bonneville, J. M. Sanfaçon, H., Pisan, B., Penswick, J., and Hohn, T. (1988). The leading sequence of caulimovirus large RNA can be folded into a large stem–loop structure. *Nucleic Acids Res.* **16**, 8377–8390.

Fütterer, J., Gordon, K., Pfeffer, P., Sanfaçon, H., Pisan, B., Bonneville, J. M., and Hohn, T. (1989). Differential inhibition of downstream gene expression by the cauliflower mosaic virus 35S RNA leader. *Virus Genes* **3**, 45–55.

Geballe, A. P., and Mocarski, E. S. (1988). Translational control of cytomegalovirus gene expression is mediated by upstream AUG codons. *J. Virol.* **62**, 3334–3340.

Goodchild, J., Fellner, P., and Porter, A. G. (1975). The determination of secondary structure in the poly(C) tract of encephalomyocarditis virus RNA with sodium bisulphite. *Nucleic Acids Res.* **2**, 887–895.

Gorbalenya, A. E., Donchenko, A. P., and Blinov, V. M. (1986). A possible common origin of poliovirus proteins with different functions. *Mol. Genet. Mikrobiol. Virusol.* **1**, 36–41.

Greif, C., Hemmer, O., and Fritsch, C. (1988). Nucleotide sequence of tomato black ring virus RNA-1. *J. Gen. Virol.* **69**, 1517–1529.

Grifo, J. A., Tahara, S. M., Morgan, M. A., Shatkin, A. J., and Merrick, W. C. (1983). New initiation factor activity required for globin mRNA translation. *J. Biol. Chem.* **258**, 5804–5810.

Grubman, M. J., and Bachrach, H. L. (1979). Isolation of foot-and-mouth disease virus messenger RNA from membrane-bound polyribosomes and characterization of its 5′ and 3′ termini. *Virology* **98**, 466–470.

Guan, K. L., and Weiner, H. (1989). Influence of the 5′-end region of aldehyde dehydrogenase mRNA on translational efficiency. Potential secondary structure inhibition of translation *in vitro*. *J. Biol. Chem.* **264**, 17764–17769.

Gupta, N. K., Ahmad, M. F., Chakrabarti, D., and Nasrin, N. (1987). Roles of eukaryotic initiation factor 2 and eukaryotic initiation factor 2 ancillary protein factors in eukaryotic protein synthesis initiation. *In* "Translational Regulation of Gene Expression" (J. Ilan, ed.), pp. 287–334. Plenum, New York.

Hackett, P. B., Petersen, R. B., Hensel, C. H., Albericio, F., Gunderson, S. I., Palmenberg, A. C., and Barany, G. (1986). Synthesis *in vitro* of a seven amino acid peptide encoded in the leader RNA of Rous sarcoma virus. *J. Mol. Biol.* **190**, 45–57.

Hagenbüchle, O., Santer, M., and Steitz, J. A. (1978). Conservation of the primary structure at the 3′ end of 18S rRNA from eucaryotic cells. *Cell* **13**, 551–563.

Haile, D. J., Hentze, M. W., Rouault, T. A., Harford, J. B., and Klausner, R. D. (1989). Regulation of interaction of the iron-responsive element binding protein with iron-responsive RNA elements. *Mol. Cell. Biol.* **9**, 5055–5061.

Hannig, E. M., and Hinnebusch, A. G. (1988). Molecular analysis of *GCN3*, a translational activator of *GCN4*: Evidence for posttranslational control of *GCN3* regulatory function. *Mol. Cell. Biol.* **8**, 4808–4820.

Harashima, S., and Hinnebusch, A. G. (1986). Multiple *GCD* genes required for repression of *GCN4*, a transcriptional activator of amino acid biosynthetic genes in *Saccharomyces cerevisiae*. *Mol. Cell. Biol.* **6**, 3990–3998.

Harris, T. J. R. (1979). The nucleotide sequence at the 5′ end of foot and mouth disease virus RNA. *Nucleic Acids Res.* **7**, 1765–1785.

Harris, T. J. R. (1980). Comparison of the nucleotide sequence at the 5' end of RNAs from nine aphthoviruses, including representatives of the seven serotypes. *J. Virol*, **36**, 659–664.

Harris, T. J. R., and Brown, F. (1976). The location of the poly(C) tract in the RNA of foot-and-mouth disease virus. *J. Gen. Virol.* **33**, 493–501.

Harris, T. J. R., and Brown, F. (1977). Biochemical analysis of a virulent and an avirulent strain of foot-and-mouth disease virus. *J. Gen. Virol.* **34**, 87–105.

Haselman, T., Camp, D. G., and Fox, G. E. (1989). Phylogenetic evidence for tertiary interactions in 16S-like ribosomal RNA. *Nucleic Acids Res.* **17**, 2215–2221.

Helentjaris, T., and Ehrenfeld, E. (1978). Control of protein synthesis in extracts from poliovirus-infected cells. I. mRNA discrimination by crude initiation factors. *J. Virol.* **2b**, 510–521.

Helentjaris, T., Ehrenfeld, E., Brown-Luedi, M. L., and Hershey, J. W. B. (1979). Alterations in initiation factor activity from poliovirus-infected HeLa cells. *J. Biol. Chem.* **254**, 10973–10978.

Hensel, C. H., Petersen, R. B., and Hackett, P. B. (1989). Effects of alterations in the leader sequence of Rous sarcoma virus RNA on initiation of translation. *J. Virol.* **63**, 4986–4990.

Hentze, M. W., Rouault, T. A., Caughman, S. W., Dancis, A., Harford, J. B., and Klausner, R. D. (1987). A cis-acting element is necessary and sufficient for translational regulation of human ferritin expression in response to iron. *Proc. Natl. Acad. Sci. U.S.A.* **84**, 6730–6734.

Herman, R. C. (1986). Internal initiation of translation on the vesicular stomatitis virus phosphoprotein mRNA yields a second protein. *J. Virol.* **58**, 797–804.

Herman, R. C. (1987). Characterization of the internal initiation of translation on the vesicular stomatitis virus phosphoprotein mRNA. *Biochemistry* **26**, 8346–8350.

Herman, R. C. (1989). Alternatives for the initiation of translation. *Trends Biochem. Sci. (Pers. Ed.)* **14**, 219–222.

Hershey, J. W. B. (1989). Protein phosphorylation controls translation rates. *J. Biol. Chem.* **264**, 20823–20826.

Hewlett, M. J., Rose, J. K., and Baltimore, D. (1976). 5'-Terminal structure of poliovirus polyribosomal RNA is pUp. *Proc. Natl. Acad. Sci. U.S.A.* **73**, 327–330.

Hinnebusch, A. G. (1984). Evidence for translational regulation of the activator of general amino acid control in yeast. *Proc. Natl. Acad. Sci. U.S.A.* **81**, 6442–6446.

Hinnebusch, A. G. (1985). A hierarchy of *trans*-acting factors modulates translation of an activator of amino acid biosynthetic genes in *Saccharomyces cerevisiae*. *Mol. Cell. Biol.* **5**, 2349–2360.

Hinnebusch, A. G. (1988). Mechanisms of gene regulation in the general control of amino acid biosynthesis in *Saccharomyces cerevisiae*. *Microbiol. Rev.* **52**, 248–273.

Hinnebusch, A. G. (1990). Involvement of an initiation factor and protein phosphorylation in translational control of *GCN4* mRNA. *Trends Biochem. Sci. (Pers. Ed.)* **15**, 148–152.

Hinnebusch, A. G., and Mueller, P. P. (1987). Translational control of a transcriptional activator in the regulation of amino acid biosynthesis in yeast. *In* "Translational Regulation of Gene Expression" (J. Ilan, ed.), pp. 397–412. Plenum, New York.

Horwitz, M. S. (1985). Adenoviruses and their replication. *In* "Virology" (B. N. Fields, ed.), pp. 433–476. Raven, New York.

Hovanessian, A. G., Galabru, J., Meurs, E., Buffet-Janvresse, C., Svab, J., and Robert, N. (1987). Rapid decrease in the levels of the double-stranded RNA-dependent protein kinase during virus infections. *Virology* **159**, 126–136.

Howell, M. T., Kaminski, A., and Jackson, R. J. (1990). Unique features of initiation of picornavirus RNA translation. *In* "New Aspects of Positive-Strand RNA Viruses" (M. A. Brinton and F. X. Heinz, eds.), pp. 144–151. Am. Soc. Microbiol., Washington, D.C.

Huang, W. M., Ao, S. Z., Casjens, S., Orlandi, R., Zeikus R., Weiss, R., Winge, D., and Fang, M. (1988). A persistent untranslated sequence within bacteriophage T4 DNA topoisomerase gene 60. *Science* **239**, 1005–1012.

Hughes, P. J., Evans, D. M. A., Minor, P. D., Schild, G. C., Almond, J. W., and Stanway, G. (1986). The nucleotide sequence of a type 3 poliovirus isolated during a recent outbreak of poliomyelitis in Finland. *J. Gen. Virol.* **67**, 2093–2102.

Hughes, P. J., North, C., Jellis, C. H., Minor, P. D., and Stanway, G. (1988). The nucleotide sequence of human rhinovirus 1B: Molecular relationships within the rhinovirus genus. *J. Gen. Virol.* **69**, 49–58.

Hughes, P. J., North, C., Minor, P. D., and Stanway, G. (1989). The complete nucleotide sequence of coxsackievirus A21. *J. Gen. Virol.* **70**, 2943–2952.

Hunt, T. (1985). False starts in translational control of gene expression. *Nature (London)* **316**, 580–581.

Hyypiä, T., Auvinen, P., and Maaronen, M. (1989). Polymerase chain reaction for human picornaviruses. *J. Gen. Virol.* **70**, 3261–3268.

Iizuka, N., Kuge, S., and Nomoto, A. (1987). Complete nucleotide sequence of the genome of coxsackievirus B1. *Virology* **156**, 64–73.

Iizuka, N., Kohara, M., Hagino-Yamagishi, K., Abe, S., Komatsu, T., Tago, K., Arita, M., and Nomoto, A. (1989). Construction of less neurovirulent polioviruses by introducing deletions into the 5' noncoding sequence of the genome. *J. Virol.* **63**, 5354–5365.

Iizuka, N., Yonekawa, H., and Nomoto, A. (1990). RNAs of less virulent CVB1 constructed *in vitro* are less efficient mRNAs. *Rinshoken Int. Conf., 5th* p. 66.

Inoue, T., Suzuki, T., and Sekiguchi, K. (1989). The complete nucleotide sequence of swine vesicular disease virus. *J. Gen. Virol.* **70**, 919–934.

Jackson, R. J. (1982). The cytoplasmic control of protein synthesis. *In* "Protein Biosynthesis in Eucaryotes" (R. Perez-Bercoff, ed.), pp. 363–417. Plenum, New York.

Jackson, R. J. (1989). Comparison of encephalomyocarditis virus and poliovirus with respect to translation initiation and processing *in vitro*. *In* "Molecular Aspects of Picornavirus Infection and Detection" (B. L. Semler and E. Ehrenfeld, eds.), pp. 51–71. Am. Soc. Microbiol., Washington, D.C.

Jackson, R. J., Howell, M. T., and Kaminski, A. (1990). Initiation of translation of picornaviral RNAs. *Rinshoken Int. Conf., 5th* p. 18.

Jacobson, M. F., and Baltimore, D. (1968). Polypeptide cleavages in the formation of poliovirus proteins. *Proc. Natl. Acad. Sci. U.S.A.* **61**, 77–84.

James, B. D., Olsen, G. J., Liu, J., and Pace, N. R. (1988). The secondary structure of ribonuclease P RNA, the catalytic element of a ribonucleoprotein enzyme. *Cell* **52**, 19–26.

Jang, S. K., and Wimmer, E. (1990). Cap-independent translation of encephalomyocarditis virus RNA: Structural elements of the internal ribosomal entry site and essential binding of a cellular 57 kD protein. *Genes Dev.* **4**, 1560–1572.

Jang, S. K., Kräusslich, H.-G., Nicklin, M. J. H., Duke, G. M., Palmenberg, A. C., and Wimmer, E. (1988). A segment of the 5' nontranslated region of encephalomyocarditis virus RNA directs internal entry of ribosomes during in vitro translation. *J. Virol.* **62**, 2636–2643.

Jang, S. K., Davies, M. V., Kaufman, R. J., and Wimmer, E. (1989). Initiation of protein synthesis by internal entry of ribosomes into the 5' nontranslated region of encephalomyocarditis virus RNA in vivo. *J. Virol.* **63**, 1651–1660.

Jansen, R. W., Newbold, J. E., and Lemon, S. M. (1988). Complete nucleotide sequence of cell culture-adapted variant of hepatitis A virus: Comparison with wild-type virus with restricted capacity for *in vitro* replication. *Virology* **163**, 299–307.

Jean-Jean, O., Weimer, T., De Recondo, A. M., Will, H., and Rossignol, J. M. (1989). Internal entry of ribosomes and ribosomal scanning involved in hepatitis B virus P gene expression. J. Virol. 63, 5451–5454.

Jenkins, O., Booth, J. D., Minor, P. D., and Almond, J. W. (1987). The complete nucleotide sequence of coxsackievirus B4 and its comparison to other members of the picornaviridae. J. Gen. Virol. 68, 1835–1848.

Johansen, H., Schumperli, D., and Rosenberg, M. (1984). Affecting gene expression by altering the length and sequence of the 5' leader. Proc. Natl. Acad. Sci. U.S.A. 81, 7698–7702.

Johnson, V. H., and Semler, B. L. (1988). Defined recombinants of poliovirus and coxsackievirus: Sequence-specific deletions and functional substitutions in the 5'-noncoding region of viral RNAs. Virology 162, 47–57.

Kaempfer, R. (1984). Regulation of eukaryotic translation. Compr. Virol. 19, 99–175.

Kaempfer, R., van Emmelo, J., and Fiers, W. (1981). Specific binding of eukaryotic initiation factor 2 to satellite tobacco necrosis virus RNA at a 5'-terminal sequence comprising the ribosome binding site. Proc. Natl. Acad. Sci. U.S.A. 78, 1542–1546.

Kahana, C., and Nathans, D. (1985). Nucleotide sequence of murine ornithine decarboxylase mRNA. Proc. Natl. Acad. Sci. U.S.A. 82, 1673–1677.

Kaufmann, Y., Goldstein, E., and Penman, S. (1976). Poliovirus-induced inhibition of polypeptide initiation in vitro on native polyribosomes. Proc. Natl. Acad. Sci. U.S.A. 73, 1834–1838.

Kawamura, N., Kohara, M., Abe, S., Komatsu, T., Tago, K., Arita, M., and Nomoto, A. (1989). Determinants in the 5' noncoding region of poliovirus Sabin 1 RNA that influence the attenuation phenotype. J. Virol. 63, 1302–1309.

Khalili, K., Brady, J., and Khoury, G. (1987). Translational regulation of SV40 early mRNA defines a new viral protein. Cell 48, 639–645.

Kirkegaard, K., and Baltimore, D. (1986). The mechanism of RNA recombination in poliovirus. Cell 47, 433–443.

Kitamura, N., Semler, B. L., Rothberg, P. G., Larsen, G. R., Adler, C. J., Corner, A. J., Emini, E. A., Hanecak, R., Lee, J. J., van der Werf, S., Anderson, C. W., and Wimmer, E. (1981). Primary structure, gene organization and polypeptide expression of poliovirus RNA. Nature (London) 291, 547–553.

Koch, F., and Koch, G. (1985). "The Molecular Biology of Poliovirus." Springer-Verlag, Berlin.

Kozak, M. (1978). How do eucaryotic ribosomes select initiation regions in messenger RNA? Cell 15, 1109–1123.

Kozak, M. (1981). Mechanism of mRNA recognition by eukaryotic ribosomes during initiation of protein synthesis. Curr. Top. Microbiol. Immunol. 93, 81–123.

Kozak, M. (1984a). Compilation and analysis of sequences upstream from the translational start site in eukaryotic mRNAs. Nucleic Acids Res. 12, 857–871.

Kozak, M. (1984b). Selection of initiation sites by eucaryotic ribosomes: Effect of inserting AUG triplets upstream from the coding sequence for preproinsulin. Nucleic Acids Res. 12, 3873–3893.

Kozak, M. (1986a). Point mutations define a sequence flanking the AUG initiator codon that modulates translation by eukaryotic ribosomes. Cell 44, 283–292.

Kozak, M. (1986b). Influences of mRNA secondary structure on initiation by eukaryotic ribosomes. Proc. Natl. Acad. Sci. U.S.A. 83, 2850–2854.

Kozak, M. (1986c). Bifunctional messenger RNAs in eukaryotes. Cell 47, 481–483.

Kozak, M. (1986d). Regulation of protein synthesis in virus-infected animal cells. Adv. Virus Res. 31, 229–292.

Kozak, M. (1987a). At least six nucleotides preceding the AUG initiator codon enhance translation in mammalian cells. J. Mol. Biol. 196, 947–950.

Kozak, M. (1987b). Effects of intercistronic length on the efficiency of reinitiation by eucaryotic ribosomes. *Mol. Cell. Biol.* **7**, 3438–3445.

Kozak, M. (1987c). An analysis of 5'-noncoding sequences from 699 vertebrate messenger RNAs. *Nucleic Acids Res.* **15**, 8125–8133.

Kozak, M. (1988). Leader length and secondary structure modulate mRNA function under conditions of stress. *Mol. Cell. Biol.* **8**, 2737–2744.

Kozak, M. (1989a). The scanning model for translation: An update. *J. Cell Biol.* **108**, 229–241.

Kozak, M. (1989b). Context effects and inefficient initiation at non-AUG codons in eucaryotic cell-free translation systems. *Mol. Cell. Biol.* **9**, 5073–5080.

Kozak, M. (1989c). Circumstances and mechanisms of inhibition of translation by secondary structure in eucaryotic mRNAs. *Mol. Cell. Biol.* **9**, 5134–5142.

Kräusslich, H.-G., Nicklin, M. J. H., Toyoda, H., Etchison, D., and Wimmer, E. (1987). Poliovirus proteinase 2A induces cleavage of eucaryotic initiation factor 4F polypeptide p220. *J. Virol.* **61**, 2711–2718.

Kuge, S., and Nomoto, A. (1987). Construction of viable deletion and insertion mutants of the Sabin strain of type 1 poliovirus: Function of the 5' noncoding sequence in viral replication. *J. Virol.* **61**, 1478–1487.

Kuge, S., Saito, I., and Nomoto, A. (1986). Primary structure of poliovirus defective-interfering particle genomes and possible generation mechanism of particles. *J. Mol. Biol.* **192**, 473–487.

Kuge, S., Kawamura, N., and Nomoto, A. (1989a). Strong inclination toward transition mutation in nucleotide substitutions by poliovirus replicase. *J. Mol. Biol.* **207**, 175–182.

Kuge, S., Kawamura, N., and Nomoto, A. (1989b). Genetic variation occurring on the genome of an *in vitro* insertion mutant of poliovirus type 1. *J. Virol.* **63**, 1069–1075.

La Monica, N., and Racaniello, V. R. (1989). Differences in replication of attenuated and neurovirulent polioviruses in human neuroblastoma cell line SH-SY5Y. *J. Virol.* **63**, 2357–2360.

La Monica, N., Meriam, C., and Racaniello, V. R. (1986). Mapping of sequences required for mouse neurovirulence of poliovirus type 2 Lansing. *J. Virol.* **57**, 515–525.

Larsen, G. R., Semler, B. L., and Wimmer, E. (1981). Stable hairpin structure within the 5'-terminal 85 nucleotides of poliovirus RNA. *J. Virol.* **37**, 328–335.

Lawson, T. G., Lee, K. A., Maimone, M. M., Abramson, R. D., Dever, T. E., Merrick, W. C., and Thach, R. E. (1989). Dissociation of double-stranded polynucleotide helical structures by eukaryotic initiation factors, as revealed by a novel assay. *Biochemistry* **28**, 4729–4734.

Laz, T., Clements, J., and Sherman, F. (1987). The role of messenger RNA sequences and structures in eukaryotic translation. *In* "Translational Regulation of Gene Expression" (J. Ilan, ed.), pp. 413–429. Plenum, New York.

Lazarus, P., Parkin, N., and Sonenberg, N. (1988). Developmental regulation by the 5' noncoding region of murine c-*myc* mRNA in *Xenopus laevis. Oncogene* **3**, 517–521.

Lee, K. A. W., Edery, I., and Sonenberg, N. (1985). Isolation and structural characterization of cap-binding proteins from poliovirus-infected HeLa cells. *J. Virol.* **54**, 515–524.

Lee, Y. F., Nomoto, A., and Wimmer, E. (1976). The genome of poliovirus is an exceptional eukaryotic mRNA. *Prog. Nucleic Acid Res. Mol. Biol.* **19**, 89–96.

Lee, Y. F., Nomoto, A., Detjen, M. B., and Wimmer, E. (1977). A protein covalently linked to poliovirus genome RNA. *Proc. Natl. Acad. Sci. U.S.A.* **74**, 59–63.

Leibold, E. A., and Munro, H. N. (1988). Cytoplasmic protein binds in vitro to a highly conserved sequence in the 5' untranslated region of ferritin heavy- and light-subunit mRNAs. *Proc. Natl. Acad. Sci. U.S.A.* **85**, 2171–2175.

Lindberg, A. M., Stålhandske, P. O. K., and Pettersson, U. (1987). Genome of coxsackievirus B3, *Virology* **156,** 50–63.

Lindquist, S. (1987). Translational regulation in the heat-shock response of *Drosophila* cells. *In* "Translational Regulation of Gene Expression" (J. Ilan, ed.), pp. 187–207. Plenum, New York.

Liu, C. C., Simonsen, C. C., and Levinson, A. D. (1984). Initiation of translation at internal AUG codons in mammalian cells. *Nature (London)* **309,** 82–85.

Lloyd, R. E., Grubman, M. J., and Ehrenfeld, E. (1988). Relationship of p220 cleavage during picornavirus infection to 2A proteinase sequencing. *J. Virol.* **62,** 4216–4223.

Logan, J., and Shenk, T. (1984). Adenovirus tripartite leader sequence enhances translation of mRNAs late after infection. *Proc. Natl. Acad. Sci. U.S.A.* **81,** 3655–3659.

London, I. M., Levin, D. H., Matts, R. L., Thomas, N. S. B., Petryshyn, R., and Chen, J. J. (1987). Regulation of protein synthesis. *Enzymes* **18,** 360–380.

López-Guerrero, J. A., Carrasco, L., Martinez-Abaraca, F., Frezno, M., and Alonso, M. A. (1989). Restriction of poliovirus RNA translation in a human monocytic cell line. *Eur. J. Biochem.* **186,** 571–582.

Macejak, D. G., Hambidge, S. J., Najita, L., and Sarnov, P. (1990). EIf-4-independent translation of poliovirus RNA and cellular mRNA encoding glucose-regulated protein 78/immunoglobulin heavy-chain binding protein. *In* "New Aspects of Positive-Strand RNA Viruses" (M. A. Brinton and F. X. Heinz, eds.), pp. 152–157. Am. Soc. Microbiol., Washington, D.C.

Marth, J. D., Overell, R. W., Meier, K. E., Krebs, E. G., and Perlmutter, R. M. (1988). Translation activation of the *lck* proto-oncogene. *Nature (London)* **332,** 171–173.

Matthews, R. E. F. (1982). Classification and nomenclature of viruses. *Intervirology* **71,** 1–199.

McGarry, T. J., and Lindquist, S. (1985). The preferential translation of Drosophila hsp70 mRNA requires sequences in the untranslated leader. *Cell* **42,** 903–911.

Meerovitch, K., Pelletier, J., and Sonenberg, N. (1989). A cellular protein that binds to the 5'-noncoding region of poliovirus RNA: Implications for internal translation initiation. *Genes Dev.* **3,** 1026–1034.

Mellor, E. J. C., Brown, F., and Harris, T. J. R. (1985). Analysis of the secondary structure of the poly(C) tract in foot-and-mouth disease virus RNAs. *J. Gen. Virol.* **66,** 1919–1929.

Melnick, J. L. (1985). Enteroviruses: Polioviruses, coxsackieviruses, echoviruses, and newer enteroviruses. *In* "Virology" (B. N. Fields, ed.), pp. 739–794. Raven, New York.

Miller, P. F., and Hinnebusch, A. G. (1989). Sequences that surround the stop codons of upstream open reading frames in *GCN4* mRNA determine their distinct functions in translational control. *Genes Dev.* **3,** 1217–1225.

Minor, P. D., and Dunn, G. (1988). The effect of sequences in the 5' non-coding region on the replication of polioviruses in the human gut. *J. Gen. Virol.* **69,** 1091–1096.

Moldave, K. (1985). Eukaryotic protein synthesis. *Annu. Rev. Biochem.* **54,** 1109–1149.

Mosenkis, J., McQueen, S. D., Janovec, S., Duncan, R., Hershey, J. W. B., Grifo, J. A., Merrick, W. C., and Thach, R. E. (1985). Shutoff of host translation by encephalomyocarditis virus infection does not involve cleavage of the eucaryotic initiation factor 4F polypeptide that accompanies poliovirus infection. *J. Virol.* **54,** 643–645.

Moss, E. G., O'Neill, R. E., and Racaniello, V. R. (1989). Mapping of attenuating sequences of an avirulent poliovirus type 2 strain. *J. Virol.* **63,** 1884–1890.

Mueller, P. P., and Hinnebusch, A. G. (1986). Multiple upstream AUG codons mediate translational control of GCN4. *Cell* **45,** 201–207.

Mueller, P. P., Harashima, S., and Hinnebusch, A. G. (1987). A segment of GCN4 mRNA containing the upstream AUG codons confers translational control upon a heterologous yeast transcript. *Proc. Natl. Acad. Sci. U.S.A.* **84,** 2863–2867.

Mueller, P. P., Jackson, B. M., Miller, P. F., and Hinnebusch, A. G. (1988). The first and fourth upstream open reading frames in *GCN4* mRNA have similar initiation efficiencies but respond differently in translational control to changes in length and sequence. *Mol. Cell. Biol.* **8,** 5439–5447.

Muller, A. J., and Witte, O. N. (1989). The 5' noncoding region of the human leukemia-associated oncogene *BCR/ABL* is a potent inhibitor of in vitro translation. *Mol. Cell. Biol.* **9,** 5234–5238.

Muñoz, A., Alonso, M. A., and Carrasco, L. (1984). Synthesis of heat-shock proteins in HeLa cells: Inhibition by virus infection. *Virology* **137,** 150–159.

Muzychenko, A. R., Lipskaya, G. Y., Maslova, S. V., Svitkin, Y. V., Pilipenko, E. V., Nottay, B. K., Kew, O. M., and Agol, V. I. (1991). Coupled mutations in the 5'-untranslated region of the Sabin poliovirus strains during in vivo passages: Structural and functional implications. *Virus Res.* (in press).

Najarian, R., Caput, D., Gee, W., Potter, S. J., Renard, A., Merryweather, J., Nest, G. V., and Dina, D. (1985). Primary structure and gene organization of human hepatitis A virus. *Proc. Natl. Acad. Sci. U.S.A.* **82,** 2627–2631.

Najita, L., and Sarnow, P. (1990). Oxidation–reduction sensitive interaction of a cellular 50 kDa protein with an RNA hairpin in the 5' noncoding region of the poliovirus genome. *Proc. Natl. Acad. Sci. U.S.A.* **87,** 5846–5850.

Newton, S. E., Carroll, A. R., Campbell, R. O., Clarke, B. E., and Rowlands, D. J. (1985). The sequence of foot-and-mouth disease virus RNA to the 5' side of the poly(C) tract. *Gene* **40,** 331–336.

Nicklin, M. J. H., Kräusslich, H. G., Toyoda, H., Dunn, J. J., and Wimmer, E. (1987). Poliovirus polypeptide precursors: Expression *in vitro* and processing by exogenous 3C and 2A proteinases. *Proc. Natl. Acad. Sci. U.S.A.* **84,** 4002–4006.

Noller, H. F., and Woese, C. R. (1981). Secondary structure of 16S ribosomal RNA. *Science* **212,** 403–411.

Nomoto, A., and Wimmer, E. (1987). Genetic studies of the antigenicity and the attenuation phenotype of poliovirus. *Soc. Gen. Microbiol. Symp.* **40,** 107–134.

Nomoto, A., Lee, Y. F., and Wimmer, E. (1976). The 5' end of poliovirus mRNA is not capped with m7G(5')ppp(5')Np. *Proc. Natl. Acad. Sci. U.S.A.* **73,** 375–380.

Nomoto, A., Omata, T., Toyoda, H., Kuge, S., Horie, H., Kataoka, Y., Genba, Y., Nakano, Y., and Imura, N. (1982). Complete nucleotide sequence of the attenuated poliovirus Sabin 1 strain genome. *Proc. Natl. Acad. Sci. U.S.A.* **79,** 5793–5797.

Nuss, D. L., Oppermann, H., and Koch, G. (1975). Selective blockage of initiation of host protein synthesis in RNA-virus-infected cells. *Proc. Natl. Acad. Sci. U.S.A.* **72,** 1258–1262.

Oh, S.-K., Hambidge, S., Najita, L., Scott, M. P., and Sarnow, P. (1990). Translation initiation by internal ribosome binding of viral and cellular mRNA molecules. *Rinshoken Int. Conf., 5th* p. 20.

Ohara, Y., Stein, S., Fu, J., Stillman, L., Klaman, L., and Roos, R. P. (1988). Molecular cloning and sequence determination of DA strain of Theiler's murine encephalomyelitis viruses. *Virology* **164,** 245–255.

O'Neill, R. E., and Racaniello, V. R. (1989). Inhibition of translation in cells infected with a poliovirus 2A^pro mutant correlates with phosphorylation of the alpha subunit of eucaryotic initiation factor 2. *J. Virol.* **63,** 5069–5075.

Ozawa, K., Ayub, J., and Young, N. (1988). Translational regulation of B19 parvovirus capsid protein production by multiple upstream AUG triplets. *J. Biol. Chem.* **263,** 10922–10926.

Palmenberg, A. C., Kirby, E. M., Janda, M. R., Drake, N. L., Duke, G. M., Potratz, K. F., and Collett, M. S. (1984). The nucleotide and deduced amino acid sequences of the encephalomyocarditis viral polyprotein coding region. *Nucleic Acids Res.* **12**, 2969–2985.

Pani, A., Julian, M., and Lucas-Lenard, J. (1986). A kinase able to phosphorylate exogenous protein synthesis initiation factor eIF-2a is present in lysate of mengovirus-infected L cells. *J. Virol.* **60**, 1012–1017.

Panniers, R., Stewart, E. B., Merrick, W. C., and Henshaw, E. C. (1985). Mechanism of inhibition of polypeptide chain initiation in heat-shocked Ehrlich cells involves reduction of eukaryotic initiation factor 4F activity. *J. Biol. Chem.* **260**, 9648–9653.

Parkin, N., Darveau, A., Nicholson, R., and Sonenberg, N. (1988a). cis-Acting translational effects of the 5' noncoding region of c-*myc* mRNA. *Mol. Cell. Biol.* **8**, 2875–2883.

Parkin, N. T., Cohen, E. A., Darveau, A., Rosen, C., Haseltine, W., and Sonenberg, N. (1988b). Mutational analysis of the 5' non-coding region of human immunodeficiency virus type 1: Effects of secondary structure on translation. *EMBO J.* **7**, 2831–2837.

Parks, G. D., Duke, G. M., and Palmenberg, A. C. (1986). Encephalomyocarditis virus 3C protease: Efficient cell-free expression from clones which link viral 5' noncoding sequences to the P3 region. *J. Virol.* **60**, 376–384.

Paul, A. V., Tada, H., von der Helm, K., Wissel, T., Kiehn, R., Wimmer, E., and Deinhardt, F. (1987). The entire nucleotide sequence of the genome of human hepatitis A virus (isolate MBB). *Virus Res.* **8**, 153–171.

Peabody, D. S., and Berg, P. (1986). Termination–reinitiation occurs in the translation of mammalian cell mRNAs. *Mol. Cell. Biol.* **6**, 2695–2703.

Peabody, D. S., Subramani, S., and Berg, P. (1986). Effect of upstream reading frames on translation efficiency in simian virus 40 recombinants. *Mol. Cell. Biol.* **6**, 2704–2711.

Pelletier, J., and Sonenberg, N. (1985). Insertion mutagenesis to increase secondary structure within the 5' noncoding region of a eucaryotic mRNA reduces translational efficiency. *Cell* **40**, 515–526.

Pelletier, J., and Sonenberg, N. (1988). Internal initiation of translation of eukaryotic mRNA directed by a sequence derived from poliovirus RNA. *Nature (London)* **334**, 320–325.

Pelletier, J., and Sonenberg, N. (1989). Internal binding of eucaryotic ribosomes on poliovirus RNA: Translation in HeLa cell extracts. *J. Virol.* **63**, 441–444.

Pelletier, J., Flynn, M. E., Kaplan, G., Racaniello, V., and Sonenberg, N. (1988a). Mutational analysis of upstream AUG codons of poliovirus RNA. *J. Virol.* **62**, 4486–4492.

Pelletier, J., Kaplan, G., Racaniello, V. R., and Sonenberg, N. (1988b). Cap-independent translation of poliovirus mRNA is conferred by sequence elements within the 5' noncoding region. *Mol. Cell. Biol.* **8**, 1103–1112.

Pelletier, J., Kaplan, G., Racaniello, V. R., and Sonenberg, N. (1988c). Translational efficiency of poliovirus mRNA: Mapping inhibitory cis-acting elements within the 5' noncoding region. *J. Virol.* **62**, 2219–2227.

Penswick, J., Hubler, R., and Hohn, T. (1988). A viable mutation in cauliflower mosaic virus, a retroviruslike plant virus, separates its capsid protein and polymerase genes. *J. Virol.* **62**, 1460–1463.

Perez-Bercoff, R., and Gander, M. (1977). The genomic RNA of mengovirus. I. Location of the poly(C) tract. *Virology* **80**, 426–429.

Perez-Bercoff, R., and Gander, M. (1978). In vitro translation of mengovirus RNA deprived of the terminally-linked (capping?) protein. *FEBS Lett.* **96**, 306–311.

Perez-Bercoff, R., and Kaempfer, R. (1982). Genomic RNA of mengovirus. V. Recognition of common features by ribosomes and eukaryotic initiation factor 2. *J. Virol.* **41**, 30–41.

Pestova, T. V., Maslova, S. V., Potapov, V. K., and Agol, V. I. (1989). Distinct modes of poliovirus polyprotein initiation *in vitro*. *Virus Res.* **14**, 107–118.

Petersen, R. B., Moustakas, A., and Hackett, P. B. (1989). A mutation in the short 5'-proximal open reading frame on Rous sarcoma virus RNA alters virus production. *J. Virol.* **63**, 4787–4796.

Pevear, D. C., Calenoff, M., Rozhon, E., and Lipton, H. L. (1987). Analysis of the complete nucleotide sequence of the picornavirus Theiler's murine encephalomyelitis virus indicates that it is closely related to cardioviruses. *J. Virol.* **61**, 1507–1516.

Pevear, D. C., Borkowski, J., Calenoff, M., Oh, C. K., Ostrowski, B., and Lipton, H. L. (1988). Insights into Theiler's virus neurovirulence based on a genomic comparison of the neurovirulent GDVII and less virulent BeAn strains. *Virology* **164**, 1–12.

Pevear, D. C., Oh, C. K., Cunningham, L. L., Calenoff, M., and Jubelt, B. (1990). Localization of genomic regions specific for the attenuated, mouse-adapted poliovirus type 2 strain W-2. *J. Gen. Virol.* **71**, 43–52.

Phillips, B. A., and Emmert, A. (1986). Modulation of the expression of poliovirus proteins in reticulocyte lysates. *Virology* **148**, 255–267.

Pilipenko, E. V., Blinov, V. M., Romanova, L. I., Sinyakov, A. N., Maslova, S. V., and Agol, V. I. (1989a). Conserved structural domains in the 5'-untranslated region of picornaviral genomes: An analysis of the segment controlling translation and neurovirulence. *Virology* **168**, 201–209.

Pilipenko, E. V., Blinov, V. M., Chernov, B. K., Dmitrieva, T. M., and Agol, V. I. (1989b). Conservation of the secondary structure elements of the 5'-untranslated region of cardio- and aphthovirus RNAs. *Nucleic Acids Res.* **17**, 5701–5711.

Pilipenko, E. V., Blinov, V. M., and Agol, V. I. (1990). Gross rearrangements within the 5'-untranslated region of the picornaviral genomes. *Nucleic Acids Res.* **18**, 3371–3375.

Pilipenko, E. V., Maslova, S. V., and Agol, V. I. (1991). Manuscript in preparation.

Pollard, S. R., Dunn, G., Cammack, N., Minor, P. D., and Almond, J. W. (1989). Nucleotide sequence of a neurovirulent variant of the type 2 oral poliovirus vaccine. *J. Virol.* **63**, 4949–4951.

Porter, A., Carey, N., and Fellner, P. (1974). Presence of a large poly(rC) tract within the RNA of encephalomyocarditis virus. *Nature (London)* **248**, 675–678.

Proud, C. G. (1986). Guanine nucleotides, protein phosphorylation and the control of translation. *Trends Biochem. Sci. (Pers. Ed.)* **11**, 73–77.

Racaniello, V. R. (1988). Poliovirus neurovirulence. *Adv. Virus Res.* **34**, 217–246.

Rancaniello, V. R., and Baltimore, D. (1981). Molecular cloning of poliovirus DNA and determination of the complete nucleotide sequence of the viral genome. *Proc. Natl. Acad. Sci. U.S.A.* **78**, 4887–4891.

Racaniello, V. R., and Meriam, C. (1986). Poliovirus temperature-sensitive mutant containing a single nucleotide deletion in the 5'-noncoding region of the viral RNA. *Virology* **155**, 498–507.

Ransone, L. J., and Dasgupta, A. (1987). Activation of double-stranded RNA-activated protein kinase in HeLa cells after poliovirus-infection does not result in increased phosphorylation of eIF-2. *J. Virol.* **61**, 1781–1787.

Ransone, L. J., and Dasgupta, A. (1988). A heat-sensitive inhibitor in poliovirus-infected cells which selectively blocks phosphorylation of the α subunit of eucaryotic initiation factor 2 by the double-stranded RNA-activated protein kinase. *J. Virol.* **62**, 3551–3557.

Rao, C. D., Pech, M., Robbins, K. C., and Aaronson, S. A. (1988). The 5' untranslated sequence of the c-*sis*/platelet-derived growth factor 2 transcript is a potent translational inhibitor. *Mol. Cell. Biol.* **8**, 284–292.

Ratner, L., Thielan, B., and Collins, T. (1987). Sequence of the 5' portion of the human cis-gene: Characterization of the transcriptional promoter and regulation of expression of the protein product by 5' untranslated mRNA sequences. *Nucleic Acids Res.* 15, 6017–6036.

Ray, B. K., Lawson, T. G., Kramer, J. C., Cladaras, M. H., Grifo, J. A., Abramson, R. D., Merrick, W. C., and Thach, R. E. (1985). ATP-dependent unwinding of messenger RNA structure by eukaryotic initiation factors. *J. Biol. Chem.* 260, 7651–7658.

Reynolds, G. A., Basu, S. K., Osborne, T. F., Chin, D. J., Gil, G., Brown, M. S., Goldstein, J. L., and Luskey, K. L., (1984). HMG CoA reductase: A negatively regulated gene with unusual promoter and 5' untranslated regions. *Cell* 38, 275–285.

Rhoads, R. E. (1988). Cap recognition and the entry of mRNA into the protein synthesis initiation cycle. *Trends Biochem. Sci. (Pers. Ed.)* 13, 52–56.

Rivera, V. M., Welsh, D., and Maizel, J. V., Jr. (1988). Comparative sequence analysis of the 5' noncoding region of the enteroviruses and rhinoviruses. *Virology* 165, 42–50.

Robertson, B. H., Grubman, M. J., Weddell, G. N., Moore, D. M., Welsh, J. D., Fischer, T., Dowbenko, D. J., Yansura, D. G., Small, B., and Kleid, D. G. (1985). Nucleotide and amino acid sequence coding for polypeptides of foot-and-mouth disease virus type A12. *J. Virol.* 54, 651–660.

Romanova, L. I., Blinov, V. M., Tolskaya, E. A., Viktorova, E. G., Kolesnikova, M. S., Guseva, E. A., and Agol, V. I. (1986). The primary structure of crossover regions of intertypic poliovirus recombinants: A model of recombination between RNA genomes. *Virology* 155, 202–213.

Rose, J. K., Trachsel, H., Leong, K., and Baltimore, D. (1978). Inhibition of transaltion by poliovirus: Inactivation of a specific initiation factor. *Proc. Natl. Acad. Sci. U.S.A.* 75, 2732–2736.

Rosen, H., Di Segni, G., and Kaempfer, R. (1982). Translational control by messenger RNA competition for eukaryotic initiation factor 2. *J. Biol. Chem.* 257, 946–952.

Ross, B. C., Anderson, B. N., Edwards, P. C., and Gust, I. D. (1989). Nucleotide sequence of high passage hepatitis A virus strain HM175: Comparison with wild-type and cell culture-adapted strains. *J. Gen. Virol.* 70, 2805–2810.

Rothberg, P. G., Harris, T. J. R., Nomoto, A., and Wimmer, E. (1978). O4-(5'-Uridylyl)tyrosine is the bond between the genome-linked protein and the RNA of poliovirus. *Proc. Natl. Acad. Sci. U.S.A.* 75, 4868–4872.

Rouault, T. A., Hentze, M. W., Caughman, S. W., Harford, J. B., and Klausner, R. D. (1988). Binding of a cytosolic protein to the iron-responsive element of human ferritin messenger RNA. *Science* 241, 1207–1210.

Rouault, T. A., Hentze, M. W., Haile, D. J., Harford, J. B., and Klausner, R. D. (1989). The iron-responsive element binding protein: A method for the affinity purification of a regulatory RNA-binding protein. *Proc. Natl. Acad. Sci. U.S.A.* 86, 5768–5772.

Roussou, I., Thireos, G., and Hauge, B. M. (1988). Transcriptional–translational regulatory circuit in *Saccharomyces cerevisiae* which involves the GCN4 transcriptional activator and the GCN2 protein kinase. *Mol. Cell. Biol.* 8, 2132–2139.

Rowlands, D. J., Harris, T. J. R., and Brown, F. (1978). More precise location of the polycytidylic acid tract in foot and mouth disease virus RNA. *J. Virol.* 26, 335–343.

Rueckert, R. R. (1985). Picornaviruses and their replication. *In* "Virology" (B. N. Fields, ed.), pp. 705–738. Raven, New York.

Saito, H., Hayday, A. C., Wiman, K., Hayward, W. S., and Tonegawa, S. (1983). Activation of the c-*myc* gene by translocation: A model for translational control. *Proc. Natl. Acad. Sci. U.S.A.* 80, 7476–7480.

Sangar, D. V., Rowlands, D. J., Harris, T. J. R., and Brown, F. (1977). Protein covalently linked to foot-and-mouth disease virus RNA. *Nature (London)* 268, 648–650.

Sangar, D. V., Newton, S. E., Rowlands, D. J., and Clarke, B. E. (1987). All foot and mouth disease virus serotypes initiate protein synthesis at two separate AUGs. *Nucleic Acids Res.* **15,** 3305–3314.

Sarnow, P. (1989). Translation of glucose-regulated protein 78/immunoglobulin heavy-chain binding protein mRNA is increased in poliovirus-infected cells at a time when cap-dependent translation of cellular mRNAs is inhibited. *Proc. Natl. Acad. Sci. U.S.A.* **86,** 5795–5799.

Sarre, T. F. (1989). The phosphorylation of eukaryotic initiation factor 2: A principle of translational control in mammalian cells. *BioSystems* **22,** 311–325.

Schlicht, H. J., Radziwill, G., and Schaller, H. (1989). Synthesis and encapsidation of duck hepatitis B virus reverse transcriptase do not require formation of core–polymerase fusion proteins. *Cell* **56,** 85–92.

Schultze, M., Hahn, T., and Jiricny, J. (1990). The reverse transcriptase gene of cauliflower mosaic virus is translated separately from the capsid gene. *EMBO J.* **9,** 1177–1185.

Sedman, S. A., Good, P. J., and Mertz, J. E. (1989). Leader-encoded open reading frames modulate both the absolute and relative rates of synthesis of the virion proteins of simian virus 40. *J. Virol.* **63,** 3884–3893.

Sedman, S. A., Gelembiuk, G. W., and Mertz, J. E. (1990). Translation initiation at a downstream AUG occurs with increased efficiency when the upstream AUG is located very close to the 5' cap. *J. Virol.* **64,** 453–457.

Seechurn, P., Knowles, N. J., and McCauley, J. W. (1990). The complete nucleotide sequence of a pathogenic swine vesicular disease virus. *Virus Res.* **16,** 255–274.

Semler, B. L., Johnson, V. H., and Tracy, S. (1986). A chimeric plasmid from cDNA clones of poliovirus and coxsackievirus produces a recombinant virus that is temperature-sensitive. *Proc. Natl. Acad. Sci. U.S.A.* **83,** 1777–1781.

Shih, D. S., Shih, C. T., Kew, O., Pallansch, M., Rueckert, R., and Kaesberg, P. (1978). Cell-free synthesis and processing of the proteins of poliovirus. *Proc. Natl. Acad. Sci. U.S.A.* **75,** 5807–5811.

Shih, D. S., Park, I. W., Evans, C. L., Jaynes, J. M., and Palmenberg, A C. (1987). Effects of cDNA hybridization on translation of encephalomyocarditis virus RNA. *J. Virol.* **61,** 2033–2037.

Skern, T., Sommergruber, W., Blaas, D., Druengler, P., Fraundorfer, F., Pieler, C., Fogy, I., and Kuechler, E. (1985). Human rhinovirus 2: Complete nucleotide sequence and proteolytic processing signals in the capsid protein region. *Nucleic Acids Res.* **13,** 2111–2126.

Skinner, M. A., Racaniello, V. R., Dunn, G., Cooper, J., Minor, P. D., and Almond, J. W. (1989). New model for the secondary structure of the 5' non-coding RNA of poliovirus is supported by biochemical and genetic data that also show that RNA secondary structure is important in neurovirulence. *J. Mol. Biol.* **207,** 376–392.

Smith, R. E., and Clark, J. M., Jr. (1979). Effect of capping upon the mRNA properties of satellite tobacco necrosis virus ribonucleic acid. *Biochemistry* **18,** 1366–1371.

Soe, L. H., Shieh, C. K., Baker, S. C., Chang, M. F., and Lai, M. M. C. (1987). Sequence and translation of the murine coronavirus 5'-end genomic RNA reveals the N-terminal structure of the putative RNA polymerase. *J. Virol.* **61,** 3968–3976.

Sonenberg, N. (1987). Regulation of translation by poliovirus. *Adv. Virus Res.* **33,** 175–204.

Sonenberg, N. (1988). Cap-binding proteins of eukaryotic messenger RNA: Functions in initiation and control of translation. *Prog. Nucleic Acid Res. Mol. Biol.* **35,** 173–207.

Sonenberg, N. (1990). Poliovirus translation. *Curr. Top. Microbiol. Immunol.* **161,** 23–47.

Sonenberg, N., and Meerovitch, K. (1990). Internal initiation of translation on poliovirus RNA. *Rinshoken Int. Conf., 5th* p. 22.

Sonenberg, N., and Pelletier, J. (1989). Poliovirus translation: A paradigm for a novel initiation mechanism. *BioEssays* **11**, 128–132.

Sonenberg, N., Guertin, D., and Lee, K. A. W. (1982). Capped mRNAs with reduced secondary structure can function in extracts from poliovirus-infected cells. *Mol. Cell. Biol.* **2**, 1633–1638.

Stanway, G., Cann, A. J., Hauptmann, R., Hughes, P., Clarke, L. D., Mountford, R. C., Minor, P. D., Schild, G. C., and Almond, J. W. (1983). The nucleotide sequence of poliovirus type 3 Leon 12 a_1b: Comparison with poliovirus type 1. *Nucleic Acids Res.* **11**, 5629–5643.

Stanway, G., Hughes, P. J., Mountford, R. C., Reeve, P., Minor, P. D., Schild, G. C., and Almond, J. W. (1984a). Comparison of the complete nucleotide sequences of the genomes of the neurovirulent poliovirus P3/Leon/37 and its attenuated Sabin vaccine derivative P3/Leon $12a_1b$. *Proc. Natl. Acad. Sci. U.S.A.* **81**, 1539–1543.

Stanway, G., Hughes, P. J., Mountford, R. C., Minor, P. D., and Almond, J. W. (1984b). The complete nucleotide sequence of a common cold virus: Human rhinovirus 14. *Nucleic Acids Res.* **12**, 7859–7875.

Strick, C. A., and Fox, T. D. (1987). *Saccharomyces cerevisiae* positive regulatory gene *PET111* encodes a mitochondrial protein that is translated from an mRNA with a long 5′ leader. *Mol. Cell. Biol.* **7**, 2728–2734.

Stroeher, V. L., Jorgensen, E. M., and Garber, R. L. (1986). Multiple transcripts from the Antennapedia gene of *Drosophila melanogaster. Mol. Cell. Biol.* **6**, 4667–4675.

Svitkin, Y. V., Krupp, G., and Gross, H. J. (1983). Discontinuity of the poly(C) tract of encephalomyocarditis virus RNA. *Bioorgan. Khim.* **9**, 1638–1643.

Svitkin, Y. V., Maslova, S. V., and Agol, V. I. (1985). The genomes of attenuated and virulent poliovirus strains differ in their *in vitro* translation efficiencies. *Virology* **147**, 243–252.

Svitkin, Y. V., Lyapustin, V. N., Lashkevich, V. A., and Agol, V. I. (1986). Synthesis and membrane-dependent processing of a precursor of the structural proteins of tick-borne encephalitis virus (flavivirus) in cell-free system. *Mol. Biol. (Moscow)* **20**, 1251–1263.

Svitkin, Y. V., Pestova, T. V., Maslova, S. V., and Agol, V. I. (1988). Point mutations modify the response of poliovirus RNA to a translation initiation factor: A comparison of neurovirulent and attenuated strains. *Virology* **166**, 394–404.

Svitkin, Y. V., Cammack, N., Minor, P. D., and Almond, J. W. (1990). Translation deficiency of the Sabin type 3 poliovirus genome: Association with an attenuating mutation $C_{472} \rightarrow U$. *Virology* **175**, 103–109.

Tahara, S. M., Morgan, M. A., and Shatkin, A. J. (1981). Two forms of purified m^7G-cap binding protein with different effects on capped mRNA translation in extracts of uninfected and poliovirus-infected HeLa cells. *J. Biol. Chem.* **256**, 7691–7694.

Theil, E. C. (1987). Storage and translation of ferritin messenger RNA. *In* "Translational Regulation of Gene Expression" (J. Ilan, ed.), pp. 141–163. Plenum, New York.

Thireos, G., Penn, M. D., and Greer, H. (1984). 5′ Untranslated sequences are required for the translational control of a yeast regulatory gene. *Proc. Natl. Acad. Sci. U.S.A.* **81**, 5096–5100.

Ting, J., and Lee, A. S. (1988). Human gene encoding the 78,000 dalton glucose-regulated protein and its pseudogene: Structure, conservation, and regulation. *DNA* **7**, 275–286.

Toyoda, H., Kohara, M., Kataoka, Y., Suganuma, T., Omata, T., Imura, N., and Nomoto, A. (1984). Complete nucleotide sequences of all three poliovirus serotype genomes. Implications for genetic relationship, gene function and antigenic determinants. *J. Mol. Biol.* **174**, 561–585.

Tracy, S., Liu, H. L., and Chapman, N. M. (1985). Coxsackievirus B3: Primary structure of the 5' non-coding and capsid protein-coding regions of the genome. *Virus Res.* **3**, 263–270.

Trono, D., Andino, R., and Baltimore, D. (1988a). An RNA sequence of hundreds of nucleotides at the 5' end of poliovirus RNA is involved in allowing viral protein synthesis. *J. Virol.* **62**, 2292–2299.

Trono, D., Pelletier, J., Sonenberg, N., and Baltimore, D. (1988b). Translation in mammalian cells of a gene linked to the poliovirus 5' noncoding region. *Science* **241**, 445–448.

Tzamarias, D., and Thireos, G. (1988). Evidence that the *GCN2* protein kinase regulates reinitiation by yeast ribosomes. *EMBO J.* **7**, 3547–3551.

Tzamarias, D., Roussou, I., and Thireos, G. (1989). Coupling of *GCN4* mRNA translational activation with decreased rates of polypeptide chain initiation. *Cell* **57**, 947–954.

van Duijn, L. P., Holsappel, S., Kasperaitis, M., Bunschoten, H., Konings, D., and Voorma, H. O. (1988). Secondary structure and expression *in vivo* of messenger RNAs into which upstream AUG codons have been inserted. *Eur. J. Biochem.* **172**, 59–66.

Vartapetian, A. B., and Bogdanov, A. A. (1987). Proteins covalently linked to viral genomes. *Prog. Nucleic Acid Res. Mol. Biol.* **34**, 209–251.

Vartapetian, A. B., Drygin, Y. F., Chumakov, K. M., and Bogdanov, A. A. (1980). The structure of the covalent linkage between proteins and RNA in encephalomyocarditis virus. *Nucleic Acids Res.* **8**, 3729–3742.

Vartapetian, A. B., Mankin, A. S., Skripkin, E. A., Chumakov, K. M., Smirnov, V. D., and Bogdanov, A. A. (1983). The primary and secondary structure of the 5'-end region of encephalomyocarditis virus RNA. A novel approach to sequencing long RNA molecules. *Gene* **26**, 189–195.

Walden, W. E., Daniels-McQueen, S., Brown, P. H., Gaffield, L., Russell, D. A., Bielser, D., Bailey, L. C., and Thach, R. E. (1988). Translational repression in eukaryotes: Partial purification and characterization of a repressor of ferritin mRNA translation. *Proc. Natl. Acad. Sci. U.S.A.* **85**, 9503–9507.

Walden, W. E., Patino, M. M., and Gaffield, L. (1989). Purification of a specific repressor of ferritin mRNA translation from rabbit liver. *J. Biol. Chem.* **264**, 13765–13769.

Watt, R., Stanton, L. W., Marcu, K. B., Gallo, R. C., Croce, C. M., and Rovera, G. (1983). Nucleotide sequence of cloned cDNA of human c-*myc* oncogene. *Nature (London)* **303**, 725–728.

Wek, R. C., Jackson, B. M., and Hinnebusch, A. G. (1989). Juxtaposition of domains homologous to protein kinases and histidyl-tRNA synthetases in GCN2 protein suggests a mechanism for coupling *GCN4* expression to amino acid availability. *Proc. Natl. Acad. Sci. U.S.A.* **86**, 4579–4583.

Westrop, G. D., Wareham, K. A., Evans, D. M. A., Dunn, G., Minor, P. D., Magrath, D. I., Taffs, F., Marsden, S., Skinner, M. A., Schild, G. C., and Almond, J. W. (1989). Genetic basis of attenuation of the Sabin type 3 oral poliovirus vaccine. *J. Virol.* **63**, 1338–1344.

Williams, M. A., and Lamb, R. A. (1989). Effect of mutations and deletions in a bicistronic mRNA on the synthesis of influenza B virus NB and NA glycoproteins. *J. Virol.* **63**, 28–35.

Williams, N. P., Mueller, P. P., and Hinnebusch, A. G. (1988). The positive regulatory function of the 5'-proximal open reading frames in *GCN4* mRNA can be mimicked by heterologous, short coding sequence. *Mol. Cell. Biol.* **8**, 3827–3836.

Wimmer, E. (1982). Genome-linked proteins of viruses. *Cell* **28**, 199–201.

Wimmer, E., Chang, A. Y., Clark, J. M., Jr., and Reichmann, M. E. (1968). Sequence studies of satellite tobacco necrosis virus RNA. *J. Mol. Biol.* **38**, 59–73.

Woese, C. R., and Gutell, R. R. (1989). Evidence for several higher order structural elements in ribosomal RNA. *Proc. Natl. Acad. Sci. U.S.A.* **86**, 3119–3122.

Zhang, Y., Dolph, P. J., and Schneider, R. J. (1989). Secondary structure analysis of adenovirus tripartite leader. *J. Biol. Chem.* **264**, 10679–10684.

ADVANCES IN VIRUS RESEARCH, VOL. 40

EFFECTS OF DEFECTIVE INTERFERING VIRUSES ON VIRUS REPLICATION AND PATHOGENESIS
in Vitro AND in Vivo

Laurent Roux,* Anne E. Simon,† and John J. Holland‡

*Département de Microbiologie
CMU
CH-1211 Geneva 4, Switzerland
†Program in Molecular and Cellular Biology and Department of Biochemistry
University of Massachusetts
Amherst, Massachusetts 01003
‡Center for Molecular Genetics and Department of Biology
University of California–San Diego
La Jolla, California 92093

I. INTRODUCTION

*Molecular Nature and Origin of Defective Interfering (DI) Particles
and Satellites—Recombination, Rearrangements,
and Biological Activities*

DI particles are subgenomic deletion mutants generated from infectious virus genomes, generally by replicase errors. DI particles and related satellite genomes of plant RNA viruses are generated by a

181

wide variety of animal, plant, and fungal viruses. For reviews see
Blumberg and Kolakofsky (1983), Barrett and Dimmock (1986), Holland (1987, 1990), Huang (1988), Schlesinger (1988), Nuss (1988), Kaper and Collmer (1988), and Simon (1988). This chapter focuses on
material not already covered in these earlier reviews. The ubiquity of
DI viruses was first clearly recognized by Huang and Baltimore
(1970). They proposed and defined the term "DI particle" to include
defective viruses containing only some portion of the infectious virus
genome, requiring homologous parental virus as helper for replication,
containing virus structural proteins and antigens, and exhibiting the
capacity to replicate preferentially at the expense of infectious helper
virus in cells infected by both. In most cases DI particle interference
results from competition for helper virus-encoded replication–encapsidation proteins. Because DI particles lack (greater or lesser) fractions of their parental virus genomes, they sometimes have mature
virions which are detectably (or markedly) smaller in size than the
parental infectious virions which generate them. In their original definition of DI particles, Huang and Baltimore did not include defective
virus particles which are not subgenomic, for example, full-sized virus
genomes which are defective because one or more mutations have
closed essential open reading frames (ORFs). There is reason to sus-
pect that such defective full-size virus genomes may sometimes exert
interfering and other important biological effects, and this is discussed
in Section II,B,3.

Satellite RNAs of plants are small RNAs which usually share little
or no homology with their helper viruses, although some do exhibit
some sequence homology with their helper viruses and thus resemble
DI particles in at least a portion of their genomes. The origin of nonhomologous satellite RNAs is unclear at present, although those with
self-cleavage ribozyme activity may have evolved from primitive RNA
precursors. [See Section III and reviews by Kaper and Collmer (1988),
Simon (1988), and Bruening *et al.* (1988) for discussions of satellite
RNAs.] All have in common the capacity for encapsidation within
protein encoded by their specific helper viruses.

The molecular origin of most DI particles of RNA animal viruses is
rather well established to be a consequence of polymerase errors, as
originally suggested by Huang (1977) and Leppert *et al.* (1977). They
suggested that DI genomes arise by virus genome rearrangements or
recombinations as a result of viral replicases "leaping" or skipping
from one virus RNA template to another or from one segment of a
template to another. During this leaping (or skipping or sliding) the
RNA replicase carries the incomplete nascent strand to a new template
(or template segment), then uses this nascent strand as a primer for

resumption of chain elongation at the new template resumption site. Depending on the leaving sites and resumption sites, DI particles can be produced which are simple internal deletions of the virus genome, simple deletions with a new terminus, or complex or bizarre genomes with multiple rearrangements of virus segments. The leaping replicase mechanism for DI particle generation has been generalized by Lazzarini et al. (1981) and Perrault (1981). This type of "copy choice" recombination of virus RNA has been experimentally verified for polioviruses by Kirkegaard and Baltimore (1986) and for coronaviruses by Lai (1990) and colleagues, and was suggested for turnip crinkle virus satellite RNAs by Cascone et al. (1990).

It is obvious that promiscuous replicase leaps might insert cellular RNA into defective (or infectious) viruses, and, in fact, Monroe and Schlesinger (1983) demonstrated the incorporation of cellular transfer RNA into the 5' termini of a class of Sindbis virus DI particles (Schlesinger, 1988). Although available evidence favors replicase leaps as the major mechanism for the generation of DI particles, other mechanisms (e.g., aberrant splicing events) might rarely be involved. Because many DI particles of RNA viruses are replicative entities only, being incapable of transcription and translation of their genomes, only those genome segments necessary for efficient replication and encapsidation need be conserved. These DI particles can undergo extensive mutational (and recombinational) change, so repetitive mutational change or biased hypermutation (Cattaneo et al., 1988a; Wong et al., 1989) may frequently affect the structure of defective RNA genomes. Some hypermutation might be due to error-prone subsets of polymerase (O'Hara et al., 1984a; Steinhauer and Holland, 1987) or to RNA unwinding enzymes which convert A residues to I (Bass and Weintraub, 1988; Lamb and Dreyfuss, 1989) or to any number of other novel phenomena of RNA genetics, such as RNA editing (Simpson, 1990) or the addition of nontemplated bases (Thomas et al., 1988). The generation and recombination of plant DI and satellite RNAs are discussed in Section III.

DI particles of DNA viruses and retroviruses may also sometimes arise by replicative error, but probably most often are generated by the variety of genetic recombination mechanisms available for DNA (Kucherlapati and Smith, 1988; Berg and Howe, 1989) and retroviruses (Coffin, 1990).

1. Interference and Amplification

In general, DI particles are replicative entities. They replicate and amplify their genomes at the expense of the replication of specific helper viruses which encode replication and encapsidation proteins, and which must compete with the DI genomes for these gene products. The genomes of the DI particles are usually rearranged to enhance their ability to replicate (and to compete for replication–encapsidation proteins). This results in selective replication and maturation of DI genomes and concomitant interference with the replication of infectious helper virus. For example, simian virus 40 DI genomes generally contain reiterated viral origins of DNA replication, and vesicular stomatitis virus (VSV) DI particles usually contain rearranged termini which favor replication and preclude transcription. Therefore, while helper virus genomes become devoted mainly to transcription, DI genomes engage only in replication. However, some classes of DI genomes are transcriptionally active, and these often interfere at a level other than replication. Sometimes DI particles interfere indirectly with virus, as when they induce interferon (IFN), alter immune responses, or restrict the cell surface expression of virus proteins. Finally, the presence or absence of encapsidation or packaging (sequences) determines the relative efficiency of maturation of DI genomes. All these parameters have been reviewed or discussed by Huang (1988, Schlesinger (1988), Holland (1987, 1990), Brockman (1977), Makino et al. (1990), Sekellick and Marcus (1980a, Barrett and Dimmock (1986), and Roux et al. (1984).

2. Influence of Host Cell Type

Host cell type is a major determinant of the biological effects of DI particles and one which is frequently overlooked in attempts to assess the influence of DI genomes on virus diseases in vivo. DI particles which interfere strongly in one host cell type may interfere weakly or not at all in other cell types. In some cases this is due to poor replication and amplification of the DI particles, but in other cell types DI particles may replicate well and amplify well, but exert only weak interference with helper virus replication. Variability of DI particle effects within different cells in vivo can be a major factor confounding interpretation of DI genome influences on disease processes. For reviews or discussion of host cell effects, see Choppin (1969), Huang (1988), Brinton et al. (1984), Cave et al. (1985), Kang et al. (1981), Holland (1987), Gillies and Stollar (1980, and Barrett and Dimmock (1986).

3. Assays for DI Particles

DI particles and DI genomes are generated by nearly all types of animal viruses, and by some plant and fungal viruses. They frequently contaminate animal virus pools in vitro and in vivo. Their presence is not always easily detected, depending on the virus assay system used, the type of virus, the DI genome class, etc. There is no sensitive reliable assay system suitable for all virus and DI genome types, so a wide variety of DI genome assays have been developed. These have been reviewed by Holland (1987). They range from electron microscopy of particles or particle purification and visualization on sucrose velocity gradients, to assays for interference with virus-directed nucleic acid synthesis or virus particle synthesis, to assays for subgenomic-sized nucleic acids in particles, cells, or tissues, to various cell protection assays. Recent polymerase chain reaction (PCR) techniques for amplifying DNA and RNA segments should prove useful, but the choice of appropriate primers will be important (and difficult when the structure of the putative DI genomes present is not known).

4. Cell Protection and Persistence

Because DI particles often interfere strongly with virus replication, it was anticipated (Huang and Baltimore, 1970) that they might frequently exhibit a direct cell-sparing effect (in addition to indirect protective effects, such as induction of IFN or other cytokines, alteration of immune responses, reduction of early virus yields, or expression at the cell surface to allow immune abrogation of infection). The interfering effects of DI particles can be so profound as to reduce the virus yield to zero in many doubly infected cells (Sekellick and Marcus, 1980b). It has frequently been observed that DI particles can facilitate the establishment and maintenance of persistent infections of cells in culture infected by a wide variety of animal viruses. This has been reviewed by Holland *et al.* (1980), Holland (1987), Huang and Baltimore (1977), Schlessinger (1988), and Barrett and Dimmock (1986). It would not, therefore, be unexpected if DI genomes often played significant roles in reducing virus yield *in vivo*, thereby facilitating complete immune system elimination of virus in some cases, or triggering persistent infections in others. However, it is not a simple matter to test this possibility in any natural infection of animals or humans. This is discussed below, and by Huang and Baltimore (1977, Huang (1988), Barrett and Dimmock (1986), and Holland (1987). Of course, many other factors such as temperature-sensitive (*ts*) and other mutations can be involved in persistent infections (Youngner and Preble, 1980; Cattaneo *et al.*, 1988b; Wong *et al.*, 1989; Enami *et al.*, 1989). Even host cell mutations are sometimes involved in persistent infections (Ahmed *et al.*, 1981).

5. Cycling Phenomena and Effects on Virus Evolution

Palma and Huang (1974) reported a significant cyclical process in the interaction of helper virus and DI particles *in vitro* when they interact continuously or repeatedly during repeated passages. They observed out-of-phase cyclic yields of each. Whenever DI particle yields were sufficient to infest most cells, the yield of infectious virus dropped sharply; then there was insufficient virus helper to support significant DI particle replication; then infectious virus yields increased again, only to decline once more as DI particles were amplified, etc. Evidence for cycling has been observed in many virus–DI interactions (Kawai *et al.*, 1975; Grabau and Holland, 1982; Roux and Holland, 1980), and it obviously could create uncertainty regarding DI particle (or virus) yields during any single sampling of a passage series, or an infected animal or human. Cave *et al.* (1984, 1985) have provided strong evidence that DI particles of VSV cause virus cycling *in vivo* in mice and that lower input ratios of DI particles are more protective than high ratios because of virus–DI cycling phenomena. This has profound implications for the stochastic nature of many virus infections, and for the severity and outcome of infections. The disease process might depend on probabilities of DI genome generation early in infection or on simultaneous infection by varying ratios of virus to DI particles (Huang, 1988). Bangham and Kirkwood (1990) have presented a simple mathematical model to explain virus–DI cycling effects on the replication and persistence of viruses.

Another type of virus–DI genome cycling can be superimposed on the basic type of cycling described above. The second type of cycling occurs during prolonged virus–DI genome interactions in persistent infections or serial passages, and it is due to the periodic unpredictable generation of DI particle-resistant ($Sdi-$) mutant viruses. The appearance of virus mutants resistant to homologous DI particles has been reported for a number of different RNA and DNA viruses, including rabies (Kawai and Matsumoto, 1977), VSV (Horodyski and Holland, 1980), lymphocytic meningitis virus (Jacobsen and Pfau, 1980), Sindbis virus (Weiss and Schlesinger, 1981), West Nile virus (Brinton and Fernandez, 1983), and DNA coliphage f1 (Enea and Zinder, 1982). Giachetti and Holland, 1988) can allow the periodic escape of virus from interference effects of previously existing DI particles; then newly generated DI genomes interfere until virus mutates again, etc. (Horodyski *et al.*, 1983; DePolo *et al.*, 1987). This type of cycling can drive rapid virus evolution and lead to unpredictable periodic emergence of new virus and DI genomes (O'Hara *et al.*, 1984a,b). Obviously, if complex cycling phenomena of the types described above occur within multiple sites of virus infection in an animal or a human, then analysis of the role of defective genomes in the disease process can become enormously complicated.

6. Involvement of Defective Virus Genomes in the Biology of Virus Disease Processes

As discussed in Section III, DI and satellite RNAs of plants clearly can exert varying *in vivo* effects on plant disease symptoms, causing reduction of symptoms, no effects on symptoms, or mild to profound intensification of disease severity (see also Kaper and Collmer, 1988; Simon, 1988). The regions of these defective genomes involved in effects on plants are now being elucidated (see Section III). The situation is considerably more complicated in animals and humans because of their immune responses, neuroendocrine responses, more highly specialized tissue and organ systems, etc.

It is well established in animal models that DI genomes and other defective genomes can influence disease processes, but their possible roles in natural infections remain largely unexplored (Huang, 1988; Holland, 1987; Barrett and Dimmock, 1986). This is partly due to difficulties in obtaining suitable tissue specimens for defective virus isolation—especially from humans. It is also due to technical problems in proving or disproving a biological role for defective genomes even when they are observed in clinical material. Several examples discussed below illustrate the difficulty.

It required decades of careful research to achieve recognition that partially defective measles virus is involved in the persistent progressive fatal brain disease subacute sclerosing panencephalitis (Hall and Choppin, 1979; Cattaneo *et al.*, 1988a,b; Wong *et al.*, 1989; Enami *et al.*, 1989), even though a large proportion of central nervous system (CNS) neurons are producing measles virus antigens and virus genomes during disease progression. Multiple mutations accumulate within virus surface envelope proteins, particularly the matrix protein, yet virus genomes are somehow transmitted from cell to cell without significant production of mature infectious virus, and despite a vigorous CNS immune response. This type of defective virus is not a DI virus, because it replicates autonomously without helper virus. Whether helper-dependent DI genomes are present and replicating at the expense of these partially defective genomes is presently unknown.

A less definitive situation exists with hepatitis B virus (HBV). A number of investigators have observed inactivating mutations, particularly in the precore and core regions of virus genomes recovered from persistently infected humans (see Okamoto *et al.*, 1990; Miller *et al.*, 1990, and references therein). These mutations might prevent synthesis and secretion of hepatitis B antigen, thereby reducing the probability of immune destruction of chronically infected hepatocytes. Miller *et al.* (1990) showed that defective genomes are not necessary in a cloned recombinant inoculum to produce chronic hepadnavirus infections in woodchucks at a rate equal to, or higher than, that obtained with an inoculum containing defective genomes. This does not, of course, rule out regular participation of defective hepadnaviruses in disease processes or persistence, since defective genomes might be undergoing cyclical interactions with nondefective genomes in many liver foci of all infections, regardless of inoculum. In fact, Miller *et al.* (1990) showed that two of three recombinant DNA clones recovered from a woodchuck inoculated with a standard serum pool had mutational defects which rendered them incapable of independent replication. In a situation such as this, it is extremely difficult to prove or disprove a biological role for defective genomes even when they are observed repeatedly in clinical specimens. This also applies to hepatitis A virus (HAV) infections, in which defective subgenomic virus genomes have been observed by some investigators, but not by others (Nüesch *et al.*, 1988; Lemon *et al.*, 1985).

Despite extensive recent investigations of human immunodeficiency virus type 1 (HIV-1), it is still uncertain whether defective genomes might be involved in causing the acquired immunodeficiency syndrome (AIDS) disease process (or attenuating it). Defective genomes, of course, play a major role in most forms of animal retrovirus oncogenesis (Coffin, 1990). Huang *et al.* (1989) have shown that a strain of murine leukemia virus causing acquired immunodeficiency in mice can do so with a helper-free defective genome alone. Apparently, this defective retrovirus genome induces oligoclonal expansion of target cells to induce a paraneoplastic immunodeficiency. Myerhans *et al.* (1989) have shown that a considerable proportion of HIV-1 genomes may be defective due to inactivating mutations in the *tat, gag,* and *env* sequences. Balfe *et al.* (1990) reported a much lower level, but in neither case is it possible to affirm or negate a role of defective genomes in disease or disease attenuation. Finally, it should be recognized that the distinction between defective and nondefective genomes is not always clear. For example, it appears that there are subgenomic mRNA replicons present during the normal replication of full-sized coronavirus genomes (Sethna *et al.*, 1989, 1991). This might allow the subgenomic replicons to compete with standard genomes in a manner analogous to DI genomes, thereby potentiating the establishment of persistent coronavirus infections, as postulated by Sethna *et al.* (1989, 1991).

Even with powerful new analytical methods for virus genome analysis, it will take much time and effort to elucidate the involvement of defective genomes in virus diseases of humans and animals. Section II reviews some recent studies of defective animal and human viruses, and Section III covers recent work with DI and satellite RNAs of plant viruses.

II. DI Viruses of Animals and Humans

As emphasized above, defective genomes and DI particles have been studied widely in *in vitro* systems in which their ability to modulate the intensity as well as the course of viral infections has been extensively described for RNA and DNA viruses. Their ability to modulate infections in experimental animal models has also been documented and extensively reviewed (Barrett and Dimmock, 1986; Holland, 1987; Huang, 1988). However, despite this large body of *in vitro* and *in vivo* information from laboratory experiments, there is still no definitive confirmation (nor refutation) of their presence and involvement in natural diseases of humans and animals. This is due, partly, to lack of interest in the potent modulators of viral infections that defective particles represent, and, mainly, as emphasized in Section I, to the unpredictable complexity of DI multiplication in a noncontrolled system. Also, the inherent characteristic of defective particles which, by definition, do not efficiently replicate in low helper virus multiplicity of infection, has prevented easy isolation from natural infections. Direct analysis of naturally infected samples, on the other hand, has often suffered from the inability to distinguish components of defective viruses from those of nondefective viruses. With the availability of more sensitive techniques (e.g., PCR), this detection is bound to become more feasible, with the major drawback that PCR may allow definitive detection only of already described defective genomes. Indeed, such a sensitive technique may turn out to be misleading in cases in which the choice of the right primers cannot be backed up by previous information regarding defective genome structures (see Section II,A,9). In this section we try to give a nonexhaustive up-to-date view of certain aspects of defective virus research, focusing, when possible, on data dealing with the possible involvement of defective genomes in animal and human diseases.

A. RNA Viruses

1. DI RNA Generation

The mechanism of DI RNA generation implies the ability of the RNA replicase to jump from one template, or one portion of a template, to another, carrying the nascent RNA strand (see Section I). The ease with which DI RNAs are then generated must therefore be inversely proportional to the processivity of the enzyme, that is, its ability to stick to its first template. Variant viruses have been described which are able to generate VSV DI RNA with a much higher efficiency (De-Polo and Holland, 1986). For such variants one would postulate that the replicase has a decreased "processivity," and/or a better ability to rebind to a new template. The protein responsible for controlling one of these two steps has apparently been identified for influenza virus (Odagiri and Tobita, 1990). A mismatch of the *NS* gene between two viral strains (*NS* of A/Aichi transferred to A/WSN by reassortment) has generated a strain (Wa-182) with the ability to produce detectable DI particles very rapidly (i.e., within a single high-multiplicity infection). This characteristic of Wa-182 was cotransferred to a third strain with the *NS* gene. Three point mutations in the Wa-182 *NS* gene were identified relative to wild type, resulting in two amino acid substitutions in the NS2 protein; NS2 is therefore likely to be a component of the replicase involved in the control of its processivity or its ability to continue chain elongation on a new template.

2. Interference at the Replicating Level

In vitro systems capable of VSV DI RNA replication have been worked out using infected cell extracts (Peluso and Moyer, 1983) or programmed reticulocyte lysates (Patton *et al.*, 1984; Wertz *et al.*, 1987). These systems, apart from giving interesting information on the viral RNA replication mechanism (Peluso and Moyer, 1988; Moyer, 1989), have allowed verification of a basic postulate of the interference mechanism (Perrault 1981; Holland, 1987). Competition between nondefective and defective genomes for availability of the *L/NS* complex to support efficient DI RNA replication has been demonstrated directly *in vitro* (Giachetti and Holland, 1989). This confirms previous *in vivo* evidence that DI particles interfere mainly at the level of replication (Huang and Manders, 1972; Perrault and Holland, 1972).

3. Packaging of DI Genomes

Competition at the level of replication is certainly the major factor explaining interference and successful amplification of DI genomes. Under multiple infectious cycles, however, packaging efficiency of DI genomes into virus particles may also intervene in their ability to outgrow nondefective or other defective genomes. It has long been recognized that if VSV DI RNAs are rapidly assembled into nucleocapsids they appear to be more slowly and less efficiently matured into virus particles (Palma and Huang, 1974; Khan and Lazzarini, 1977;

Moyer and Gatchell, 1979). This question has been reinvestigated (Von Laer *et al.*, 1988). Under low interference conditions [DI/ST standard virus = 0.125 in the inoculum] DI particles were, in fact, found to be delayed in their production relative to nondefective particles. Under higher interference conditions (DI/ST > 2), however, both ST and DI particles were found to be delayed in their production, with no particular handicap for DI nucleocapsids. This delay could then be correlated with retarded, but more prolonged, viral protein synthesis. By quantitative comparison of intracellular and extracellular Sendai virus copy-back DI nucleocapsids, restriction of DI nucleocapsid budding relative to ST nucleocapsid budding was observed, but only above a certain level of DI RNA replication (Mottet and Roux, 1990). These two examples show that DI nucleocapsid packaging can be modulated either temporally or quantitatively, depending on the conditions of interference. This may, in turn, influence the course of infections by modulation of the ratios of DI/ST in successive inocula. Restriction of DI nucleocapsid budding during excessive intracellular DI replication may prevent dying out of the infection due to lack of helper virus in the yields. Not all types of DI nucleocapsids are restricted in their packaging in the same way. Larger nucleocapsids appear to be less restricted than smaller ones (Mottet and Roux, 1990; Re and Kingsbury, 1988). This may be indicative of as yet unknown rules directing nucleocapsid packaging.

4. A Model to Account for DI Cell-Sparing Effects

Under conditions of high interference, decreased amounts of intracellular VSV M and G proteins have been observed together with prolonged synthesis of host proteins (Von Laer *et al.*, 1988). These conditions correspond to the cell-sparing effects following Sendai virus DI particle infections. Cell-sparing is observed together with a strong restriction of total virus particle budding (Tuffereau and Roux, 1988). Thus, it was proposed that negative-strand virus DI allows cell survival and the establishment of persistent infections by promoting reduced efficiency of viral particle assembly (Tuffereau and Roux, 1988). In this model it is the efficient and stable formation of virus prebudding structures (glycoproteins /M/ST nucleocapsids) at the plasma membrane which would otherwise lead to cell death. DI nucleocapsids would be unable to enter, or stabilize, these prebudding structures. Virus particle budding is then decreased (Tuffereau and Roux, 1988), and M. G, or HN (for Sendai virus) are decreased in amount (Tuffereau and Roux, 1988; Von Laer *et al.*, 1988) due to faster degradation (Tuffereau and Roux, 1988; Roux *et al.*, 1984). Any alterations (e.g., mutation or deletion) of proteins involved in the stable formation of these prebudding structures would have similar cell-sparing effects on infected cells.

5. DI Particles and Interferon Induction

Cell-sparing effects produced by DI particles can be observed in the absence of any IFN production. IFN induction may, however, be another way by which DI particles can modulate the course of a viral infection. The ability of DI particle-enriched VSV inocula to promote IFN production has been recognized previously, but with ambiguity due to the ability of wild-type virus to induce IFN, on the one hand, and to inhibit host cell protein synthesis, on the other (reviewed by Holland, 1987). This complex question has recently been reexamined for VSV (Marcus and Gaccione, 1989). DI containing self-complementary (snapback) RNA was shown to induce 20- to 30-fold higher amounts of IFN than plaque-derived VSV. This effect is heat resistant and thus likely due to the double-stranded nature of infecting DI RNA. In the absence of heat treatment, strong induction of IFN by DI was observed only at low multiplicity of infection (1–2 DI particles and 0.1–0.2 plaque-forming units). Higher multiplicities caused marked reduction in IFN yield. Since this reduction was not observed after heat treatment of DI stocks, the authors postulate that IFN reduction at higher multiplicity of infection results from the inhibitory effect of VSV on cell protein synthesis via small leader RNA synthesis (Grinnell and Wagner, 1984). What is noteworthy here, in the context of possible effects of DI on the course of viral infections in animals, is the minimal multiplicity of DI infection required to promote maximal effects on IFN induction. Thus, potential biological effects of DI in animal infections would not necessarily require massive doses.

6. DI Particles and Immune Response Modulation

Another effect of DI particles on animal infections is their puzzling ability to modulate the specific humoral immune response (McLain and Dimmock, 1989). Mice infected with lethal doses of influenza viruses produced high levels of antineuraminidase antibodies in the lungs. Coinfection with a lifesaving dose of DI viruses (but not with β-propriolactone-inactivated DIs) was accompanied by the appearance of not only antineuraminidase, but also hemagglutination-inhibiting antibodies. The latter show protective ability when passively transferred to naive mice before infection. These antibodies, entirely of the immunoglobulin G types, are nevertheless nonneutralizing antibodies and are found uniquely in the lungs (from which they disappear within 15 days). Their protective activity was postulated to be due to opsonization of virus or infected cells.

7. DI Particles in Natural Infections

A correlation was observed between high virulence of influenza virus epidemics in chickens, with the absence of DI particles produced by the viral strains implicated (Bean et al., 1985; Webster et al., 1986; Chambers and Webster, 1987). Involvement of DI particles in natural infections has also been supported by Northern blot analysis of RNAs extracted from specimens (feces and blood) of humans suffering from HAV infections. The presence of subgenomic viral RNA molecules was observed (Nüesch et al., 1989). Interestingly, the identification of the deletion end points in these RNAs, by RNase protection experiments, is compatible with two of the three deletions identified in RNAs of HAV particles grown in vitro and shown to interfere with standard virus replication (Nüesch et al., 1989). These HAV DI RNAs have been identified in all HAV-infected cell culture systems reported to date. The propensity of most of the in vitro infection systems (cited by Nüesch et al., 1988) to readily evolve to persistent infections, the extended incubation times of HAV in vivo (i.e., up to 36 days), and the detection of DI RNAs in both these situations might be related phenomena.

8. DI Particles and Viral Attenuated Vaccines

Attenuated viral vaccines have classically been produced by serially passaging virus in nonnatural hosts or host cells. These multiple passages could have allowed the emergence of DI particles in at least some cases. Recently, measles-attenuated vaccine preparations have been shown to contain DI particles and/or subgenomic viral RNAs as well (Calain and Roux, 1988; Bellocq et al., 1990). Whether these DI particles can replicate on vaccination and whether they participate in vaccine attenuation are open questions. Measles virus DI genomes were shown to survive 3 days in infected cells in the absence of helper virus before being rescued (Mottet et al., 1990). In theory, this would allow vaccine DI genomes to be amplified in the body under conditions of low multiplicity of infection, through rescue by standard virus produced in adjacent cells. Likewise, the ability to detect rubella virus DI particles after three undiluted passages (Frey and Hemphill, 1988) makes likely the presence of DI particles in attenuated rubella vaccines (which are prepared following 27–80 passages).

9. Toward More Sensitive DI Detection

More sensitive detection of DI particles is likely to be needed to show their involvement in natural infections. This higher sensitivity might come from classical techniques such as the cytopathology inhibition test of McLain et al. (1988), which appears to detect influenza DI particles with a 320,000-fold higher sensitivity than the previous infectious center inhibition test. PCR, however, would appear to be a method of choice to directly demonstrate the presence of defective viral genomes in specimens derived from natural infections. PCR may, however, suffer from its extreme sensitivity when attempting the detection of rare molecules or molecules with structures which are unfavorable for the amplification reaction. For example, amplification of copy-back RNA molecules has been found to be surprisingly difficult (P. Calain and L. Roux, unpublished observations). This is likely due to competition of the primers with the complementary RNA termini in the hybridization reactions (inter- versus intramolecular reactions). On the other hand, incorrect short-deletion DI DNAs have been amplified due to rare nonspecific priming when the appropriate DI RNAs were too long to be amplified (P. Calain and L. Roux, unpublished observations). Previous knowledge of the DI genome structures being sought may therefore be needed to exclude reverse transcriptase and Taq polymerase artifacts.

B. Retroviruses

Defective retroviruses have long been known to be potent biological modulators (Coffin, 1990) and, obviously, the most widely studied so far are the "acutely transforming defectives," which have replaced part of their genomes with an oncogene. They thereby induce cell-proliferative syndromes at high frequency in animals, and the basis for this increased pathogenicity is now reasonably well understood.

1. Feline Leukemia Virus (FelV)

Different classes of retroviruses have been shown to induce immunodeficiency diseases in various animal species (Desrosiers and Letvin, 1987), and this is rather common for FelVs (references cited by Overbraugh et al., 1988). In viral DNA isolated from cat tissues infected with a natural isolate of FelV which consistently induces fatal immunodeficiency syndrome (FelV-FAIDS), variant forms of FelV genomes were identified and these became the predominant forms just prior to the onset of disease (Overbraugh et al., 1988). By direct cloning from the infected tissues, a replication-defective variant (61C) was isolated and shown to be responsible for this increased pathogenicity. Variant 61C, in association with the common form (61E), induced the immunodeficiency syndrome, and transfer of the 3' end of the 61C genome (env long terminal repeat) onto the 61E genome conferred pathogenicity to the chimera. Amino acid substitutions, small deletions and insertions in the gp70 coding region, as well as three nucleotide changes in the long terminal repeat were identified in 61C. Therefore, variant 61C, a replication-defective virus which can outgrow and interfere with its helper virus, appears to drastically modulate infections of cats.

2. Murine Leukemia Virus (MulV)

Similarly, a replication-defective variant of MulVs has been shown to be associated with severe immunodeficiency in mice (Chattopadhyay *et al.*, 1989) and to induce this syndrome (Aziz *et al.*, 1990). This variant, DUH5, was characterized by a deleted genome (4.8 kb versus 8.8kb for the helper virus) lacking most of the pol–env coding regions. The gag region, on the other hand, was relatively well conserved (Aziz *et al.*, 1990). DUH5 was shown to code *in vivo* for a 60-kDa protein representing a modified gag protein with a unique p12 domain (Chattopadhyay *et al.*, 1989; Huang and Jolicoeur, 1990). This Pr60^{gag} was normally myristylated, phosphorylated, and associated with the plasma membrane. However, it was not efficiently processed and could not be released into the medium in the absence of helper Pr65^{gag} (Huang and Jolicoeur, 1990). It appeared to interfere with efficient processing of Pr65^{gag}. DUH5 itself, without any detected helper function, was shown to induce immunodeficiency in correlation with oligoclonal proliferation of infected cells (Huang *et al.*, 1989). Immunodeficiency is thus postulated to derive from this primary neoplasia. These replication-defective interfering retroviruses were identified and found to be closely associated with immunodeficiency by *direct analysis* of infected tissues. This stresses the need for a similar strategy in searching for defective particles in virus infections in nature, because defective viruses are likely to be lost by usual methods of virus isolation.

3. HIV

A similar situation has not yet been described for human AIDS, caused by HIV. What is known, however, is that defective HIV viruses do exist in nature. For example, a highly defective strain with a nonfunctional tat protein was isolated from a healthy individual (Huet *et al.*, 1989). Comparison of HIV isolated in tissue culture from an individual at different times of infection, with endogenous viruses characterized by direct analysis at the time of isolation, showed marked discrepancies between the two populations of viruses (Myerhans *et al.*, 1989). This study, which was confined to the *tat* gene, inferred that many or most of the proviruses *in vivo* may be defective. This contrasted with a rather low frequency of inactivating mutations found in viral genome segments directly amplified and sequenced from peripheral blood mononuclear lymphocytes of HIV patients (Balfe et al., 1990). Defective HIV viruses containing defects in tat, rev, or gag proteins have been experimentally produced (Gorelick et al., 1990; Trono et al., 1989; Green et al., 1989; Malim et al., 1989). These lethal variants expressed a dominant negative phenotype over the infectious virus and, although they are not DI viruses stricto sensu, they can interfere with efficient wild-type virus multiplication. The possible presence of such defective genomes in natural HIV infections is still open to question. Once again, such defective viruses would escape any detection method which involves preliminary amplification steps in tissue culture.

C. DNA Viruses

1. Epstein–Barr Virus (EBV)

EBV is a human herpesvirus that establishes a persistent replicative infection in oropharyngeal and genital sites. This contrasts with the stringently latent life cycle observed in circulating B lymphocytes. These different types of infections have been attributed to variations in host cell regulation. However, in vitro EBV replicative systems have been shown to contain a defective EBV form (see references cited by Patton et al., 1990). This defective EBV contains a deleted rearranged DNA (het-DNA, for heterogeneous DNA) which appears to replicate preferentially over the standard EBV DNA (Miller et al., 1985). When added to latently infected cells, het-DNA activates expression of endogenous EBV genomes and causes disruption of latency (Fresen et al., 1978; Miller et al., 1984). het-DNA can be transmitted horizontally from cell to cell, and therefore appears to be a defective form of EBV capable of modulating infections. PCR amplification of a sequence specific for the rearranged het-DNA has led to identification of this DNA segment in human epithelial lesions (oral hairy leukoplakia), known to contain abundant replicative EBV (Patton et al., 1990). The possibility of participation of this defective form of EBV DNA in the modulation of natural human infections must therefore be considered.

2. Herpesvirus

Herpesvirus DI is known to promote establishment of persistent infections and to inhibit infected cell cytolysis (Frenkel, 1981). The interfering mechanisms are not understood. Competition of strong promoters present on the reiterated DI DNA with promoters of the helper virus DNA, leading to down-regulation of the helper functions involved in cytolysis, has been proposed (Frenkel, 1981). More recently, a modified ORF was identified on an equine herpes type 1 DI genome. This modified ORF, created by rearrangement, results from the replacement of the last 97 amino acids, or an ORF of 469, by a sequence of 68 different residues (Gray *et al.*, 1989; Yalamanchili *et al.*, 1990). This has opened the possibility that this new function, if expressed by the DI genome, would be able to interfere with helper virus functions. It is tempting to speculate that the rearrangements involved in DI genome generation could create trans-dominant negative regulators of the type engineered by truncation of the transcriptional activator VP16 (see Section I; Friedman *et al.*, 1988).

3. Hepadnavirus

As discussed in Section I, a high incidence of defective HBV genomes containing a disruption of the precore coding sequence has been identified in patients having undergone anti-HBe seroconversion (Carman *et al.*, 1989). For three of these patients, exclusive appearance of such defective HBV genomes could be exactly correlated with seroconversion (Okamoto *et al.*, 1990). This was observed under conditions in which no genomes carrying defects in the precore region could be detected in three other patients who lacked anti-HBe antibodies (Okamoto *et al.*, 1990). Since anti-HBe seroconversion is generally tied to evolution toward chronic asymptomatic infections, it is tempting to speculate that the course of HBV infections is often modulated by defective viruses. A study by Miller *et al.* (1990), on the other hand, suggests that this may not be necessary. Infections of woodchucks with a serum pool containing defective particles and with a recombinant-derived virus pool (equivalent to a plaque-derived inoculum) led to 65% and 80% of chronic carriers, respectively, suggesting that defective virus in the inoculum is not essential for the establishment of persistent hepadnavirus infection (Miller *et al.*, 1990). These results were discussed in Section I.

D. Satellite Virus (Delta Agent)

Viroids, virusoids, and satellite viruses or satellite RNAs are commonly involved in plant diseases (see Section III). A thus far unique equivalent of such plant genomes in the reign of animal viruses is the human delta agent (reviewed by Taylor, 1990). Hepatitis delta virus (HDV) infection is closely associated with HBV because HDV uses HBV surface antigen to form enveloped particles capable of cell-to-cell transmission; hence, its definition as a satellite virus. HDV, however, is capable of autonomous RNA replication in the absence of HBV (Kuo *et al.*, 1991). Coinfection with HDV often results in exacerbation of the underlying HBV hepatitis (Rizzetto *et al.*, 1988). The molecular basis of this modulation of infection is not understood. Restricted levels of HBV replication have been reported in some cases, but high levels of both nucleic acids in the serum have been documented as well (Monjardino and Saldanha, 1990). HDV genome is transcribed into an mRNA which encodes the delta antigen (Hsieh *et al.*, 1990). Actually, two forms of the delta antigen have been detected in infected tissues, corresponding to two different RNA genomes with an ORF corresponding to proteins of 195 and 214 residues, respectively. The additional 19 amino acids consist of a carboxy-terminal extension of the small antigen resulting from a transition in the stop codon (Luo *et al.*, 1990). Remarkably, this variant genome accumulates on replication of the small protein-encoding genome. Whereas the small delta antigen strongly enhances genome replication, the large form acts as a dominant negative repressor of such replication (Chao *et al.*, 1990). Therefore, while the delta agent is itself a modulator of HBV infection, its own replication is, in turn, regulated by the generation of a defective mutant genome able to exert significant interference with the delta agent.

III. DI Viruses of Plants

Plant viruses can be associated with a variety of small subviral RNAs such as satellites, DIs, and chimeric molecules with properties of both satellites and DIs. This section reviews our current knowledge of the replication and symptom modulation ability of these small RNAs, focusing primarily on systems in which both satellites and DIs have been found.

As summarized in Section I, satellites are defined as RNAs which require a helper virus for infectivity, yet share little sequence similarity with the viral genomic RNA(s) (Murant and Mayo, 1982). Satellites that encode structural proteins which compose the satellite capsid are referred to as satellite viruses, whereas satellites that are encapsidated in helper virus particles are called satellite RNAs. Satellites can be either linear or circular; circular satellites are commonly known as virusoids. One of the intriguing properties of satellites is their ability to modulate the symptoms of the helper virus, either by attenuating symptoms (analogous to many animal virus DIs, as described in Sections I and II) or by intensifying symptoms. The replication strategies of these molecules are also fascinating, since the helper viral replicase (which, presumably, is required for subviral RNA replication) must, in many cases, recognize sequences and structures not present in the viral genome.

Only recently have DI RNAs been found to be associated with plant viruses. Two viruses shown to have DIs are members of the tombusvirus group: the cherry strain of tomato bushy stunt virus (TBSV, also known as petunia asteroid mosaic virus) (Hillman *et al.*, 1987) and cymbidium ringspot virus (CyRSV) (Burgyan *et al.*, 1989; Rubino *et al.*, 1990). A third virus associated with DI RNAs, turnip crinkle virus (TCV) (Li *et al.*, 1989), is a member of the structurally related carmoviruses. All three viruses support the replication of satellite RNAs. Both TBSV and CyRSV DIs ameliorate symptoms of the helper virus, while TCV DIs dramatically intensify viral symptoms on a variety of hosts. In this section we first discuss plant virus systems in which both DI and satellite RNAs have been well characterized, and then describe the effects of the subviral RNAs on replication and symptom production of their helper viruses. In Sections III, E and F we present information on the generation of DI and other discontinuous RNAs in plants and possible future uses of DIs in controlling disease.

A. TCV

TCV is a 30nm icosahedral plant virus with a single genomic RNA of 4051 nucleotides and a coat consisting of 180 subunits of a 38-kDa protein (Morris and Carrington, 1988). Although TCV is related to TBSV and CyRSV both structurally and at the nucleotide sequence level, the genomic organization of TCV differs markedly from the two tombusviruses; TBSV and CyRSV contain two additional ORFs beyond the coat protein ORF, and TCV has a small ORF upstream of the coat protein-coding sequence (Carrington *et al.*, 1989; Grieco *et al.*, 1989; Hearne *et al.*, 1990). TCV is associated with a variety of subviral RNAs including satellite RNAs, DI RNAs, and molecules with characteristics of both satellites and DIs (Fig. 1). All these subviral RNAs accumulate in such vast quantities that they are clearly visible in ethidium bromide-stained gels. Two isolates of TCV have been characterized: TCV-B and TCV-M. The TCV-B strain has mild symptoms on turnip and is naturally associated with a DI RNA, DI RNA G, and a satellite RNA, sat-RNA D (Li *et al.*, 1989). DI RNA G is a mosaic molecule composed of a sequence near the 5' end of TCV as well as viral terminal 3' -end sequences. The first 10 nucleotides at the 5' end of DI RNA G, however, are not derived from TCV genomic RNA, but rather share complete or nearly complete homology with the 5' ends of the TCV satellite RNAs. The next 10 nucleotides of DI RNA G are of unknown origin (or have diverged significantly from a genomic or satellite sequence).

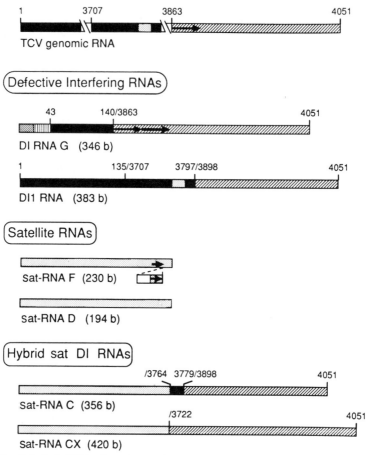

FIG. 1.   Subviral RNA of turnip crinkle virus (TCV). Sequences in common are shaded alike.

A TCV-similar sequence begins with base 21 of DI RNA G, which corresponds to base 43 of TCV. Sequences of six DI RNA G cDNA clones reveal heterogeneity among numerous bases, single-nucleotide insertions and deletions, and a direct repeat of 36 bases (Li *et al.*, 1989).

TCV-B genomic RNA has been cloned and sequenced (Carrington *et al.*, 1989), and transcripts synthesized *in vitro* are infectious (Heaton *et al.*, 1989). Inoculation of turnip with *in vitro*-synthesized viral transcripts can result in the accumulation of DI RNAs *de novo* (Li *et al.*, 1989). Characterization of one *de novo*-generated DI RNA (DI1 RNA) revealed a molecule containing both the exact 5' and 3' ends of TCV flanking one interior segment.

The TCV-M isolate produces a severe disease on turnip and other hosts (Li and Simon, 1990) and is naturally found with two satellite RNAs, sat-RNAs D and F, and one RNA which is a hybrid between a satellite and DI RNA, sat-RNA C (Simon and Howell, 1986). sat-RNA C, which historically has been referred to as a satellite, has the complete sequence of sat-RNA D at the 5′ end and two regions of TCV genomic RNA at the 3′ end. Recently, a second hybrid satellite has been identified which contains a sat-RNA D sequence at the 5′ end and a single segment of TCV genomic sequence at the 3′ end (Zhang et al., 1991).

All junctions of TCV discontinuous RNAs [DIs, hybrid satellite/DI RNAs and recombinant RNAs (see below)] have right-side junction sequences which can be classified as one of three dissimilar motifs of approximately 20 nucleotides each (Cascone et al., 1990). One of the motifs is found near the 5′ end of TCV (beginning at base 10), while a second is located at the extreme 5′ end of the satellite RNAs and DI RNA G. The third motif, which is located at the right side of several junctions, is also located immediately downstream of the 5′ end of the TCV subgenomic RNA which encodes the capsid polypeptide.

The satellite and DI RNAs naturally associated with TCV are efficiently packaged in the helper virus capsid (Altenbach and Howell, 1981; Heaton et al., 1989). Using assays involving the protection of nucleic acid from RNase treatment by tightly bound coat protein, Wei et al. (1990) have identified several high-affinity coat protein binding sites on the RNA genome. None of these sequences, however, is present in the DI RNAs. Studies with purified DI RNA G suggest that an additional high-affinity site is located in the DI RNA sequence which is not protected in assays using genomic RNA. These authors propose that differential folding of the two molecules may occlude or expose this coat protein binding site. Both sat-RNA C and the DI RNAs also contain this coat protein-protected sequence, which may explain the efficient accumulation of these molecules in planta. Since sat-RNAs D and F lack sequence similarity with the helper virus, excluding 7 nucleotides at the 3′ ends, and are efficiently packaged, other sequences with affinity for TCV coat protein remain to be discovered.

B. TBSV

TBSV, the type member of plant tombusviruses, is a 30-nm icosahedral virus consisting of 180 copies of a 41-kDa structural protein subunit in the capsid (Morris and Carrington, 1988). TBSV has a single-stranded positive-sense RNA genome of 4776 nucleotides (Hearne et *al.*, 1990) and can support the replication of a 0.7-kb satellite RNA which was originally isolated from plants infected with CyRSV (Hillman, 1986). A 0.4-kb DI RNA associated with a laboratory stock of TBSV markedly attenuates the symptoms expressed by the virus on tobacco [*Nicotiana clevelandii* (Hillman *et al.*, 1987)]. The 396-base RNA is a complex mosaic molecule composed of five noncontiguous segments of viral homology. Three nucleotides at the extreme 5' terminus of this DI RNA differ from the viral 5'-end sequence, and homology begins 13 nucleotides into the genomic RNA sequence. Computer analysis of individual junctions reveals no obvious sequence or structural similarities. Unlike the satellite RNA, which is efficiently encapsidated by helper virus encoded coat protein, the DI RNA is inefficiently packaged, composing less than 30% of the total encapsidated RNA on a molar basis (Morris and Hillman, 1989). DI RNAs have also been generated following high-multiplicity passaging of a clonally pure TBSV inculum (Morris and Hillman, 1989). These DI RNAs, presumably generated *de novo*, vary in size (< 0.7 kb) and degree of symptom attenuation.

C. CyRSV

CyRSV contains a single-stranded positive-sense genomic RNA of 4733 bases and shares nearly 70% sequence similarity with TBSV (Grieco *et al.*, 1989; Hearne *et al.*, 1990). Isolates of CyRSV can contain either a 621-base satellite RNA or a 499-base DI RNA (Burgyan *et al.*, 1989; Rubino *et al.*, 1990). Approximately 17% of the satellite primary sequence is similar to several interspersed regions of CyRSV genomic RNA. The mosaic DI RNA is composed of six discontinguous fragments of CyRSV genomic sequence. Unlike the DI RNA of TBSV and DI RNA G of TCV, CyRSV DI begins at the 5' end of the helper virus genomic RNA. The right sides of the junction sequences share some sequence similarity with the very 5' end of the genomic RNA. Unlike the satellite RNA, CyRSV DI RNA is not efficiently packaged, which suggests that common sequences between the satellite and the genomic RNA, not present in the DI RNA, may be candidates for encapsidation signals (Burgyan *et al.*, 1989).

Full-length transcripts synthesized *in vitro* from cloned CyRSV genomic RNA are infectious (Burgyan *et al.*, 1990). DI RNAs were found to accumulate in *N. clevelandii* previously inoculated with sap from CyRSV transcript-infected plants (Burgyan *et al.*, 1991). The *de novo*-generated DI RNAs were similar to the DI RNA in the original inoculum in having sequences entirely derived from the helper virus genome. RNA samples taken from younger and older leaves contained different DI RNA sizes, with larger DIs present in older tissue. Sequence analysis revealed that the smaller DIs, which were stable through later passages, were nearly always completely contained within the larger molecules.

D. Attenuation and Intensification of Symptoms by Plant Subviral RNAs

DI RNAs associated with TBSV and CyRSV dramatically attenuate the symptoms of their helper viruses. Inoculation of *N. clevelandii* with TBSV genomic RNA purified over a sucrose gradient results in severe systemic necrosis and plant death within 2 weeks (Hillman *et al.*, 1987). Plants infected with an inoculum containing purified virus and DI RNA exhibit local necrotic lesions only on the inoculated leaves followed by a persistent infection. Symptom attenuation is correlated with both a decrease in the amount of virus recovered from plants and an increase in the accumulation of DI RNA. Studies using tobacco protoplasts to analyze the influence of a 0.7-kb DI RNA on TBSV replication indicate that the presence of the DI results in a 65% reduction in the rate of helper virus genomic RNA synthesis (Jones *et al.*, 1990). By analyzing the rate of genomic RNA synthesis, Jones *et al.* were able to discount the possibility that the DI RNA affects viral accumulation due to enhanced degradation of the genomic RNA or selective suppression of subgenomic RNA synthesis. The 0.7-kb CyRSV satellite also has an effect on symptoms produced by TBSV. When the satellite is inoculated with TBSV on *N. clevelandii*, symptoms are slightly less severe and delayed by about 1 week (Hillman, 1986). The level of accumulation of viral genomic RNA in the presence of the satellite is substantially reduced in whole plants.

The satellite and DI RNAs associated with CyRSV also ameliorate the symptoms of the virus on *N. clevelandii* (Burgyan *et al.*, 1989). Plants inoculated with genomic RNA alone exhibit apical necrosis followed by plant death. Infection of plants with viral genomic and satellite RNAs reduces symptoms to stunting with no necrosis. Plants infected with CyRSV and the DI RNA produce mild symptoms visible only on younger leaves. Although the effect of DI or satellite RNA on virus levels has not been rigorously examined, CyRSV genomic RNA levels are decreased in plants infected with both genomic and DI RNAs (Burgyan *et al.*, 1989).

Unlike DI RNAs of other plant or animal viruses, both the hybrid satellite/DI and DI RNAs of TCV markedly intensify the symptoms of the helper virus (Altenbach and Howell, 1981; Li *et al.*, 1989). Most host plants infected with helper virus with or without the two small satellites (sat-RNAs D and F) exhibit mild symptoms such as stunting, slight leaf crinkling, and mosaic coloring (Simon *et al.*, 1989). Addition of either sat-RNA C or DI RNA G results in stunting and tightly crinkled leaves with a dark green pigment (Li *et al.*, 1989; Li and Simon, 1990). For one plant, *Arabidopsis thaliana*, addition of sat-RNA C results in the death of most ecotypes (Li and Simon, 1990; X. H. Li and A. E. Simon, unpublished observations). Curiously, symptoms are intensified even though sat-RNA C and DI RNA G interfere with the accumulation of TCV (Li *et al.*, 1989; A. E. Simon, unpublished observations). Virus levels are 5-fold lower when DI RNA G is included in the inoculum and about 2-fold lower in the presence of sat-RNA C.

All the TCV subviral RNAs which increase symptoms have 153 bases of genomic RNA 3'-end sequence in common (see Fig. 1). Studies using cloned transcripts of sat-RNA C have indicated that this region is indeed responsible for symptom production (Simon *et al.*, 1988). Attempts to define precise sequences responsible for increased symptoms have been unsuccessful due to the limited infectivity of sat-RNA C (Simon *et al.*, 1988; C. D. Carpenter and A. E. Simon, unpublished observations) or DI RNA G (Li and Simon, 1991) transcripts with mutations near the 3' end.

The ability of subviral RNAs to produce disease symptoms is best understood for the satellite RNA associated with cucumber mosaic virus (CuMV); alterations at any of three nucleotides distinguishes necrogenic and nonnecrogenic satellites (Devic *et al.*, 1990; Sleat and Palukaitis, 1990b). Computer-generated secondary-structure predictions for necrogenic and nonnecrogenic satellites do not reveal any correlations between structure and pathogenesis (Sleat and Palakaitis, 1990b), implying that the primary nucleotide sequence is somehow responsible for disease symptoms. Furthermore, the secondary-structure model for one necrogenic satellite indicates that the necrogenic domain is in a highly nuclease-sensitive region (Hidaka *et al.*, 1988), suggesting that this region of the satellite may interact with either host- or virus-encoded factors to initiate disease production.

The intensification of symptoms by the virulent satellite or DI RNA of TCV is completely dependent on the origin of the helper virus genomic RNA (C. D. Carpenter, X. H. Li, and A. E. Simon, unpublished observations) and the ability of the genomic RNA to produce symptoms on a host plant (Li and Simon, 1990). Plants display increased symptoms only if inoculated with subviral RNA and the genomic RNA from the TCV-M isolate. Similar results have been obtained for several CuMV satellites; systemic chlorotic symptoms are produced on tobacco only when the satellites are associated with one subgroup of the CuMV helper virus (Sleat and Palukaitis, 1990a). These results suggest that an interaction between satellite RNAs or DI RNAs and the viral genomic RNA or encoded product is necessary for symptom production.

E. Generation of Plant DI RNAs

A template-switching model in which specific sequences are targeted by the replicase during reinitiation of synthesis has been proposed for the generation of DI RNAs and the satellite/DI hybrid RNAs in the TCV system (Cascone et al., 1990). This model is based on the similarity among right-side junction sequences and one of three motifs (Cascone et al., 1990; Zhang et al., 1991). Since the motifs are also found at the 5' ends of the TCV genomic RNA and subviral RNAs as well as near the initiation site of the 1.45-kb subgenomic RNA, template switching by the replicase has been suggested to occur during replication of minus-strand RNA. This model is similar to one proposed by Re et al. (1985) for the generation of Sendai virus DI RNAs. Pyrimidine-rich putative promoter signals are found at one side of Sendai virus DI junctions which resemble a sequence located 5–16 nucleotides from the 3' end of the genomic RNA. A similar mechanism may also be responsible for the generation of CyRSV DI RNAs, since there is sequence similarity at the right side of all junctions (Rubino et al., 1990). Recently, recombination between TCV satellites, and between satellites and the genomic RNA, has been detected (Cascone et al., 1990; Zhang et al., 1991). The right-side junction sequences of the single and double recombinant RNAs are all similar to one of the three motifs mentioned above. The importance of specific nucleotides within one motif was recently demonstrated by the elimination of recombination between satellite RNAs following single-base alterations within the motif present in sat-RNA C, which is always found at the right side of satellite recombinant junctions (P. J. Cascone and A. E. Simon, unpublished observations).

After initial generation of a DI RNA, further accumulation depends on the ability of the molecule to be replicated and to move along with the helper virus through the plant. The observation that most subviral RNAs of TCV range in size from 346 to 420 bases indicates that the size of the DI RNA may play an important role in efficient accumulation. Further, more conclusive evidence for the importance of size comes from recent studies using an infectious clone of TCV DI RNA G (Li and Simon, 1991). Plants inoculated with transcripts of the DI containing 30 base deletions near the 5' end do not accumulate any detectable DI RNA. However, if the deleted sequence is replaced with a similar-sized fragment derived from a bacterial plasmid, the DI RNA accumulates at normal levels. Replacement of the 30-base deletion with 60 unrelated nucleotides also produces a viable molecule. There are at least two possible explanations for the size requirement of the DI RNA: (1) Length of the molecule may be important in encapsidation into the TCV isocahedral capsid; and (2) distance between the 3' and 5' ends may play a role in replication processes. Further work using plant protoplasts which are competent to replicate the viral and subviral RNAs should distinguish between these two possibilities.

F. Future Uses of Plant DI RNAs

Viruses are responsible for major crop losses worldwide, and ways of improving plant resistance to viruses are currently being pursued in a number of laboratories. Since DI RNAs (in general) interfere with virus accumulation and disease production, DI RNAs associated with viruses are actively being sought. However, with the exception of the plant rhabdovirus Sonchus yellow net virus (Ismail and Milner, 1988) and the reovirus wound tumor virus (Anzola et al., 1987), DI RNAs have been found only for the related viruses described above. The creation of artificial DI RNAs may be a viable alternative to reliance on naturally formed DIs. Although these artificial DIs may not be packaged for transport through the plant, possibly due to a lack of packaging signals and/or size constraints, the ability to transform plants such that each cell produces DI RNA transcripts should alleviate the need for natural movement. The DIs would still require the cis-acting signals necessary for replication by the helper virus, and identification of such replication signals is an active area of plant virus research. Similar experiments using satellite RNAs to control the pathogenesis of CuMV proved quite successful; tobacco transformed with cDNA to a CuMV satellite which ameliorates CuMV symptoms, under transcriptional control of a con-stitutive plant promoter, produces low levels of satellite RNA which are greatly amplified following infection with CuMV. This results in disease attenuation (Harrison et al., 1987).

IV. SUMMARY

DI viruses and defective viruses generally are widespread in nature. Laboratory studies show that they can sometimes exert powerful disease-modulating effects (either attenuation or intensification of symptoms). Their role in nature remains largely unexplored, despite recent suggestive evidence for their importance in a number of systems.

ACKNOWLEDGMENTS

This work was supported by grants from the Swiss National Foundation for Scientific Research, the National Science Foundation (DMV 8704124 and 8803853), and the National Institutes of Health (AI 14627).

REFERENCES

Ahmed, R., Canning, W. M., Kaufmann, R. S., Sharp, A. H., Hallum, J. V., and Fields, B. W. (1981). *Cell* **25,** 325–332.
Altenbach, S. B., and Howell, S. H. (1981). *Virology* **112,** 25–33.
Anzola, J. V., Xu, Z., Asamizu, T., and Nuss, D. L. (1987). *Proc. Natl. Sci. U.S.A.* **84,** 8301–8305.
Aziz, D. C., Hanna, Z., and Jolicoeur, P. (1990). *Nature (London)* **338,** 505–508.
Balfe, P., Simmonds, P., Ludlum, C. A., Bishop, J. O., and Leigh Brown, A. J. (1990). *J. Virol.* **64,** 6621–6233.
Bangham, C. R., and Kirkwood, T. B. L. (1990). *Virology* **179,** 821–826.
Barrett, A. D., and Dimmock, N. J. (1986). *Curr. Top. Microbiol. Immunol.* **128,** 55–84.
Bass, B. L., and Weintraub, H. (1988). *Cell* **55,** 1089–1098.
Bean, W. J., Kawaoka, Y., Wood, J. M., Pearson, J. E., and Webster, R. G. (1985). *J. Virol.* **54,** 151–160.
Bellocq, C., Mottet, G., and Roux, L. (1990). *Biologicals* **18,** 337–343.
Berg, D. E., and Howe, M. H., eds. (1989). "Mobile DNA." Am. Soc. Microbiol., Washington, D.C.
Blumberg, B. M., and Kolakofsky, D. (1983). *J. Gen. Virol.* **64,** 1839–1847.
Brinton, M. A., and Fernandez, A. V. (1983). *Virology* **129,** 107–115.
Brinton, M. A., Blank, K. J., and Nathanson, N. (1984). *In* "Concepts in Viral Pathogenesis" (A. L. Notkins and M. B. A. Oldstone, eds.), pp. 71–78. Springer-Verlag, New York.
Brockman, W. W. (1977). *Prog. Med. Virol.* **23,** 69–95.
Bruening, G., Buzayan, J. M., Hampel, A., and Gerlach, W. L. (1988). *In* "RNA Genetics" (E. Domingo, J. J. Holland, and P. Ahlquist, eds.), Vol. 2, pp. 127–145. CRC Press, Boca Raton, Florida.
Burgyan, J., Grieco, F., and Russo, M. (1989). *J. Gen. Virol.* **70,** 235–239.
Burgyan, J., Nagy, P. D., and Russo, M. (1990). *J. Gen. Virol.* **71,** 1857–1860.
Burgyan, J., Rubino, L., and Russo, M. (1991). *J. Gen. Virol.* **72,** 505–509.
Calain, P., and Roux, L. (1988). *J. Virol.* **62,** 2859–2866.
Carman, W. F., Jacyna, M. R., Hadziyannis, S., Karayiannis, P., McGarvey, M. J., Markis, A., and Thomas, H. C. (1989). *Lancet* **2,** 588–591.
Carrington, J. C., Heaton, L. A., Zuidema, D., Hillman, B. I., and Morris, T. J. (1989). *Virology* **170,** 219–226.
Cascone, P. J., Carpenter, C. D., Li, X. H., and Simon, A. E. (1990). *EMBO J.* **9,** 1709–1715.
Cattaneo, R., Schmid, A., Billeter, M. A., Sheppard, R. D., and Udem, S. A. (1988a). *J. Virol.* **62,** 1388–1397.
Cattaneo, R., Schmid, A., Eschle, D., Baczko, K., ter Meulen, V., and Billeter, M. A. (1988b). *Cell* **55,** 255–265.
Cave, D. R., Hagen, F. S., Palma, E. L., and Huang, A. S. (1984). *J. Virol.* **50,** 86–91.
Cave, D. R., Hendrickson, F. M., and Huang, A. (1985). *J. Virol.* **55,** 366–373.
Chambers, T. M., and Webster, R. G. (1987). *J. Virol.* **61,** 1517–1523.

Chao, M., Hsieh, S.-Y., and Taylor, J. (1990). J. Virol. 64, 5066–5069.

Chattopadhyay, S. K., Morse, H. C., III, Makino, M., Ruscetti, S. K., and Hartley, J. W. (1989). Proc. Natl. Acad. Sci. U.S.A. 86, 3862–3866.

Choppin, P. W. (1969). Virology 39, 130–134.

Coffin, J. (1990). In "Virology" (B. N. Fields, D. M. Knipe, R. M. Chanock, J. L. Melnick, M. S. Hirsh, T. P. Monath, and B. Roizman, eds.), pp. 1437–1500. Raven, New York.

DePolo, N., and Holland, J. J. (1986). Virology 151, 371–378.

DePolo, N. J., Giachetti, C., and Holland, J. J. (1987). J. Virol. 61, 454–464.

Desrosiers, R. C., and Letvin, N. L. (1987). Rev. Infect. Dis. 9, 438–446.

Devic, M., Jaegle, M., and Baulcombe, D. (1990). J. Gen. Virol. 71, 1443–1449.

Enami, M., Sato, T. A., and Sugiura, A. (1989). J. Gen. Virol. 70, 2191–2196.

Enea, V., and Zinder, N. D. (1982). Virology 122, 222–226.

Frenkel, N. (1981). In "The Human Herpesviruses. An Interdisciplinary Perspective" (A. J. Nahmias, W. R. Dowdle, and R. F. Schinzai, eds.), pp. 91–120. Elsevier, New York.

Fresen, K.-O., Cho, M.-S., and zur Hasen, H. (1978). Int. J. Cancer 22, 378–383.

Frey, T. K., and Hemphill, M. L. (1988). Virology 164, 22–29.

Friedman, S. J., Triezenberg, S. J., and McKnight, S. L. (1988). Nature (London) 335, 452–454.

Giachetti, C., and Holland, J. J. (1988). J. Virol. 62, 3614–3621.

Giachetti, C., and Holland, J. J. (1989). Virology 170, 264–267.

Gillies, S., and Stollar, J. (1980). Virology 107, 509–513.

Gorelick, R. J., Nigida, S. M., Bess, J. W., Arthus, L. O., Henderson, L. E., and Rein, A. (1990). J. Virol. 64, 3207–3211.

Grabau, E. A., and Holland, J. J. (1982). J. Gen. Virol. 60, 87–97.

Gray, W. L., Yalamanchili, R., Raengsakulrach, B., Baumann, R. P. Staczeck, J., and O'Callaghan, D. J. (1989). Virology 172, 1–10.

Green, M., Ishino, M., and Loewenstein, P. M. (1989). Cell 58, 215–223.

Grieco, F., Burgyan, J., and Russo, M. (1989). Nucleic Acids Res. 17, 6383.

Grinnell, B. W., and Wagner, R. R. (1984). Cell 36, 533–543.

Hall, W. W., and Choppin, P. W. (1979). Virology 99, 443–447.

Harrison, B. D., Mayo, M. A., and Baulcombe, D. C. (1987). Nature (London) 328, 799–802.

Hearne, P. Q., Knorr, D. A., Hillman, B. I., and Morris, T. J. (1990). Virology 177, 141–151.

Heaton, L. A., Carrington, J. C., and Morris, T. J. (1989). Virology 170, 214–218.

Hidaka, S., Hanada, K., Ishikawa, K., and Miura, K.-I. (1988). Virology 164, 326–333.

Hillman, B. I. (1986). Ph.D. thesis, University of California, Berkeley, California.

Hillman, B. I., Carrington, J. C., and Morris, T. J. (1987). Cell 51, 427–433.

Holland, J. J. (1987). In "The Rhabdoviruses" (R. R. Wagner, ed.), pp. 297–360. Plenum, New York.

Holland, J. J. (1990). In "Virology" (B. N. Fields, D. M. Knipe, R. M. Chanock, J. L. Melnick, M. S. Hirsh, T. P. Monath, and B. Roizman, eds.), pp. 151–165. Raven, New York.

Holland, J. J., Kennedy, S. I. T., Semler, B. L., Jones, C. L., Roux, L., and Grabau, E. A. (1980). Compr. Virol. 16, 137–192.

Horodyski, F. M., and Holland, J. J. (1980). J. Virol. 36, 627–631.

Horodyski, F. M., Nichol, S. T., Spindler, K. R., and Holland, J. J. (1983). Cell 33, 801–810.

Hsieh, S.-Y., Chao, M., Coates, L., and Taylor, J. (1990). J. Virol. 64, 3192–3198.

Huang, A. S. (1977). Bacteriol. Rev. 41, 811–821.

Huang, A. S. (1988). In "RNA Genetics" (E. Domingo, J. J. Holland, and P. Ahlquist, eds.), Vol. 3, pp. 195–208. CRC Press, Boca Raton, Florida.

Huang, A. S., and Baltimore, D. (1970). *J. Mol. Biol.* **47**, 275–291.
Huang, A. S., and Baltimore, D. (1977). *Compr. Virol.* **10**, 73–116.
Huang, M., and Jolicoeur, P. (1990). *J. Virol.* **64**, 5764–5772.
Huang, A. S., and Manders, E. K. (1972). *J. Virol.* **9**, 909–916.
Huang, M., Sinard, C., and Jolicoeur, P. (1989). *Science* **246**, 1614–1617.
Huet, T., Dazza, M.-C., Brun-Vézinet, F., Roelands, G. E., and Wain-Hobson, S. (1989). *AIDS* **3**, 707–715.
Ismail, I. D., and Milner, J. J. (1988). *J. Gen. Virol.* **69**, 999–1006.
Jacobsen, S., and Pfau, C. J. (1980). *Nature (London)* **283**, 311–313.
Jones, R. W., Jackson, A. O., and Morris, T. J. (1990). *Virology* **176**, 539–545.
Kang, C. Y., Weide, L. G., and Tischfield, J. A. (1981). *J. Virol.* **40**, 946–952.
Kaper, J. M., and Collmer, C. W. (1988). *In* "RNA Genetics" (E. Domingo, J. J. Holland, and P. Ahlquist, eds.), Vol. 3, pp. 171–194. CRC Press, Boca Raton, Florida.
Kawai, A., and Matsumoto. S. (1977). *Virology* **76**, 60–71.
Kawai, A., Matsumoto, S., and Tanabe, K. (1975). *Virology* **67**, 520–533.
Khan, S. R., and Lazzarini, R. A. (1977). *Virology* **77**, 109–206.
Kirkegaard, K., and Baltimore, D. (1986). *Cell* **47**, 433–443.
Kucherlapati, R. S., and Smith, G. R., eds. (1988). "Genetic Recombination." Am. Soc. Microbiol., Washington, D.C.
Kuo, M. Y.-P., Chao, M., and Taylor, J. (1991). *J. Virol.* **63**, 1946–1950.
Lai, M. M. C. (1990). *Annu. Rev. Microbiol.* **44**, 303–333.
Lamb, R. A., and Dreyfuss, G. (1989). *Nature (London)* **337**, 19–20.
Lazzarini, R. A., Keene, J. D., and Schubert, M. (1981). *Cell* **26**, 145–154.
Lemon, S. M., Jansen, R. W., and Newbold, J. E. (1985). *J. Virol.* **54**, 78–85.
Leppert, M., Kort, L., and Kolakofsky, D. (1977). *Cell* **12**, 539–552.
Li, X. H., and Simon, A. E. (1990). *Phytopathology* **80**, 238–242.
Li, X. H., and Simon, A. E. (1991). *J. Virol.*, in press.
Li, X. H., Heaton, L. A., Morris, T. J., and Simon, A. E. (1989). *Proc. Natl. Acad. Sci. U.S.A.* **86**, 9173–9177.
Luo, G., Chao, M., Hsieh, S.-T., Sureau, C., Nishikuba, K., and Taylor, J. (1990). *J. Virol.* **64**, 1021–1027.
Makino, S., Yokomori, K., and Lai, M. M. C. (1990). *J. Virol.* **64**, 6045–6053.
Malim, M. H., Böhnlein, S., Hauber, J., and Cullen, B. R. (1989). *Cell* **58**, 205–214.
Marcus, P. I., and Gaccione, C. (1989). *Virology* **171**, 630–633.
McLain, L., and Dimmock, N. J. (1989). *J. Gen. Virol.* **70**, 2615–2624.
McLain, L., Armstrong, S. J., and Dimmock, N. J. (1988). *J. Gen. Virol.* **69**, 1415–1419.
Miller, G., Rabson, M., and Heston, L. (1984). *J. Virol.* **50**, 174–182.
Miller, G., Heston, L., and Countryman, J. (1985). *J. Virol.* **54**, 45–52.
Miller, R. H., Girones, R., Cote, P. J., Hornbuckle, W. E., Chestnut, T., Balwin, B. H., Korba, B. E., Tennant, B. C., Gerin, J. L., and Purcell, R. H. (1990). *Proc. Natl. Acad. Sci. U.S.A.* **87**, 9329–9332.
Monjardino, J. P., and Saldanha, J. A. (1990). *Br. Med. Bull.* **42**, 399–407.
Monroe, S. S., and Schlesinger, S. (1983). *Proc. Natl. Acad. Sci. U.S.A.* **80**, 3279–3283.
Morris, T. J., and Carrington, J. C. (1988). *In* "The Plant Viruses" (R. Koenig, ed.), Vol. 3, pp. 73–112. Plenum, New York.
Morris, T. J., and Hillman, B. I. (1989). *UCLA Symp. MOl. Cell. Biol.* **101**, 185–197.
Mottet, G., and Roux, L. (1990). *Virus Res.* **14**, 175–188.
Mottet, G., Curran, J., and Roux, L. (1990). *Virology* **176**, 1–7.
Moyer, S. A. (1989). *Virology* **172**, 341–345.
Moyer, S. A., and Gatchell, S. H. (1979). *Virology* **92**, 168–179.
Murant, A. F., and Mayo, M. A. (1982). *Annu. Rev. Phytopathol.* **20**, 49–70.

Myerhans, A., Cheynier, R., Albert, J., Seth, M., Kwok, S., Sninsky, J., Morfeldt-Manson, L., Asjo, B., and Wain-Hobson, S. (1989). *Cell* **58**, 901–910.

Nüesch, J. P. F., Krech, S., and Siegl, G. (1988). *Virology* **165**, 419–427.

Nüesch, J. P. F., de Chastonay, J., and Siegl, G. (1989). *J. Gen. Virol.* **70**, 3475–3480.

Nuss, D. L. (1988). *In* "RNA Genetics" (E. Domingo, J. J. Holland, and P. Ahlquist, eds.), Vol. 2, pp. 187–210. CRC Press, Boca Raton, Florida.

Odagiri, T., and Tobita, K. (1990). *Proc. Natl. Acad. Sci. U.S.A.* **87**, 5988–5992.

O'Hara, P. J., Horodyski, F. M., Nichol, S. T., and Holland, J. J. (1984a). *J. Virol.* **49**, 793–798.

O'Hara, P. J., Nichol, S. T., Horodyski, F. M., and Holland, J. J. (1984b). *Cell* **36**, 915–924.

Okamoto, H., Yotsumoto, S., Akahane, Y., Yamanaka, T., Miyazaki, Y., Sugai, Y., Tsuda, F., Tanaka, T., Miyakawa, Y., and Mayumi, M. (1990). *J. Virol.* **64**, 1298–1303.

Overbraugh, J., Donahue, P. R., Quackenbush, S. L., Hoover, E. A., and Mullins, J. I. (1988). *Science* **239**, 906–910.

Palma, E. L., and Huang, A. S. (1974). *J. Infect. Dis.* **129**, 402–410.

Patton, D. F., Shirley, P., Raab-Traub, N., Resnick, L., and Sixbey, J. W. (1990). *J. Virol.* **64**, 397–400.

Patton, J. T., Davis, N. L., and Wertz, G. W. (1984). *J. Virol.* **49**, 303–309.

Peluso, R. W., and Moyer, S. A. (1983). *Proc. Natl. Acad. Sci. U.S.A.* **80**, 3198–3202.

Peluso, R. W., and Moyer, S. A. (1988). *Virology* **162**, 369–376.

Perrault, J. (1981). *Curr. Top. Microbiol. Immunol.* **93**, 151–207.

Perrault, J., and Holland, J. J. (1972). *Virology* **50**, 159–170.

Re, G. G., and Kingsbury, D. W. (1988). *Virology* **165**, 331–337.

Re, G. G., Morgan, E. M., and Kingsbury, D. W. (1985). *Virology* **46**, 27–37.

Rizzetto, M., Ponzetto, A., Bonino, F., and Smedile, A. (1988). *In* "Viral Hepatitis and Liver Disease" (A. J. Zuckerman, ed.), pp. 389–394. Liss, New York.

Roux, L., and Holland, J. J. (1980). *Virology* **100**, 53–64.

Roux, L., Beffy, P., and Portner, A. (1984). *Virology* **138**, 118–128.

Rubino, L., Burgyan, J., Grieco, F., and Russo, M. (1990). *J. Gen. Virol.* **71**, 1655–1660.

Schlesinger, S. (1988). *In* "RNA Genetics" (E. Domingo, J. J. Holland, and P. Ahlquist, eds.), Vol. 2, pp. 167–185. CRC Press, Boca Raton, Florida.

Sekellick, M. J., and Marcus, P. I. (1980a). *Ann. N.Y. Acad. Sci.* **350**, 545–557.

Sekellick, M. J., and Marcus, P. I. (1980b). *Virology* **104**, 247–252.

Sethna, P. B., Hung, S.-L., and Brian, D. A. (1989). *Proc. Natl. Acad. Sci. U.S.A.* **86**, 5626–5630.

Sethna, P. B., Hoffman, M. A., and Brian, D. A. (1991). *J. Virol.* **65**, 320–325.

Simon, A. E. (1988). *Plant Mol. Biol. Rep.* **6**, 240–252.

Simon, A. E., and Howell, S. H. (1986). *EMBO J.* **5**, 3423–3428.

Simon, A. E., Engel, H., Johnson, R., and Howell, S. H. (1988). *EMBO J.* **7**, 2645–2651.

Simon, A. E., Engel, H., and Howell, S. H. (1989). *UCLA Symp. Mol. Cell. Biol.* **101**, 217–227.

Simpson, L. (1990). *Science* **250**, 512–513.

Sleat, D. E., and Palukaitis, P. (1990a). *Virology* **176**, 292–295.

Sleat, D. E., and Palukaitis, P. (1990b). *Proc. Natl. Acad. Sci. U.S.A.* **87**, 2946–2950.

Steinhauer, D. A., and Holland, J. J. (1987). *Annu. Rev. Microbiol.* **41**, 409–433.

Taylor, J. M. (1990). *Cell* **61**, 371–373.

Thomas, S. M., Lamb, R. A., and Paterson, R. G. (1988). *Cell* **54**, 891–902.

Trono, D., Feinberg, M. B., and Baltimore, D. (1989). *Cell* **59**, 113–120.

Tuffereau, C., and Roux, L. (1988). *Virology* **162**, 417–426.

Von Laer, D. M., Mack, D., and Kruppa, J. (1988). *J. Virol.* **62**, 1323–1329.
Webster, R. G., Kawaoka, Y., and Bean, W. J. (1986). *Virology* **149**, 165–173.
Wei, N., Heaton, L. A., and Morris, T. J. (1990). *J. Mol. Biol.* **214**, 85–95.
Weiss, B., and Schlesinger, S. (1981). *J. Virol.* **37**, 840–844.
Wertz, G. W., Davis, N. L., and Patton, J. (1987). *In* "The Rhabdoviruses" (R. R. Wagner, ed.), pp. 271–296. Plenum, New York.
Wong, T. C., Ayata, M., Hirano, A., Yoshikawa, Y., Tsuruoka, H., and Yamanouchi, K. (1989). *J. Virol.* **63**, 5464–5468.
Yalamanchili, R. R., Raengsakulrach, B., Baumann, R. P., and O'Callaghan, D. J. (1990). *Virology* **175**, 448–455.
Youngner, J. S., and Preble, O. T. (1980). *Compr. Virol.* **16**, 73–135.
Zhang, C., Carcone, P. J., and Simon, A. E. (1991). *Virology,* in press.

ADVANCES IN VIRUS RESEARCH, VOL. 40

STRUCTURE AND FUNCTION OF THE HEF GLYCOPROTEIN OF INFLUENZA C VIRUS

Georg Herrler and Hans-Dieter Klenk

Institut für Virologie
Philipps-Universität Marburg
D-3550 Marburg, Germany

I. Introduction

Soon after the first isolation of an influenza C virus from a patient (Taylor, 1949), it became obvious that this virus differs from other myxoviruses in several aspects. Pronounced differences have been observed in the interactions between the virus and cell surfaces, suggesting that influenza C virus attaches to receptors different from those recognized by other myxoviruses. While influenza A and B viruses agglutinate erythrocytes from many species, including humans, the spectrum of erythrocytes agglutinated by influenza C virus is much more restricted. Erythrocytes from rats, mice, and adult chickens are suitable for hemagglutination and hemadsorption tests; cells from other species, however, react not at all or only poorly with influenza C virus (Hirst, 1950; Minuse *et al.*, 1954; Chakraverty, 1974; Ohuchi *et al.*, 1978). Differences are also observed so far as hemagglutination inhibitors are concerned. A variety of glycoproteins have been shown to prevent influenza A and B viruses from agglutinating erythrocytes. In the case of influenza C virus, rat serum was for a long time the only known hemagglutination inhibitor (Styk, 1955; O'Callaghan *et al.*, 1980).

A difference in the receptors for influenza C virus and other myxo-viruses was also suggested by studies on the receptor-destroying enzyme. The ability of influenza C virus to inactivate its own receptors was reported soon after the first isolation of this virus from a patient (Hirst, 1950). However, the influenza C enzyme did not affect the receptors of other myxoviruses and, conversely, the receptor-destroying enzyme of either of the latter viruses was unable to inactivate the receptors for influenza C virus on erythrocytes. While the enzyme of influenza A and B virus was characterized as a neuraminidase in the 1950s (Klenk et al., 1955), even with refined methodology no such activity was detectable with influenza C virus (Kendal, 1975; Nerome, et al., 1976).

It is now known that both the receptor-binding and receptor-destroying activities as well as the fusion activity of influenza C virus are mediated by the only glycoprotein present on the surface of the virus particle. The structure and functions of this protein, which is designated HEF, are reviewed in the following sections.

II. STRUCTURE

A. Primary Structure

For two strains of influenza C virus, the RNA segment containing the genetic information for HEF has been cloned and sequenced (Nakada et al., 1984; Pfeifer and Compans, 1984). Sequence data for several strains have been obtained by direct sequencing of the viral RNA (Buonagurio et al., 1985; Adachi et al., 1989). The gene is 2070–2075 nucleotides in length and can code for a polypeptide of 654–655 amino acids (Fig. 1). The predicted polypeptide has a molecular weight of about 72,000. At the amino terminus there is a stretch of 12 hydrophobic amino acids, which may represent the signal sequence. Cleavage of this sequence results in a polypeptide with a molecular weight of about 70,500. Two additional hydrophobic sequences are located at positions 447–463 and 627–652. The former is probably involved in the fusion activity, as discussed in Section III,C. The hydrophobic amino acid sequence at the carboxy-terminal end is assumed to function as a membrane anchor, which is followed by a cytoplasmic tail of only three amino acids. While a homology of 30% has been observed between the hemagglutinins of influenza A and B viruses (Krystal et al., 1982), no significant values of homology were found when these glycoproteins were compared with the HEF protein of influenza C virus. The similarity between the HEF sequence and the HA sequence is restricted to the presence of the three hydrophobic domains mentioned above. Using this criterion, sequence alignments have been reported, with six to nine cysteines being conserved in the glycoproteins of the three types of influenza virus (Nakada et al., 1984; Pfeifer and Compans, 1984). Comparison of the other influenza C proteins with their influenza A and B counterparts also revealed only a very low degree of sequence similarity (Yamashita et al., 1989). Together, the sequence data suggest that influenza A and B viruses are more closely related to one another than they are to influenza C virus.

B. Co- and Posttranslational Modifications

Among the modifications of the HEF polypeptide, gycosylation has been studied in greatest detail. In the presence of the inhibitor tunicamycin the unglycosylated form of the protein is obtained (Nagele, 1983; Hongo et al., 1986a). This finding indicates that the native glycoprotein only contains N-linked oligosaccharides, while O-linked carbohydrate structures are absent. As indicated in Fig. 1, the amino acid sequence contains eight consensus sequences Asn–X–Ser/Thr suitable for the attachment of N-linked oligosaccharides (Nakada et al., 1984; Pfeifer and Compans, 1984). Analysis of the synthesis of the influenza C glycoprotein in the presence of limiting concentrations of glycosylation inhibitors suggested the presence of seven oligosaccharides on the native protein (Nagele, 1983): six on the HEF_1 portion and only one on HEF_2.

Three size classes of oligosaccharides—G_1, G_2, and G_3—have been resolved by gel chromatography (Nakamura et al., 1979). Oligosaccharides corresponding to the two smaller size classes (i.e., G_2 and G_3) have also been observed in influenza A virus, while G_1 is restricted to influenza C virus. G_3 appears to represent the mannose-rich type of oligosaccharides. The oligosaccharides of size classes G_1 and G_2 have both been shown to contain N-acetylneuraminic acid, indicating that they are of the complex type. Because of the presence of sialic acid on the viral surface, influenza C virus is able to inhibit the hemagglutinating activity of influenza A viruses (Nerome et al., 1976; Meier-Ewert et al., 1978). The structure of the different oligosaccharides has not been determined. It has been suggested that some HEF polypeptides contain predominantly oligosaccharides of the larger size classes, while others are glycosylated with the smaller size classes (Nagele, 1983). This would provide an explanation as to why, after sodium dodecyl sulfate–polyacrylamide gel electrophoresis (SDS–PAGE), HEF is detected as a doublet band (Herrler et al., 1979; Sugawara et al., 1981).

```
                                                                                    60
AGC AGA AGG GGG TTA ATA ATG TTT TTC TCA TTA CTC TTG GTG TTG GGC CTC ACA CAC GAG
                    Met Phe Phe Ser Leu Leu Leu Val Leu Gly Leu Thr His Glu
                                                                          △
                                                                                   120
GCT GAA AAA ATA AAG TGC CTT CAA AAG GAA GAA ACA CTA AGC CTA CAC AAT
Ala Glu Lys Ile Lys Cys Leu Gln Lys Glu Glu Thr Leu Ser Leu His Asn
                                                                    O
                                                                                   180
GGC TTC GGA GGA AAT TTG GCC ACA GAA GAA GTC GTC TTT GAG CTT GTT AAG CCC
Gly Phe Gly Gly Asn Leu Ala Thr Glu Glu Val Val Phe Glu Leu Val Lys Pro
                                                                                   240
AAA GCT GGA GCC TCT GTT GAT CAA AGT GGA GAT TCA AGG TCA AGG GAT TCA AGG ACT
Lys Ala Gly Ala Ser Val Asp Gln Ser Gly Asp Ser Arg Ser Met Asp Ser Arg Thr
                                                                                   300
GAC AAA AGC AAT AGT CCT TCT AGG GCT GAT GAT ACT GCT GAT AAG
Asp Lys Ser Asn Ser Pro Ser Arg Ala Asp Asp Thr Ala Asp Lys
                                                           □
                                                                                   360
TTT CGT TTT TTG CTT TCT GGT GGA TTG ATG AGT GGT TTT GGC CCA CCT GGG AAG GTA
Phe Arg Phe Leu Leu Ser Gly Gly Leu Met Ser Gly Phe Gly Pro Pro Gly Lys Val
                                                                                   420
GAC TAC CTT CTT TAC TGT CAA GGA ATA AAT GTT TAT TTT GAT GAT ATG AAC TGG AGT
Asp Tyr Leu Leu Tyr Cys Gln Gly Ile Asn Val Tyr Phe Asp Asp Met Asn Trp Ser
                                                                                   480
CCA CAT GCT GCT ATA AAT TGT TGT CAT TGC GCA TCA GAA ATG GCC AAT TTC CAG
Pro His Ala Ala Ile Asn Cys Cys His Cys Ala Ser Glu Met Ala Asn Phe Gln
                          ●
                                                                                   540
AAA ACT ATT CCT TTA CAA GTG ACT GGG GCA AAT TGC AGC AAT TGC TGG TTG GAC
Lys Thr Ile Pro Leu Gln Val Thr Gly Ala Asn Cys Ser Asn Cys Trp Leu Asp
                                                                                   600
AAA AAT CCA GCA TTG CTT ACA CAA GAA GTC TCA AAC AAC GAA TGT GGG AAA GAA
Lys Asn Pro Ala Leu Leu Thr Gln Glu Val Ser Asn Asn Glu Cys Gly Lys Glu
                                                                                   660
AAT CTT GCT TTC TTC TAT ACC ACC CCA CCA TTT GGA GAG TAT AAT GGA TAC AGG
Asn Leu Ala Phe Phe Tyr Thr Thr Pro Pro Phe Gly Glu Tyr Asn Gly Tyr Arg
                                                                                   720
GTG GCT TCT TGC TAT TAT GAT TCA AGT GAA AAA AGA GGA CTA GAT TGT GAC
Val Ala Ser Cys Tyr Tyr Asp Ser Ser Glu Lys Arg Gly Leu Asp Cys Asp
                          ■
                                                                                   780
AAC TAC TTT CAA GTG ATC GTC GTT GGA AAA GGA GGA CTA TTA GAT AAC AGG
Asn Tyr Phe Gln Val Ile Val Val Gly Lys Gly Gly Leu Leu Asp Asn Arg
                                                                                   840
GTA TCA CCT TAC TAC AAT GGG AAT TCT GGA GAC ACC TGT CAA ATG CTC CAG
Val Ser Pro Tyr Tyr Asn Gly Asn Ser Gly Asp Thr Cys Gln Met Leu Gln
                                                                                   900
CTG AAA CCT GGA AGA TAT TCA GTA AGA AGA TCA ACA TTA CCT GAA AGA
Leu Lys Pro Gly Arg Tyr Ser Val Arg Arg Ser Thr Leu Pro Glu Arg
                                                                                   960
AGT TAT TTT GAC ATG ATA AAA ATG AAA TCC ATT GTC ATT TGG GGA
Ser Tyr Phe Asp Met Ile Lys Met Lys Ser Ile Val Ile Trp Gly
                                                                                  1020
AAA AGC AGA GAA TAT GCA GAA AAA ACT TCC CAA GTC CCA ATT TGG GGA
Lys Ser Arg Glu Tyr Ala Glu Lys Thr Ser Gln Val Pro Ile Trp Gly
                                                                                  1080
AAA AAT GCA GAA GAA TCT GAC GTG GAT GCT GCT CCA AGC ACT GGG TGC ATG
Lys Asn Ala Glu Glu Ser Asp Val Asp Ala Ala Pro Ser Thr Gly Cys Met

                                                                                  1140
TTG ATC CAA AAG CCA TAC CAA AAG CCA GAT CAC ATT GGA GAA GCT GAT CAA CAA
Leu Ile Gln Lys Pro Tyr Gln Lys Pro Asp His Ile Gly Glu Ala Asp Gln Gln
                                                                                  1200
ATG AGG GAG TTG TCA GGA CTG GAC CTG TAT GAA GCT ATA TCA CAA TCA GGG TGG
Met Arg Glu Leu Ser Gly Leu Asp Leu Tyr Glu Ala Ile Ser Gln Ser Gly Trp
                                                                                  1260
GGC TTC GGA GGG AAT ACC AGT TTT GAG ACG TTT ACG GAG AAA TTT GGA AGA TGC
Gly Phe Gly Gly Asn Thr Ser Phe Glu Thr Phe Thr Glu Lys Phe Gly Arg Cys
                                                                               O
                                                                                  1320
CCT TTG GCT GCA ATC AAA ATC CCA AAA ATC CCA CTT CTA ATT CCC ACC
Pro Leu Ala Ala Ile Lys Ile Pro Lys Ile Pro Leu Leu Ile Pro Thr
                                                                                  1380
AGT GGA ACC AGC TCT TTT GGA ATT TTT GGA ATC GAT ATC GAT GAC CTC
Ser Gly Thr Ser Ser Phe Gly Ile Phe Gly Ile Asp Ile Asp Asp Leu
                                                                  ←
                                                                                  1440
ATT ATT GGT TTG GTT GTT GCA ATC GTT GAA ACA GAA ATT GGA GGC TAT CTG CTT GGA
Ile Ile Gly Leu Val Val Ala Ile Val Glu Thr Glu Ile Gly Gly Tyr Leu Leu Gly
                                                                                  1500
AGT AGA AAA TCA GCA GGT GGT GTG ACA GAA ACA GAA AAA GGG TTT GAG AAA
Ser Arg Lys Ser Ala Gly Gly Val Thr Glu Thr Glu Lys Gly Phe Glu Lys
                                                                                  1560
ATT GGA AAT GAC ATA TCT TTA AAA GCC ATC GCA ATA ATC GAC AAA AAT CTA CTA AAC
Ile Gly Asn Asp Ile Ser Leu Lys Ala Ile Ala Ile Ile Asp Lys Asn Leu Leu Asn
                                                                                  1620
GAC AGA ATT GAG GAT CAA GCA ATC ACT ATA GAA ATT GAA AAT GCA
Asp Arg Ile Glu Asp Gln Ala Ile Thr Ile Glu Ile Glu Asn Ala
                                                                                  1680
AGA GAG GCA TTG GTG CAA TTG TTA GGA GCC TTG GTA TTG ATA GGA ATA
Arg Glu Ala Leu Val Gln Leu Leu Gly Ala Leu Val Leu Ile Gly Ile
                                                                                  1740
AGC AGT GCA ATT TCT CTA AAT GTC GCA GAA CTA ACA ATT AAC AGA GCA GGA
Ser Ser Ala Ile Ser Leu Asn Val Ala Glu Leu Thr Ile Asn Arg Ala Gly
                                                                               ●
                                                                                  1800
GAT CTA GCA TTT TTC TCA CCA GGT TGC GGT GAC ATT GAC TGG TTT TAC GAT GAT CAA
Asp Leu Ala Phe Phe Ser Pro Gly Cys Gly Asp Ile Asp Trp Phe Tyr Asp Asp Gln
                                                                                  1860
AGC TGT TGC CAA AAT TTC AAG GAT CTG GAA TTG TGG GAT CCA ACC ATT CCC CCT
Ser Cys Cys Gln Asn Phe Lys Asp Leu Glu Leu Trp Asp Pro Thr Ile Pro Pro
                              ●
                                                                                  1920
CTT GAC ACA TTT CAA GAT CTG CAG TCA CTA TCA CAT GCT GGA AGC TTG GGC TTA
Leu Asp Thr Phe Gln Asp Leu Gln Ser Leu Ser His Ala Gly Ser Leu Gly Leu
                                                                                  1980
GCA ATA ACT ACT ACA AGA GCA ATG ATG AAA AAA TAT TGC AGC ATC GCC ATC TGC AGA
Ala Ile Thr Thr Thr Arg Ala Met Met Lys Lys Tyr Cys Ser Ile Ala Ile Cys Arg
                                                                                  2040
ACT AAA TGA TTG AGA CAA AAA AAT CCC CTT GCT ACT GCT
Thr Lys End

TTT ATA AAA AAC AAC CCT GTC AAT GCT GCT ACT GCT
```

As discussed in Section II,D, glycosylation of HEF is important for the presentation of the antigenic epitopes. Furthermore, the carbohydrate side chains are crucial for the stability of the glycoprotein by protecting it from proteolytic degradation. In the presence of tunicamycin, virions are released from the infected cells; however, the virus particles are lacking surface proteins (Hongo et al., 1986a).

Another posttranslational modification of HEF is the proteolytic cleavage of the precursor polypeptide HEF_0 into the cleavage products HEF_1 and HEF_2. As discussed in Section III,C, this modification is required for viral fusion activity. Cleavage is caused by a cellular protease. Some cultured cells [e.g., chick embryo fibroblasts, LLC-MK2, or Madin–Darby canine kidney (MDCK) cells] are lacking an appropriate enzyme or have only low amounts of it. On the surface of virions released by such cells, the glycoprotein HEF is found predominantly in the uncleaved form, which can be cleaved in vitro by incubation with trypsin and elastase (Compans et al., 1977; Herrler et al., 1979; Sugawara et al., 1981). Influenza C viruses grown in embryonated eggs or primary chick kidney cells contain most of their glycoprotein molecules in the cleaved form. The cleavage products are detected after SDS–PAGE only in the presence of reducing agents (Herrler et al., 1979).

This observation indicates that the two polypeptides are held together by disulfide bonds, as observed with several viral surface glycoproteins which are proteolytically cleaved. The disulfide bonds contribute to a unique electrophoretic behavior of HEF which is not observed with the glycoproteins of other influenza viruses. Under nonreducing conditions the electrophoretic mobility of HEF_0 suggests a molecular weight of about 100,000, which is not in accord with the size deduced from the sequence data. In the presence of reducing agents, the electrophoretic migration of the uncleaved glycoprotein suggests a molecular weight of about 80,000, which is in the size range expected for the glycosylated HEF_0. A shift from the 100K form to the 80K form is also observed under nonreducing conditions after proteolytic cleavage of HEF_0 into the disulfide-bonded products HEF_1 and HEF_2

←──────────────────────────────

Fig. 1.   DNA sequence of gene segment 4 of influenza C/JHB/1/66 and its translation in open reading frame 1 (Pfeifer and Compans, 1984). The sequence is written in message sense. Hydrophobic sequences are marked with wavy lines. The predicted HEF_1–HEF_2 cleavage site is indicated by an arrow. The predicted cleavage site of the leader peptide is indicated by an open triangle. Solid circles indicate potential glycosylation sites. Open circles indicate cysteine residues conserved among hemagglutinins of influenza A, B, and C viruses. The active-site serine (amino acid 71) is indicated by an open square. The mutation site of a mutant with increased receptor-binding efficiency is marked with a solid square (Thr 284).

(designated $HEF_{1,2}$). No evidence for the release of a polypeptide could be obtained, which would explain the shift in the molecular weight by about 20,000 (Meier-Ewert *et al.*, 1980, 1981a,b).

Therefore, it is assumed that the uncleaved glycoprotein has a peculiar conformation which is maintained by disulfide bonds. This conformation may allow only association with a reduced amount of SDS, thereby causing aberrant electrophoretic migration behavior. The conformational constraint is released either by abolishing the disulfide bonds or by proteolytic cleavage of HEF_0 into HEF_1 and HEF_2. It is not known whether the formation of disulfide bonds is a co- or posttranslational modification of the glycoprotein. The proteolytic cleavage was found to be a late modification. In pulse–chase experiments hardly any cleaved glycoprotein was detectable in infected chick kidney cells. Therefore, the proteolytic cleavage may occur only shortly before virus particles are released by budding.

A modification of the influenza C glycoprotein, which has been described only recently, is the acylation with fatty acids (Veit *et al.*, 1990). The acyl chains are attached presumably to cysteine residues, as indicated by their release after treatment with either hydroxylamine or mercaptoethanol. Such a labile thioester-type linkage has been found on many acylated glycoproteins of both viral and cellular origin. In all cases tested palmitic acid was the predominant fatty acid. The HEF glycoprotein was unique in this respect, because stearic acid was detected as the prevailing fatty acid. The reason for this difference in the acylation is unknown. Cysteine residues in the cytoplasmic tail have been identified as fatty acid attachment sites for several glycoproteins. The cytoplasmic domain of HEF is very short. It contains only a single cysteine, which, therefore, is the candidate for attachment of stearic acid. So far, no biological function can be attributed to the fatty acid of the influenza C glycoprotein.

C. Supramolecular Structure of the HEF Spike

A characteristic feature of influenza C virus was revealed by electron microscopy long before the proteins had been analyzed. The surface projections are usually arranged in a reticular structure consisting mainly of hexagons (Flewett and Apostolov, 1967), which can be seen on both filamentous and spherical particles (Fig. 2). A single spike protein is observed on each of the six vertices of the hexagons (Herrler *et al.*, 1981). Values for the length of individual spikes are in the range of 8–15 nm. The low-resolution structure of the spikes determined by electron microscopy indicated that the influenza C glycoprotein is a trimer (Hewat *et al.*, 1984). The trimeric structure was confirmed when the sedimentation of the glycoprotein in sucrose

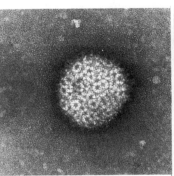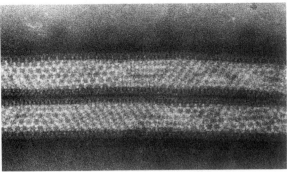

FIG. 2.   Electron micrograph of influenza C virions. Both a spherical and a filamentous particle are shown. (Adapted from Herrler *et al.*, 1981.)

gradients was analyzed (Formanowski and Meier-Ewert, 1988; Formanowski *et al.*, 1989). Glycoprotein, which has been released from the viral surface by bromelain treatment, is still found as a trimer, indicating that the membrane anchor and the cytoplasmic tail are not essential for maintaining this structure. Calcium ions, however, appear to play an important role. On sucrose gradient centrifugation of bromelain-released HEF, trimers were detected only in the presence of calcium ions. When calcium-deficient buffers were used, the glycoprotein dissociated into monomers (Formanowski and Meier-Ewert, 1988; Formanowski *et al.*, 1989).

Lateral interactions between trimeric spike glycoproteins are probably involved in the formation of the hexagonal array on the viral surface. This is suggested by the finding that the reticular arrangement is sometimes maintained after removal of the spikes from the viral membrane by either protease treatment or spontaneous release (Herrler *et al.*, 1981). Lateral interactions are also suggested by electron micrographs of detergent-isolated spikes. On removal of the detergent, membrane glycoproteins (e.g., the influenza A hemagglutinin) form rosettes, where the proteins are connected at a central point via their hydrophobic membrane anchor. In contrast, the influenza C glycoproteins are arranged in an elongated beetlelike structure with individual spikes standing side by side (Formanowski *et al.*, 1989, 1990). The lateral interactions are not dependent on the proteolytic cleavage of the glycoprotein. The hexagonal array is observed with virus containing predominantly HEF_0 as well as with virus containing $HEF_{1,2}$. Glycoprotein in the uncleaved form also maintains the reticular pattern at pH 5.0. Cleaved glycoprotein, however, undergoes a major conformational change at low pH, resulting in the loss of the regular hexagonal arrangement of the spikes (Hewat *et al.*, 1984; For-manowski *et al.*, 1990).

It has been reported that crystals of bromelain-released HEF have been obtained, which are suitable for X-ray diffraction studies (Rosenthal *et al.*, 1990). Thus, there is hope that in the near future the three-dimensional structure of the influenza C glycoprotein will be known, which would represent major progress toward understanding the structure–function relationship.

D. Antigenic Epitopes

Among monoclonal antibodies directed against the HE protein, two groups have been distinguished. Group A antibodies inhibited the hemagglutinating and hemolytic activities and neutralized the infectivity of influenza C virus, whereas group B antibodies lacked any of these reactivities (Sugawara *et al.*, 1986). Analysis of antigenic variants selected for resistance against either of these monoclonal antibodies suggested the presence of four antigenic epitopes: A-1, A-2, B-1, and B-2 (Sugawara *et al.*, 1988). Competitive binding assays indicated that sites A-1 and A-2 may be located close to one another (Sugawara *et al.*, 1988). Both A epitopes were shown to be sensitive to denaturing conditions (e.g., treatment with SDS). Therefore, on Western blots, HEF was detected only by group B antibodies. These results indicate that sites B-1 and B-2 are sequence-dependent epitopes, whereas sites A-1 and A-2 are conformation-dependent epitopes. The conformation of both A epitopes was found to be dependent on the glycosylation of HEF. The nonglycosylated form of the protein synthesized in the presence of tunicamycin was recognized by group B antibodies, while group A antibodies reacted only poorly or not at all (Hongo *et al.*, 1986b; Sugawara *et al.*, 1988). The antigenic sites are presumably different from the functional epitopes of the receptor-binding and receptor-destroying activities. The ability of several monoclonal antibodies to inhibit the hemagglutinating activity of influenza C virus (Sugawara *et al.*, 1986, 1988; Vlasak *et al.*, 1987; Herrler *et al.*, 1988a) may be due to steric hindrance. Some of the antibodies caused partial inhibition of the receptor-destroying enzyme, when the esterase activity was determined with large substrates, but no inhibitory effect was observed when small substrates were used (Vlasak *et al.*, 1987; Herrler *et al.*, 1988a; Hachinohe *et al.*, 1989).

The antigenic variation among different strains of influenza C virus is less pronounced than in the case of influenza A viruses. Several reports revealed a high degree of cross-reactivity between different strains, irrespective of the time and place of isolation (Czekalowski and Prasad, 1973; Chakraverty, 1974, 1978; Meier-Ewert *et al.*, 1981c;

Kawamura *et al.*, 1986). Using monoclonal antibodies, it was possible, however, to demonstrate antigenic variation (Sugawara *et al.*, 1986, 1988; Adachi *et al.*, 1989). The low extent of variation is not due to a low capacity to produce antigenic variants. Escape mutants resistant against monoclonal antibodies have been obtained with a frequency similar to values reported for influenza A virus (Sugawara *et al.*, 1988). Maybe the immune selection is less pronounced in the case of influenza C virus. This may also explain why no antigenic drift has been observed with this group of viruses. Among the antigenic variants arising within influenza A viruses, one usually becomes dominant and replaces the older ones. In contrast, analyses of different influenza C strains indicate that several antigenic variants cocirculate (Adachi *et al.*, 1989). This conclusion is supported by studies on the genetic variation in the HEF as well as the *NS* gene of influenza C virus (Buonagurio *et al.*, 1985, 1986; Kawamura *et al.*, 1986; Adachi *et al.*, 1989).

III. Functions

A. Receptor-Destroying Enzyme

Although the receptor-destroying enzyme of influenza C virus was described by Hirst in 1950, more than 30 years passed before its specificity was elucidated. Identification of the enzyme activity was accomplished by analyzing its effect on hemagglutination inhibitors. Rat serum has long been known for its inhibitory activity (Styk, 1955; O'Callaghan *et al.*, 1980). Two components of rat serum have been shown to account for most of the hemagglutination inhibition activity: murinoglobulin and α_1-macroglobulin (Herrler *et al.*, 1985b; Kitame *et al.*, 1985). The carbohydrate portion of the latter compound was found to consist primarily of N-linked biantennary oligosaccharides (Herrler *et al.*, 1985b). The only effect of the influenza C enzyme on these oligosaccharides was a change in the terminal sialic acid residue. While the native macroglobulin contained 40% of its sialic acid as *N*-acetyl-9-*O*-acetylneuraminic acid, this amount was reduced to 10% after treatment with the receptor-destroying enzyme.

Concomitant with the decrease of the 9-*O*-acetylated sialic acid, an increase of *N*-acetylneuraminic acid was observed (Herrler *et al.*, 1985c). The same effect was obtained with bovine submandibulary mucin, another hemagglutination inhibitor. In both cases the change in the sialic acid was paralleled by loss of the inhibitory activity, indicating that the receptor-destroying enzyme of influenza C virus is a sialate 9-*O*-acetylesterase (Fig. 3). The enzyme has been shown to be

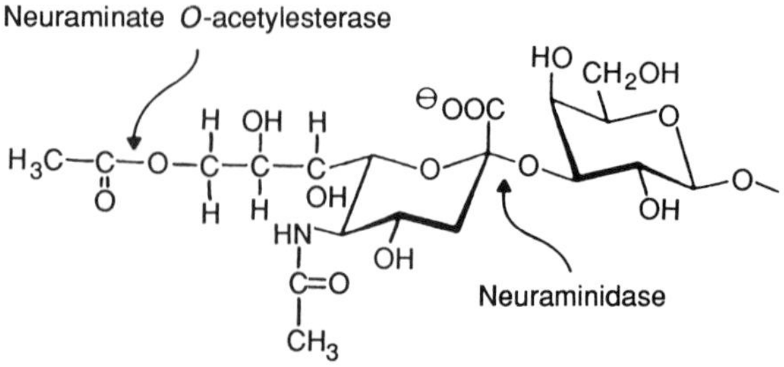

Neuraminate *O*-acetylesterase

FIG. 3.   Structure of *N*-acetyl-9-*O*-acetylneuraminic acid connected to galactose via an α2,3-linkage. The sites of action of the acetylesterase of influenza C virus and the neuraminidase of influenza A and B viruses are shown.

a function of HEF by several approaches: expression of the cloned HEF gene in vertebrate cells (Vlasak *et al.*, 1987) and analysis of the purified protein after isolation by detergent (Herrler *et al.*, 1988a) or protease treatment (Formanowski and Meier-Ewert, 1988).

The influenza C esterase belongs to the class of serine hydrolases which are inhibited by diisopropyl fluorophosphate (DFP). The inhibitor abolishes the enzyme activity without affecting the hemagglutinating activity (Muchmore and Varki, 1987). This finding suggests that the active site of the esterase and the receptor-binding site are different epitopes on the influenza C glycoprotein. There is some information on the amino acids which are crucial for the formation of the active site of the esterase. From the knowledge about other serine hydrolases (e.g., trypsin and chymotrypsin), it is expected that the enzyme mechanism involves a charge relay system, which is accomplished by a catalytic triad composed of the amino acids serine, histidine, and aspartic acid (Kraut, 1977). Taking advantage of the fact that DFP binds covalently to the serine in the active site of serine hydrolases, amino acid 71 of HEF has been identified as active-site serine (Herrler *et al.*, 1988b; Vlasak *et al.*, 1989). This amino acid is part of the sequence Phe–Gly–Asp–Ser–Arg (Fig. 1). While the motif Gly–Asp–Ser is found in the active site of many serine hydrolases, including trypsin and chymotrypsin, the following arginine residue has been detected so far only in the active site of the acetylesterases of influenza C virus and coronaviruses (see Section IV and Fig. 5). From inhibition studies with arginine-specific modifying reagents, it has been suggested that this arginine residue may be important for substrate recognition, possibly interacting with the carboxyl group of *N*-acetyl-9-*O*-acetylneuraminic acid (Neu5,9Ac$_2$) (Hayes and Varki, 1989).

Analysis of a series of compounds revealed that the esterase of influenza C virus has a high specificity for O-acetyl groups, Neu5,9Ac$_2$

TABLE I

Substrate Specificity of the Acetylesterase of Influenza
C/JHB/1/66[a]

Substrate	Relative cleavage rate (%)
N-Acetyl-9-O-acetylneuraminic acid	100
N-Acetyl-4-O-acetylneuraminic acid	3
N-Glycolyl-9-O-acetylneuraminic acid	33
N-Acetyl-7-O-acetylneuraminic acid	—
Bovine submandibular gland mucin	30
Rat serum glycoprotein	90
Rat erythrocytes	25
Equine submandibular gland mucin	—
4-Methylumbelliferyl acetate	220
4-Methylumbelliferyl butyrate	14
4-Nitrophenyl acetate	3500
α-Naphthyl acetate	2200

[a] From Schauer et al. (1988).

being hydrolyzed at the highest rate among all natural substrates tested (Schauer et al., 1988) (see Table I). Some aromatic acetates (e.g., 4-nitrophenyl acetate or α-naphthyl acetate) are cleaved at higher rates than Neu5,9Ac$_2$ (Vlasak et al., 1987; Schauer et al., 1988; Wagaman et al., 1989). These compounds are substrates for many serine hydrolases, including proteases, and therefore are not suited for determination of the enzyme specificity. They enable, however, fast and sensitive assays. α-Naphthyl acetate has been shown to be useful for cytochemical detection of influenza C-infected cells (Wagaman et al., 1989). Treatment of erythrocytes with influenza C virus has been reported to change the reactivity of the cells with lectins specific for N-acetylgalactosamine (Luther et al., 1988). From this finding it has been inferred that the receptor-destroying enzyme is able to release the acetyl residue of N-acetylgalactosamine. However, there is no direct chemical evidence supporting this conclusion.

Apart from DFP the esterase activity of influenza C virus is also inhibited by diethyl-4-nitrophenyl phosphate and some isocoumarins (Schauer et al., 1988; Vlasak et al., 1989). Inhibition of the esterase by DFP or isocoumarins has been reported to reduce the infectivity of the virus (Muchmore and Varki, 1987; Vlasak et al., 1989). This finding may suggest that the receptor-destroying enzyme is required for virus entry into cells. However, both the hemagglutination (i.e., receptor-binding) and hemolytic (i.e., fusion) activities are not affected by the inactivation of the esterase (Muchmore and Varki, 1987; Vlasak et al., 1989). Thus, more experiments are necessary to show whether the

TABLE II

RECEPTOR SPECIFICITY OF INFLUENZA A, B, AND C VIRUSES[a]

Sialic acid on human erythrocytes	HA titer (HA units/ml)[b]		
	C/JHB/1/66	B/HK/8/73	A/PR/8/34
Native	0	64	256
Asialo	0	0	0
Neu5Ac			
α2,3Galβ1,3GalNAc	0	2	256
α2,3Galβ1,4GlcNAc	0	128	128
α2,6Galβ1,4GlcNAc	0	64	128
Neu5Gc			
α2,6Galβ1,4GlcNAc	0	2	0
Neu5,9Ac$_2$			
α2,3Galβ1,3GalNAc	128	0	0
α2,3Galβ1,4GlcNAc	128	0	0
α2,6Galβ1,4GlcNAc	128	0	0

[a] Adapted from Rogers et al. (1986).
[b] 0, HA titer <2.

reduction of the infectivity is correlated with the inhibition of the enzyme activity or whether it is due to an indirect effect of the inhibitor.

B. Receptor-Binding Activity

The identification of the receptor-destroying enzyme as a sialate 9-O-acetylesterase implied that Neu5,9Ac$_2$ (see Fig. 3) is a crucial component of the cellular receptors for influenza C virus (Herrler et al., 1985c). Direct evidence for the role of Neu5,9Ac$_2$ as a receptor determinant was provided by studies with erythrocytes which had been modified to contain only a single type of sialic acid. Influenza C virus was able to agglutinate erythrocytes which had been sialylated with Neu5,9Ac$_2$, but not cells containing N-acetyl- or N-glycolylneuraminic acid (Rogers et al., 1986) (see Table II). On the basis of these results, it was possible to explain previous observations which seemed to argue against an involvement of sialic acid in the attachment of influenza C virus to cells. The resistance of the erythrocyte receptors to periodate treatment (Ohuchi et al., 1978) is due to a greatly reduced oxidation of Neu5,9Ac$_2$ by periodate compared to Neu5Ac (Haverkamp et al., 1975). The difficulty in inactivating the influenza C receptors with viral and bacterial neuraminidases (Hirst, 1950; Kendal, 1975; Herrler et al., 1985a) is explained by the relative resistance of Neu5,9Ac$_2$ to the action of these enzymes (Corfield et al., 1981). The importance of

Neu5,9Ac$_2$ as a receptor determinant is not restricted to erythrocytes. 9-O-Acetylated sialic acid is also required for influenza C virus to initiate the infection of cultured cells (Herrler and Klenk, 1987a). In fact, lack of this type of sialic acid is a major reason for the resistance of many cell lines to influenza C virus.

Insertion of artificial receptors into the plasma membrane of cultured cells rendered several resistant cells sensitive to an influenza C infection. Moreover, an increase in the yield of virus was observed with cells which usually produce only low amounts of virus (Herrler and Klenk, 1987a). The presence of 9-O-acetylated sialic acid appears to be the major factor in determining whether a glycoconjugate can serve as a receptor for influenza C virus. Erythrocytes which have been resialylated to contain Neu5,9Ac$_2$ were agglutinated by influenza C virus regardless of whether the sialic acid molecule was attached to galactose via an α-2,3 or α-2,6 linkage (Rogers et $al.$, 1986) (see Table II). The sialyltransferase specific for the latter linkage type only acts on glycoproteins. Therefore, receptors generated by this enzyme are glycoproteins. On the other hand, it has been shown that bovine brain gangliosides can also serve as receptors for influenza C virus, although the active species among the mixture of gangliosides has not been determined (Herrler and Klenk, 1987a,b). Thus, both glycoproteins and glycolipids can be used as receptors by influenza C virus, provided they contain Neu5,9Ac$_2$. A larger number of glycoconjugates must be analyzed, however, in order to know whether factors other than the presence of Neu5,9Ac$_2$ are important for the receptor function of a glycoconjugate.

It has been suggested that, in addition to Neu5,9Ac$_2$, influenza C virus may also recognize N-acetylgalactosamine (Luther et $al.$, 1988). The conclusion is based on the finding (mentioned in Section III,A) that erythrocytes treated with influenza C virus differ from control cells in their reactivity with lectins specific for N-acetylgalactosamine. However, direct evidence for such a receptor specificity is lacking.

The amino acids involved in the receptor-binding site of HEF have not been determined. Valuable information should be obtained by the analysis of mutants with a change in the receptor-binding activity. A mutant has been described which has an expanded cell tropism due to a more efficient recognition of Neu5,9Ac$_2$-containing receptors compared to the parent virus (Szepanski et $al.$, 1989). Sequence analysis of this mutant indicated that a single point mutation (Thr 284 to isoleucine; see Fig. 1) is responsible for the change in the receptor-binding activity (Szepanski et $al.$, 1991). Interestingly, the mutation site is located next to a sequence (Gly–Asn–Ser–Gly) which, in similar form

(Gly–Gln–Ser–Gly), is also found in several subtypes of influenza A hemagglutinins (Fig. 1). The homologous sequence in the H3 subtype composing amino acids 225–228 has been shown to be part of the receptor-binding pocket (Weis *et al.*, 1988). These data suggest that the amino acids Gly 279 to Thr 284 may be constituents of the receptor-binding site of HEF and that the mutant is altered at this site. The observation that these amino acids are located on the unfolded polypeptide at a distance far from Ser 71 at the catalytic center of the esterase, together with the DFP effects (Section III,A), supports the notion that receptor binding and receptor inactivation are exerted by different structural domains of HEF.

Another example of an influenza C virus with a change in the receptor-binding activity has been reported (Camilleri and Maassab, 1988). Virus isolated from persistently infected MDCK cells was found to be more sensitive to the action of hemagglutination inhibitors than was wild-type virus. Sequence analysis of more mutants or variants of this type should help further define the receptor-destroying and receptor-binding sites of HEF. Obviously, however, final answers to these problems can be given only when the three-dimensional structure of the glycoprotein is available. The importance of individual amino acids involved in the formation of the functional epitopes can then be evaluated by site-directed mutagenesis.

The ability of the influenza C glycoprotein to attach to $Neu5,9Ac_2$-containing receptors can be used as powerful tool to detect 9-*O*-acetylated sialic acid (Muchmore and Varki, 1987). The ability of influenza C virus to agglutinate erythrocytes from an adult chicken, but not those from a 1-day-old chicken, was the basis for the discovery that $Neu5,9Ac_2$ is a differentiation marker on chicken erythrocytes, which has been confirmed by chemical analysis of the sialic acids on these cells (Herrler *et al.*, 1987). The sensitivity of the receptor recognition by influenza C virus is evident from studies with human erythrocytes. By chemical analysis only Neu5Ac has been detected, not $Neu5,9Ac_2$ (Shukla and Schauer, 1982). Agglutination and binding studies indicate, however, that erythrocytes from some individuals contain low levels of 9-*O*-acetylated sialic acid on their surface (Ohuchi *et al.*, 1978; Nishimura *et al.*, 1988).

C. Membrane Fusion

The fusion activity of influenza C virus was first demonstrated with erythrocytes. Microscopic observation of virus-induced cell fusion and photometric detection of hemolysis indicated that the virus is able to fuse with mouse and chicken erythrocytes (Ohuchi *et al.*, 1982; Kitame *et al.*, 1982). Recently, the fusion between virus membranes and artificial membranes has been monitored using a resonance energy assay (Formanowski et al., 1990). In contrast to the hemagglutinating (Herrler et al., 1979; Sugawara et al., 1981) and esterase activities of HEF (Herrler et al., 1988a), the fusion activity requires the proteolytic cleavage of HEF_0 into polypeptides HEF_1 and HEF_2 (Ohuchi et al., 1982; Kitame et al., 1982), described in Section II,B. The dependence of the influenza C virus-induced fusion on the cleavage of HEF indicated that this activity is a function of the surface glycoprotein. Virus with uncleaved glycoprotein can be rendered fusiogenic by in vitro cleavage of HEF. Both trypsin and elastase have been shown to be effective in this respect, whereas other proteases (e.g., chymotrypsin and thermolysin) were unable to activate the glycoprotein (Kitame et al., 1982; Ohuchi et al., 1982; Formanowski et al., 1990).

An additional characteristic of the fusion activity is pH dependence. Similar to influenza A and B viruses and several other viruses, influenza C virus causes fusion only at a low pH. Optimal pH values for hemolysis of erythrocytes vary between 5.0 and 5.7, depending on the virus strain. Optimal fusion between virus and unilamellar liposomes was detected in the range of 5.6–6.1. Several changes have been observed when the glycoprotein is shifted from neutral to acidic pH values: (1) The glycoprotein becomes susceptible to trypsin digestion; (2) the endogenous tryptophan fluorescence decreases; and (3) the hexagonal arrangement of the surface projections disappears (Formanowski et al., 1990). These changes, which were only observed with virus containing the cleaved HEF (i.e., $HEF_{1,2}$), suggest that exposure to a low pH results in a conformational change of the glycoprotein.

The characteristics of the influenza C virus-induced fusion described so far (i.e., a dependence on both proteolytic cleavage and low pH and a conformational change at low pH) are very similar to those reported for the fusion activity of influenza A and B viruses. It is therefore likely that fusion occurs by a similar mechanism for all influenza viruses. With influenza A virus it is widely accepted that the conformational change observed at acidic pH results in the exposure of the amino-terminal portion of the membrane-bound cleavage product (HA_2). This part of the protein is made up of a stretch of hydrophobic amino acids, which probably interact with the cellular membrane, thereby inducing fusion between the viral envelope and the membrane of the target cell. This model is also applicable to influenza C virus.

Differences between influenza A and C viruses have been observed so far as the kinetics of the fusion process are concerned. In the case of influenza A virus, the conformational change is fast and a later step is rate limiting. With influenza C virus the conformational change has been found to be a rate-limiting step (Formanowski et al., 1990). The reason for the delayed conformational change may be related to the

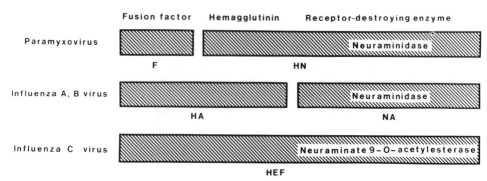

Fig. 4. The glycoproteins of paramyxoviruses (HN and F), influenza A and B viruses (HA and NA), and influenza C virus (HEF), illustrating differences in the distribution of biological activities (i.e., fusion, hemagglutination, and receptor inactivation). The sizes of the boxes representing the individual glycoproteins are not proportional to the molecular weights.

hexagonal arrangement of the spikes. The close packing of the glycoproteins might be a hindrance in adopting the conformation required for fusion.

In the course of virus infection, the viral fusion activity is crucial for the penetration of enveloped viruses. Viruses with a pH-dependent fusion activity are generally assumed to enter a cell via endosomes. The acidic pH within such vesicles triggers the fusion reaction, resulting in the release of the nucleocapsid into the cytoplasm. This may also apply to influenza C virus, although no evidence has been presented to support this assumption. In any case the fusion activity is crucial for the infectivity of the virus. Virus with uncleaved glycoprotein is lacking not only fusion activity, but also infectivity (Herrler et al., 1979; Sugawara et al., 1981). Restoration of the fusion activity in vitro by proteolytic cleavage of the glycoprotein is accompanied by restoration of the infectivity.

Due to the characteristics of the glycoprotein HEF, influenza C virus is unique among myxoviruses. Influenza A and B viruses as well as paramyxoviruses differ from influenza C virus in the specificity of the receptor-binding activity (Neu5Ac versus Neu5,9Ac$_2$) and the receptor-destroying enzyme (neuraminidase versus acetylesterase). In addition, HEF is responsible for three activities (receptor binding, receptor inactivation, and fusion), while both paramyxoviruses and influenza A and B viruses have two surface glycoproteins for these activities (Fig. 4). The unique characteristics of the influenza C glycoprotein are reflected in the designation "HEF," which has been proposed to indicate that this protein can function as a hemagglutinin, as an esterase, and as fusion factor (Herrler *et al.*, 1988a). Others have chosen the designation "HE" (Vlasak *et al.*, 1987), which ignores the fusion activity. In addition, there is an HE protein present on some coronaviruses (Cavanagh *et al.*, 1990). This protein, described in Section IV, has hemagglutinating and esterase activities, but no fusion activity. Thus, "HEF" is an appropriate designation for the influenza C glycoprotein, to distinguish it from the coronavirus glycoprotein.

IV. RELATIONSHIP BETWEEN HEF AND THE CORONAVIRUS
GLYCOPROTEIN HE

For many years ortho- and paramyxoviruses have been thought to be the only animal viruses containing receptor-destroying enzymes. Prompted by a sequence similarity between an open reading frame on the genome of mouse hepatitis virus and the HEF gene of influenza C virus (Luytjes *et al.*, 1988) it was found that bovine coronavirus (BCV) is able to inactivate its own receptors on erythrocytes (Vlasak *et al.*, 1988a). The enzyme turned out to be a sialate 9-*O*-acetylesterase similar to the receptor-destroying enzyme of influenza C virus. In fact, the coronavirus enzyme was able to inactivate the receptors for influenza C virus on erythrocytes, and the esterase of influenza C virus inactivated the receptors for BCV (Vlasak *et al.*, 1988a), indicating that both viruses use the same receptor determinant for attachment to cells (i.e., $Neu5,9Ac_2$). This conclusion was confirmed by resialylation studies with erythrocytes and has been extended to a porcine coronavirus, hemagglutinating encephalomyelitis virus (HEV) (Schultze *et al.*, 1990).

An acetylesterase activity has been reported not only for BCV but also for HEV and some strains of mouse hepatitis virus (Yokomori *et al.*, 1989; Schultze *et al.*, 1991a; Pfleiderer *et al.*, 1991). The acetylesterase activity of BCV has been shown to be a function of a surface glycoprotein which is detected as a disulfide-linked dimer with a molecular weight of about 140,000. (Vlasak *et al.*, 1988b; Schultze *et al.*, 1991a). The same protein has been identified previously as a hemagglutinin (King *et al.*, 1985), and therefore the designation "HE" has been chosen to indicate its dual function as hemagglutinin and esterase (Cavanagh *et al.*, 1990). Similar to its influenza C counterpart, the esterase of coronaviruses is a serine esterase which can be inhibited by DFP (Vlasak *et al.*, 1988b; Schultze *et al.*, 1990, 1991a). Inhibition of the enzyme activity results in a dramatic reduction of infectivity, suggesting an important role for the esterase in an early stage of the infection (Vlasak *et al.*, 1988b). The importance of the HE protein has also been demonstrated with monoclonal antibodies which

```
1  MFSSLLLVLGLTEAEKIKICLQKQVNSSFSLHNGFGGNLYATEEKRMFELVKPKAGASVL  60
                       || ||   |   |   |  |||| |    | |
2                       MGSTCIAMAPRTLLLLIGCQLVFGFNEPLNIVSHL  35

          ▽
1  NQSTWIGFGDSRTD......KSNSAFPRSADVSAK.TADKFRFLSGGSLMLSMFGPPGKV  113
   ||  || |||||        ||  |   |||     |||   ||  |
2  ND.DWFLFGDSRSDCTYVENNGHPKLDWLDLDPKLCNSGKISAKSGNSLFRSFHFTDF..  92


1  DYLYQGCGKHKVFYEGVNWSPHAAINCY....RKNWTDIKLNFQKNIYELASQSHCMSLV  169
   |  | |  | |||||||||||| ||| |    |    | |    |||   |  ||||||
2  .YNYTGEGDQIVFYEGVNFSPNHGFKCLAYGDNKRWMGNKARFYARVYEKMAQYRSLSFV  151


1  NALDKTIPLQVTAGTAGNCNNSFLKNPALYTQEVKPSENKCGKENLAFFTLPTQFGTYEC  229
   |          |||   |  |         |   |  |        |            ||
2  NVPYAYGGKAKPTSICKH.KTLTLNNPTFISKESNYVDYYYESE........ANFTLAGC  202


1  KLHLVASCYFIYDSKEVYNKRGCDNYF..QVIYDSFGKVVGGLDNRVSPYTGNSGDTPTM  287
   |||   |  |   |||      |   |   || ||   || ||| |            |
2  DEFIVPLCVFNGHSKGSSSDPANKYYMDSQSYYNMDTGVLYGFNSTLDVGNTAKDPGLDL  262


1  QCDMLQLKPGRYSVRSSPRFLLMPERSYCFDMKEK.GPVTAVQSIWGKGRESDYAVDQAC  346
   |   | ||| |    |   || |  | |||  |    ||  || | |    | || |||
2  TCRYLALTPGNYKAVSLEYLLSLPSKAICLRKPKRFMPVQVVDSRWNSTRQSDNMTAVAC  322


1  LSTPGCMLIQKQKPYIGEADDHHGDQEMRELLSGLDYEARCISQSG..WVNETSPFTEKY  404
   |   |   |  | ||  |     |||| |||| |  |  |  |         |   |
2  QLPYCFFRNTSADYSGGTHDVHHGDFHFRQLLSGLLLNVSCIAQQGAFLYNNVSSSWPAY  382


1  LLPPKFGRCPLAAKEESIPKIPDGLLIPTSGTDTTVTKPKSRIPGIDDLIIGLLFVAIVE  464
     || || ||   | | |     ||    || |                ~~~~~~~~~~~~~
2  ....GYGQCPTAANIGYMAPVCIYDPLPVVLLGVLLGIAVLIIVFLILYFMTDSGVRLHE  438


1  TGIGGYLLGSRKESGGGVTKESAEKGFEKIGNDIQILKSSINIAIEKLNDRISHDENAIR  524

2  A  439


1  DLTLEIENARSEALLGELGIIRALLVGNISIGLQESLWELASEITNRAGDLAVEVSPGCW  584


1  IIDNNICDQSCQNFIFKFNETAPVPTIPPLDTKIDLQSDPFYWGSSLGLAITATISLAAL  644
                                                   ~~~~~~~~~~~~~~~

1  VISGIAICRTK  655
   ~~~~~~
```

FIG. 5. Alignment of the amino acid sequences of the HEF protein of influenza C/JHB/1/66 (1) and the HE protein of mouse hepatitis virus, strain JHM (2). Every tenth position is indicated by a dot. Wavy lines indicate hydrophobic sequences. With HEF these compose the amino-terminal signal sequence (amino acids 1–12), the presumptive fusion peptide (amino acids 447–463), and the membrane anchor (amino acids 624–652). HE has only two hydrophobic regions, the signal peptide (amino acids 1–17) and the membrane anchor (amino acids 404–429), which do not align with the corresponding domains of HEF. The active-site serine of the acetylesterase, which is conserved in both proteins, is marked with an open triangle. Stretches of amino acids identical in both sequences are indicated by underscoring. Vertical lines indicate identical or related amino acids. The figure is based on an alignment of HEF_1 and HE. (Courtesy of S. G. Siddell.)

were shown to neutralize BCV both *in vivo* and *in vitro* (Deregt *et al.,* 1989).

When the amino acid sequence of the HE protein is aligned with the sequence of the influenza C glycoprotein HEF, homology is observed only with the HEF_1 cleavage product. There is no sequence on the HE protein which is related to the HEF_2 polypeptide (Fig. 5). This observation is not surprising, because HEF_2 is responsible for the fusion activity of influenza C virus, whereas, in the case of coronaviruses, fusion is a function not of HE, but of the S protein (reviewed by Spaan *et al.,* 1988). The homology between the amino acid sequences of HE and HEF_1 has been reported to be 30% (Luytjes *et al.,* 1988). The alignment indicates that there are many conservative substitutions. A few regions are completely identical in both sequences (Fig. 5). Among these is the sequence Phe–Gly–Asp–Ser–Arg, which, in the case of influenza C virus, has been shown to contain the active-site serine of the esterase (Herrler *et al.,* 1988b; Vlasak *et al.,* 1989).

It is interesting to note that, on the other hand, the putative constituent sequence of the HEF receptor-binding site (Gly 279–Thr 284) does not have a homologous counterpart in the HE sequence. This observation may be related to the recent finding that HE is not very efficient in agglutinating erythrocytes (Schultze *et al.,* 1991a) and that the major hemagglutinin of BCV is the peplomer glycoprotein S (Schultze *et al.,* 1991b). It has been argued that the extent of identity between HE and HEF_1 is high enough to rule out convergent evolution, and, therefore, it has been speculated that coronaviruses acquired the HE gene from influenza C virus by nonhomologous recombination between ancestors of both viruses (Luytjes *et al.,* 1988). However, acetylesterases are also found in cells. If coronaviruses actually acquired the esterase gene by a recombination event, the gene might as well be derived from a cellular gene. More information about the viral and cellular esterases is required to distinguish between these possibilities.

REFERENCES

Adachi, K., Kitame, F., Sugawara, K., Nishimura, H., and Nakamura, K. (1989). *Virology* **172,** 125–133.

Buonagurio, D. A., Nakada, S., Desselberger, U., Krystal, M., and Palese, P. (1985). *Virology* **146,** 221–232.

Buonagurio, D. A., Nakada, S., Fitch, W. M., and Palese, P. (1986). *Virology* **153,** 12–21.

Camilleri, J. J., and Maassab, H. F. (1988). *Intervirology* **29,** 178–184.

Cavanagh, D., Brian, D. A., Enjuanes, L., Holmes, K. V., Lai, M. M. C., Laude, H., Siddell, S. G., Spaan, W., Taguchi, F., and Talbot, P. J. (1990). *Virology* **176,** 306–307.

Chakraverty, P. (1974). *J. Gen. Virol.* **25,** 421–425.

Chakraverty, P. (1978). *Arch. Virol.* **58,** 341–348.

Compans, R. W., Bishop, D. H. L., and Meier-Ewert, H. (1977). *J. Virol.* **21**, 658–665.

Corfield, A. P., Michalski, J. C., and Schauer, R. (1981). In "Sialidases and Sialidoses. Perspectives in Inherited Metabolic Diseases" (G. Tettamanti, P. Durand, and S. Di Donato, eds.), Vol. 4, pp. 3–70. Edi Fermes, Milan, Italy.

Czekalowski, J. W., and Prasad, A. K. (1973). *Arch. Gesamte Virusforsch.* **42**, 215–227.

Deregt, D., Gifford, G. A., Ijaz, M. K., Watts, T. C., Gilchrist, J. E., Haines, D. M., and Babiuk, L. A. (1989). *J. Gen. Virol.* **70**, 993–998.

Flewett, T. H., and Apostolov, K. (1967). *J. Gen. Virol.* **1**, 297–304.

Formanowski, F., and Meier-Ewert, H. (1988). *Virus Res.* **10**, 177–192.

Formanowski, F., Wrigley, N. G., and Meier-Ewert, H. (1989). In "Genetics and Pathogenicity of Negative Strand Viruses" (B. W. J. Mahy and D. Kolakofsky, eds.), pp. 16–23. Elsevier, Amsterdam.

Formanowski, F., Wharton, S. A., Calder, L. J., Hofbauer, C., and Meier-Ewert, H. (1990). *J. Gen. Virol.* **71**, 1181–1188.

Hachinohe, S., Sugawara, K., Nishimura, H., Kitame, F., and Nakamura, K. (1989). *J. Gen. Virol.* **70**, 1287–1292.

Haverkamp, J., Schauer, R., Wember, M., Kamerling, J. P., and Vliegenthart, J. F. G. (1975). *Hoppe-Seyler's Z. Physiol. Chem.* **365**, 1575–1583.

Hayes, B. K., and Varki, A. (1989). *J. Biol. Chem.* **264**, 19443–19448.

Herrler, G., and Klenk, H.-D. (1987a). *Virology* **159**, 102–108.

Herrler, G., and Klenk, H.-D. (1987b). In "The Biology of Negative Strand Viruses" (B. W. J. Mahy and D. Kolakofsky, eds.), pp. 63–67. Elsevier, Amsterdam.

Herrler, G., Compans, R. W., and Meier-Ewert, H. (1979). *Virology* **99**, 49–56.

Herrler, G., Nagele, A., Meier-Ewert, H., Bhown, A. S., and Compans, R. W. (1981). *Virology* **113**, 439–451.

Herrler, G., Rott, R., and Klenk, H.-D. (1985a). *Virology* **159**, 102–108.

Herrler, G., Geyer, R., Müller, H.-P., Stirm, S., and Klenk, H.-D. (1985b). *Virus Res.* **2**, 183–192.

Herrler, G., Rott, R., Klenk, H.-D., Müller, H.-P., Shukla, A. K., and Schauer, R. (1985c). *EMBO J.* **4**, 1503–1506.

Herrler, G., Reuter, G., Rott, R., Klenk, H.-D., and Schauer, R. (1987). *Biol. Chem. Hoppe-Seyler* **368**, 451–454.

Herrler, G., Dürkop, I., Becht, H., and Klenk, H.-D. (1988a). *J. Gen. Virol.* **69**, 839–846.

Herrler, G., Multhaup, G., Beyreuther, K., and Klenk, H.-D. (1988b). *Arch. Virol.* **102**, 269–274.

Hewat, E. A., Cusack, S., and Verwey, C. (1984). *J. Mol. Biol.* **175**, 175–193.

Hirst, G. K. (1950). *J. Exp. Med.* **91**, 177–185.

Hongo, S., Sugawara, K., Homma, M., and Nakamura, K. (1986a). *Arch. Virol.* **89**, 171–187.

Hongo, S., Sugawara, K., Homma, M., and Nakamura, K. (1986b). *Arch. Virol.* **89**, 189–201.

Kawamura, H., Tashiro, M., Kitame, F., Homma, M., and Nakamura, K. (1986). *Virus Res.* **4**, 275–288.

Kendal, A. P. (1975). *Virology* **65**, 87–99.

King, B., Potts, B. J., and Brian, D. A. (1985). *Virus Res.* **2**, 53–59.

Kitame, F., Sugawara, K., Ohwada, K., and Homma, M. (1982). *Arch. Virol.* **73**, 357–361.

Kitame, F., Nakamura, K., Saito, A., Sinohara, H., and Homma, M. (1985). *Virus Res.* **3**, 231–244.

Klenk, E., Faillard, H., and Lempfrid, H. (1955). *Hoppe-Seyler's Z. Physiol. Chem.* **301**, 235–246.

Kraut, J. (1977). *Annu. Rev. Biochem.* **46**, 331–358.

Krystal, M., Elliot, R. M., Benz, E. W., Young, J. F., and Palese, P. (1982). *Proc. Natl. Acad. Sci. U.S.A.* **79**, 4800–4804.
Luther, P., Cushley, W., Hölzer, C., Desselberger, U., and Oxford, J. S. (1988). *Arch. Virol.* **101**, 247–254.
Luytjes, W., Bredenbeek, P. J., Noten, A. F., Horzinek, M. C., and Spaan, W. J. (1988). *Virology* **166**, 415–422.
Meier-Ewert, H., Compans, R. W., Bishop, D. H. L., and Herrler, G. (1978). *In* "Negative Strand Viruses and the Host Cell" (B. W. J. Mahy and R. D. Barry, eds.), pp. 127–133. Academic Press, New York.
Meier-Ewert, H., Herrler, G., Nagele, A., and Compans, R. W. (1980). *In* "Structure and Variation in Influenza Virus" (W. G. Laver and G. M. Air, eds.), pp. 357–366. Elsevier/North-Holland, Amsterdam.
Meier-Ewert, H., Nagele, A., Herrler, G., Basak, S., and Compans, R. W. (1981a). *In* "Replication of Negative Strand Viruses" (D. H. L. Bishop and R. W. Compans, eds.), pp. 173–180. Elsevier/North-Holland, Amsterdam.
Meier-Ewert, H., Nagele, A., Herrler, G., Basak, S., and Compans, R. W. (1981b). *In* "Genetic Variation among Influenza Virus" (D. P. Nayak and C. F. Fox, eds.), pp. 263–272. Academic Press, New York.
Meier-Ewert, H., Petri, G., and Bishop, D. H. L. (1981c). *Arch. Virol.* **67**, 141–147.
Minuse, E., Quilligan, J. J., and Francis, T., Jr. (1954). *J. Lab. Clin. Med.* **43**, 31–43.
Muchmore, E. A., and Varki, A. (1987). *Science* **236**, 1293–1295.
Nagele, A. (1983). Ph.D. thesis, Technical University, Munich.
Nakada, S., Creager, R. S., Krystal, M., Aaronson, R. P., and Palese, P. (1984). *J. Virol.* **50**, 118–124.
Nakamura, K., Herrler, G., Petri, T., Meier-Ewert, H., and Compans, R. W. 91979). *J. Virol.* **29**, 997–1005.
Nerome, K., Ishida, M., and Nakayama, M. (1976). *Arch. Virol.* **50**, 241–244.
Nishimura, H., Sugawara, K., Kitame, F., and Nakamura, K. (1988). *J. Gen. Virol.* **69**, 2545–2553.
O'Callaghan, R. J., Gohd, R. S., and Labat, D. D. (1980). *Infect. Immun.* **30**, 500–505.
Ohuchi, M., Homma, M., Muramatsu, M., and Ohyama, S. (1978). *Microbiol. Immunol.* **22**, 197–203.
Ohuchi, M., Ohuchi, R., and Mifune, K. (1982). *J. Virol.* **42**, 1076–1079.
Pfeifer, J. B., and Compans, R. W. (1984). *Virus Res.* **1**, 281–296.
Pfleiderer, M., Routledge, E., Herrler, G., and Siddell, S. (1991). *J. Gen. Virol.* **72**, 1309–1315.
Rogers, G. N., Herrler, G., Paulson, J. C., and Klenk, H.-D. (1986). *J. Biol. Chem.* **261**, 5947–5951.
Rosenthal, P. B., Formanowski, F., Skehel, J. J., Meier-Ewert, H., and Wiley, D. C. (1990). *Int. Congr. Virol., 8th* **P17–025**, 218.
Schauer, R., Reuter, G., Stoll, S., Posadas del Rio, F., Herrler, G., and Klenk, H.-D. (1988). *Biol. Chem. Hoppe-Seyler* **369**, 1121–1130.
Schultze, B., Gross, H.-J., Brossmer, R., Klenk, H.-D., and Herrler, G. (1990). *Virus Res.* **16**, 185–194.
Schultze, B., Wahn, K., Klenk, H.-D., and Herrler, G. (1991a). *Virology* **180**, 221–228.
Schultze, B., Gross, H. J., Brossmer, R., and Herrler, G. (1991b). Submitted for publication.
Shukla, A. K., and Schauer, R. (1982). *Hoppe-Seyler's Z. Physiol. Chem.* **363**, 255–262.
Spaan, W., Cavanagh, D., and Horzinek, M. C. (1988). *J. Gen. Virol.* **69**, 2939–2952.
Styk, B. (1955). *Folia Biol. (Prague)* **1**, 207–212.
Sugawara, K., Ohuchi, M., Nakamura, K., and Homma, M. (1981). *Arch. Virol.* **68**, 147–151.

Sugawara, K., Nishimura, H., Kitame, F., and Nakamura, K. (1986).*Virus Res.* **6,** 27–32.

Sugawara, K., Kitame, F., Nishimura, H., and Nakamura, K. (1988). *J. Gen. Virol.* **69,** 537–547.

Szepanski, S., Klenk, H.-D., and Herrler, H. (1989). *In* "Cell Biology of Virus Entry, Replication, and Pathogenesis" (R. W. Compans, A. Helenius, and M. B. A. Oldstone, eds.), pp. 125–134. Liss, New York.

Szepanski, S., Gross, H. J., Brossmer, R., Klenk, H.-D., and Herrler, G. (1991). Submitted for publication.

Taylor, R. M. (1949). *Am. J. Public Health* **39,** 171–178.

Veit, M., Herrler, G., Schmidt, M. F. G., Rott, R., and Klenk, H.-D. (1990). *Virology* **177,** 807–811.

Vlasak, R., Krystal, M., Nacht, M., and Palese, P. (1987). *Virology* **160,** 419–425.

Vlasak, R., Luytjes, W., Spaan, W., and Palese, P. (1988a). *Proc. Natl. Acad. Sci. U.S.A.* **85,** 4526–4529.

Vlasak, R., Luytjes, W., Leider, J., Spaan, W., and Palese, P. (1988b). *J. Virol.* **62,** 4686–4690.

Vlasak, R., Muster, T., Lauro, A. M., Powers, J. C., and Palese, P. (1989). *J. Virol.* **63,** 2056–2062.

Wagaman, P. C., Spence, H. A., and O'Callaghan, R. J., (1989). *J. Clin. Microbiol.* **27,** 832–836.

Weis, W., Brown, J. H., Cusack, S., Paulson, J. C., Skehel, J. J., and Wiley, D. C. (1988). *Nature (London)* **333,** 426–431.

Yamashita, M., Krystal, M., and Palese, P. (1989). *Virology* **171,** 458–466.

Yokomori, K., La Monica, M., Makino, S., Shieh, C.-K., and Lai, M. M. C. (1989). *Virology* **173,** 683–691.

ADVANCES IN VIRUS RESEARCH, VOL. 40

BUNYAVIRIDAE: GENOME ORGANIZATION AND REPLICATION STRATEGIES

Michèle Bouloy

Unité de Virologie Moléculaire and URA CNRS 545
75724 Paris Cedex 15, France

I. INTRODUCTION

When created in 1973, the family Bunyaviridae was known to comprise more than 250 viruses formerly classified into group C or supergroup Bunyamwera of arboviruses (Porterfield *et al.*, 1973/1974, 1975/1976; Bishop *et al.*, 1980). Now members of the Bunyaviridae are grouped into five genera: *Bunyavirus, Phlebovirus, Hantavirus, Nairovirus*, and *Uukuvirus*. Many viruses of the family are transmitted by arthropods and can replicate in both vertebrates and invertebrates. Most bunyaviruses are transmitted by mosquitoes and gnats, nairo- and uukuviruses by ticks, and phleboviruses by sandflies (phlebotomines) and gnats (Bishop and Shope, 1979). Nevertheless, the family is not composed exclusively of arthropod-borne viruses. It appears that Hantaan and Hantaan-related viruses, which infect rodents and humans, have no known arthropod vector. They are probably transmitted to humans by bites or aerosols (Lee *et al.*, 1981). It has been demonstrated that Bunyaviridae are not restricted to animals and insects, since they have representatives among plant viruses. Tomato spotted wilt virus shares the properties of Bunyaviridae (Milne

235

and Francki, 1984; de Haan *et al.*, 1989, 1990). It is a plant virus transmitted by thrips in which the virus also multiplies. It has been proposed that this virus should be assigned to the new genus of the Bunyaviridae, the *"Tospovirus"* (P. de Haan, personal communication).

Bunyaviridae are found in most parts of the world. Some of these viruses are associated with human or domestic animal infections (e.g., Rift Valley fever, Crimean–Congo hemorrhagic fever, California encephalitis, and Korean hemorrhagic fever with renal syndrome). On the other hand, none of the 12 uukuviruses presently isolated is known to cause human diseases (for a review see Gonzalez-Scarano and Nathanson, 1990).

Within each genus viruses are grouped into serogroups which were determined originally on the basis of complement fixation, neutralization, and hemagglutination tests. Recently, biochemical data have substantiated and given support to this classification. Bunyaviruses are classified into 16 serogroups; nairoviruses, into six serogroups. In the *Hantavirus*, *Uukuvirus*, and *Phlebovirus* genera viruses are grouped into only one serogroup (Table I).

Viruses which belong to the Bunyaviridae family share common properties: the viral particle is roughly spherical. 75–115 nm in diameter, possesses a lipid envelope with protruding glycoprotein spikes, and contains three internal circular ribonucleoprotein complexes. Each ribonucleoprotein is composed of numerous copies of the nucleoprotein N; a few copies of the L protein, which is a transcriptase; and a molecule of single-stranded RNA segment, called large (L), medium (M), or small (S), depending on its relative size class. Except for rare cases, viral morphogenesis occurs not at the plasma membrane, but in the Golgi apparatus, where budding has been observed. In addition, as expected for viruses with a segmented genome, these viruses are able to interact genetically with closely related viruses and to produce reassortants.

Understanding of the genomic organization and replication processes of Bunyaviridae has progressed extensively during the past few years, due to the development of the cDNA tool. Features and genetics of the family have been described by Obijeski and Murphy (1977), Bishop (1979), and Bishop and Shope (1979). More recently, various aspects of the biology of these viruses have been presented in several reviews (Parsonson and McPhee, 1985; Bishop, 1986, 1990a; Pettersson and Von Bonsdorff, 1987; Beaty and Bishop, 1988; Gonzalez-Scarano *et al.*, 1988; Elliott, 1990; Gonzalez-Scarano and Nathanson, 1990; Schmaljohn and Patterson, 1990). Thus, this chapter focuses on the molecular biology of these viruses, with special emphasis on the most recent data.

TABLE I

CLASSIFICATION INTO SEROGROUPS OF THE MEMBERS OF THE BUNYAVIRIDAE[a]

Genus				
Bunyavirus	Phlebovirus[b]	Nairovirus	Uukuvirus	Hantavirus
Anopheles A	Sandfly fever	Crimean–Congo	Uukuniemi	Hantaan
Anopheles B	Bujaru	hemorrhagic fever		Ungrouped
Bunyamwera	Candiru			
Bwamba		Dera Ghazi Khan		
Group C	Chilibre			
California	Frijoles	Kao Shuan		
Capim		Hughes		
Guama				
Guamboa	Punta Toro	Nairobi sheep		
		disease		
Koongol	Rift Valley fever	Qalyub		
Minatitlan	Salehabad	Sakhalin		
Olifantsvlei	Ungrouped			
Patois				
Simbu				
Tete				
Turlock				
Ungrouped				

[a] Most of the viruses were listed in the *International Catalogue of Arboviruses* (Karabatsos, 1985) as well as in several reviews (Bishop and Shope, 1979; Bishop, 1986, 1990a; Gonzalez-Scarano and Nathanson, 1990). A sixth genus, *Tospovirus*, has been proposed; it includes the Tomato spotted wilt virus.

[b] *Phlebovirus* genus comprises only one serogroup. Listed here are the eight antigenic complexes within the serogroup.

II. MORPHOLOGY AND STRUCTURAL COMPONENTS

New methods have been developed to determine the structure of virions by electron micoscopy, one of which involves ultrarapid freezing in a thin layer of vitreous ice (Lepault *et al.*, 1983). When applied to examine the *Bunyavirus* La Crosse, the vitrified hydrated particles showed uniform spherical shapes with diameters ranging from 75 to 115 nm (Talmon *et al.*, 1987). This differs from the pleiomorphism observed after the classical negative staining (Obijeski *et al.*, 1976a). This method, which preserves structural details of the virion in the native state, has also revealed 10-nm-long regular spikes and a 4-nm-thick bilayer structure. No distinct pattern was observed at the surface of bunyavirus particles (Talmon *et al.*, 1987). This observation is in contrast with the results obtained by Von Bonsdorff and Pettersson (1975) with uukuniemi virus, in which the surface glycoproteins are clustered to form hollow cylindrical units, arranged into penton–hexon clusters in a T = 12, P = 3 icosahedral surface lattice with hexon-penton distances of 17 nm. This type of structure is also present in phleboviruses (Smith and Pifat, 1982; Martin et al., 1985). A specific arrangement on the surfaces of nairoviruses and hantaviruses has also been observed (Donets et al., 1977; McCormick et al., 1982; White et al., 1982; Martin et al., 1985).

Although the virion is schematically represented as a spherical particle containing three ribonucleoproteins (one of each size), purified virus preparations do not contain equimolar ratios of L, M, and S RNA molecules (Pettersson and Kääriäinen, 1973; Bouloy et al., 1973/1974; Obijeski et al., 1976b; Pettersson et al., 1977). This led Talmon et al. (1987) to speculate that the difference in virion size observed within the preparations is related to the RNA content; the larger particles would contain duplicates (or triplicates) of at least one segment.

All Bunyaviridae members have three single-stranded RNA genome segments called L, M, and S, depending on their respective size. Table II summarizes the size of the genomic segments of representatives of each genus and within the proposed member of the plant kingdom, as determined by cloning and sequencing. For clarity, the references are listed separately (Table III). Although sequence data on the L segment are yet limited to Bunyamwera and Hantaan viruses, genome analyses in denaturing gels seem to indicate that the L segments of bunya-, hanta-, phlebo-, and uukuviruses are similar in size, whereas nairoviruses possess the largest L segment, almost twice the size of the L segments of viruses in the other genera (Clerx and Bishop, 1981; Clerx et al., 1981; Watret and Elliott, 1985). No sequence data are available yet for the M segment of nairoviruses. When analyzed in gels, the size of this segment fits within the range defined for other genera (Foulke et al., 1981; Clerx and Bishop, 1981; Clerx et al., 1981; Watret and Elliott, 1985; Watret et al., 1985).

Most of the viruses of the Bunyaviridae family are composed of four structural proteins, called L. N, G1, and G2. Proteins L and N are associated with RNA, forming nucleocapsids; G1 and G2 are located at the surface and form the spikes. The sizes of the structural proteins from viruses of each genus are reported in Table IV. In addition to their structural proteins, bunyaviruses and phleboviruses express two nonstructural proteins (NS_M and NS_S) and uukuviruses express one (NS_S). Interestingly, hantaviruses do not seem to express any of the established nonstructural proteins during their life cycle. Data on nairoviruses are presently incomplete but, so far as the S segment is concerned, there is no polypeptide corresponding to NS_S. Thus, the existence and the number of the nonstructural proteins depend on the

TABLE II

SIZE OF THE THREE RNA SEGMENTS (IN BASES)

Genus	Segment		
	L	M	S
Bunyavirus			
Bunyamwera	6875	4458	961
Germiston	—	4534	980
Maguari	—	—	945
La Crosse	—	4526	981
Snowshoe hare	—	4527	982
Aino	—	—	850
Hantavirus			
Hantaan	6530	3616	1696
Nephropathia Epidemica	—	3682	1785
Sapporo rat	—	3651	1769
Prospect Hill	—	—	1675
Nairovirus			
Dugbe	—	—	1712
Phlebovirus			
Rift Valley fever	—	3884	1690
Punta Toro	—	4330	1904
Sandfly fever			
Sicilian	—	—	1746
Toscana	—	—	1869
Uukuvirus			
Uukuniemi	—	3231	1720
Tomato spotted wilt	—	—	2916

genus (Table IV). This raises the question of their functions, which, at the present time, have not been determined for any of them. Small open reading frames (ORFs) identified in some of the L, M, and S segments by sequencing, but of unknown significance (i.e., with no corresponding mRNA or polypeptide detected in infected cells), have not been indicated in Table IV.

Unlike other enveloped RNA viruses, Bunyaviridae do not contain an internal matrix protein. This suggests that the ribonucleoproteins, which are the sole internal structures, interact directly with the envelope. In several cases, when examined under the electron microscope, the ribonucleoproteins were observed to form a circular structure (Pettersson and Von Bonsdorff, 1975; Samso *et al.*, 1975; Obijeski *et al.*, 1976b). The existence of such circular structures is explained by the presence, in each genomic RNA molecule, of 3' and 5' sequences, complementary to one another, which can form stable base-paired panhandles. This RNA–RNA structure resistant to RNase digestion exists in

TABLE III

REFERENCE LIST FOR THE CLONING AND SEQUENCING OF THE THREE RNA SEGMENTS

Genus	Segment		
	L	M	S
Bunyavirus			
Bunyamwera	Elliott (1989b)	Lees *et al.* (1986)	Elliott (1989a)
Germiston	—	Pardigon *et al.* (1988)	Gerbaud *et al.* (1987b)
Maguari	—	—	Elliott and McGregor (1989)
La Crosse	—	Grady *et al.* (1987)	Cabradilla *et al.* (1983) Akashi and Bishop (1983)
Snowshoe hare	—	Eshita and Bishop (1984)	Bishop *et al.* (1982)
Aino	—	—	Akashi *et al.* (1984)
Hantavirus			
Hantaan	Schmaljohn (1990)	Schmaljohn *et al.* (1987b)	Schmaljohn *et al.* (1986b)
Nephropathia Epidemica	—	Yoo and Kang (1987)	Stohwasser *et al.* (1990)
Sapporo rat	—	Giebel *et al.* (1989)	Arikawa *et al.* (1990)
Prospect Hill	—	Arikawa *et al.* (1990)	Parrington and Kang (1990)
Nairovirus			
Dugbe	—	—	Ward *et al.* (1990)
Phlebovirus			
Rift Valley fever	—	Collett *et al.* (1985), Takehara *et al.* (1989)	Giorgi *et al.* (1991)
Punta Toro	—	Ihara *et al.* (1985b)	Ihara *et al.* (1984)
Sandfly fever Sicilian	—	—	Marriott *et al.* (1989)
Toscana	—	—	Giorgi *et al.* (1991)

(*continued*)

TABLE III

(*Continued*)

Genus	L	M	S
		Segment	
Uukuvirus			
Uukuniemi	—	Rönnholm and Pettersson (1987)	Simons *et al.* (1990)
Tomato spotted wilt	—	—	de Haan *et al.* (1990)

deproteinized genomic RNA (Pardigon *et al.*, 1982; Patterson *et al.*, 1983) as well as within La Crosse virus nucleocapsids (Raju and Kolakofsky, 1989). From sequence data it is possible to predict the most stable structures by using computer programs. In all Bunyaviridae segments sequenced so far, the 5′- and 3′-terminal sequences are predicted to form stable base-paired structures. Depending on the segment and the virus, the length of these structures varies from some

TABLE IV

Structural and Nonstructural Proteins[a]

Genus	Structural			Nonstructural		
	L	G1	G2	N	NS$_S$	NS$_M$
Bunyavirus	180–200 (259)	100–125	29–40	19–27 (10.5–11)	7–12	11–16
Hantavirus	180–200 (246.5)	68–72	54–60	50–52 (48–49)	None	None
Nairovirus[b]	180–200	72–84	80–40	48–54 (49)	None	?
Phlebovirus	180–200	55–70	50–60	28 (27–28)	29 (29–36)	14 and 78[c] or 30[d] (17[c]–30[d])
Uukuvirus	180–200	70–75	65–70	25 (28)	30 (32)	None
Tomato spotted wilt	200	78	58	29 (29)	50 (52)	?

[a] Size is expressed in kilodaltons, estimated from analyses in sodium dodecyl sulfate–polyacrylamide gels. When available, values in parentheses repesent the size predicted from sequence data (references are cited in Table III).

[b] In the nairovirus Hazara three glycoproteins have been described (Foulke *et al.*, 1981) for Rift Valley fever virus[c] and for Punta Toro virus[d].

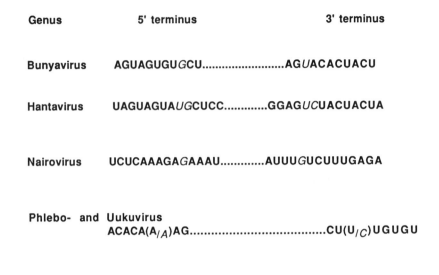

| Genus | 5' terminus | 3' terminus |

Bunyavirus   AGUAGUGU*G*CU.........................AG*U*ACACUACU

Hantavirus   UAGUAGUA*U*GCUCC..............GGAG*UC*UACUACUA

Nairovirus   UCUCAAAGA*G*AAAU..............AUUU*G*UCUUUGAGA

Phlebo- and Uukuvirus
ACACA(A$_{/A}$)AG.....................................CU(U$_{/C}$)UGUGU

tomato spotted wilt
AGAGCAAU.....................................AUUGCUCU

FIG. 1. Conservation of the terminal sequences of the three segments. Sequences are written from 5' to 3'. References are listed in Table III. Mismatches within the complementary sequences are indicated by letters in italics.

20 to 60 nucleotides. Such base-paired structures possibly play a role in transcriptase recognition and encapsidation.

Not only are the 5'- and 3'-terminal sequences complementary, but they are also conserved within the three segments. These consensus sequences are short (i.e., 8–10 bases long) and specific for each genus, except for phlebo- and uukuviruses, for which they are identical (Fig. 1). In the case of Bunyamwera virus, in which the three segments have been sequenced, the terminal 11 bases are conserved and complementary, except for positions 9 and −9 (Elliott, 1990). This mismatch is present in all bunyavirus segments sequenced so far. Surprisingly, a significant level of similarity was observed between bunya- and hantavirus consensus sequences. If such similarities are discussed in terms of evolution and phylogeny, it is worth mentioning the plant virus Rice stripe (and probably also Maize stripe virus). This virus does not belong to the Bunyaviridae family (Toriyama and Watanabe, 1989; Ishikawa *et al.*, 1989), but its RNA genome, which is composed of four segments, possesses 3'- and 5'-terminal sequences complementary to each other and significantly similar to those of phlebo- and uukuviruses (Kakutani *et al.*, 1990). Moreover, the smallest segment of Rice stripe virus shares the ambisense arrangement of phlebo- and uukuviruses.

III. Genome Organization

Like other viruses with a segmented genome, the Bunyaviridae have the ability to exchange genome segments during mixed infections. Genome reassortment has been demonstrated to occur in cell cultures infected with certain bunyaviruses of the same serogroup, but to fail with viruses from different serogroups, even when they belong to the same genus. This phenomenon, which is limited to a few viruses, can occur in mosquitoes dually infected in the laboratory (Shope et al., 1981; Beaty et al., 1981, 1982, 1985; Beaty and Bishop, 1988; Chandler et al., 1990; Turell et al., 1990) and in nature (Klimas et al., 1981; Ushijima et al., 1981). Reassortants between viruses in the Bunyavirus genus have been exploited by means of biochemical analyses to determine the coding assignments of each segment. Reassortants between viruses of the California serogroup were first obtained by Bishop and colleagues, who established the coding assignments of the S and M segments (for reviews see Bishop and Shope, 1979; Bishop, 1979). The M segment encodes three proteins: the two glycoproteins G1 and G2 and a nonstructural protein NS_M. The S segment codes for two proteins: the nucleoprotein N and a nonstructural polypeptide NS_S. These results were later confirmed by Elliott (1985), who analyzed reassortants between viruses of the Bunyamwera serogroup obtained by Pringle et al. (1984). Similar approaches using reassortants between viruses of the California serogroup (Endres et al., 1989) or between viruses of the Bunyamwera serogroup (Elliott, 1989b) finally established that the L protein is encoded by the L segment. These and other results have been collated in reviews to which the reader is referred for more detailed information on the genetics of Bunyaviridae (Bishop, 1979, 1985; Bishop and Shope, 1979).

During the past decade, cDNA cloning and sequencing have proved to be very potent tools. By using the cloned cDNA to develop other experimental approaches such as in vitro translation, expression via recombinant vectors, and protein sequencing, it has become possible to decipher the genomic segments of the five genera. The data are reported in the following section as well as in a schematic representation of the three segments and their organization, presented in Fig. 2.

A. L RNA Segment

The L segments of Bunyamwera virus (Elliott, 1989b) and Hantaan virus (Schmaljohn, 1990) are 6875 and 6530 nucleotides long, respectively. These viruses are the prototypes of the Bunyavirus and Hantavirus genera. They are the only ones, so far, whose complete genomes have been sequenced. The L segments of both viruses have a similar organization: A long ORF is identified in the viral complementary-sense RNA, potentially encoding 2238 and 2150 amino acids, respectively. When analyzed in sodium dodecyl sulfate (SDS)–polyacrylamide gels, the apparent molecular weight of the L protein was estimated to be 200,000, which is slightly smaller than expected from the theoretical sequence (Table IV). In addition to the large ORF, a small ORF consisting of 129 (Bunyamwera virus) and 221 (Hantaan virus) amino acids is present in the virus-sense L RNA. There are several reasons to question the significance of the small ORF: (1) There is no homology between the putative polypeptides encoded by the small ORFs of the two viruses, nor can the products and their mRNAs be detected in virus-infected cells; (2) interestingly, no similar small ORF is detected in the virion sense of the La Crosse virus L segment (Hacker *et al.*, 1990); (3) the ORFs coding for the L protein and the putative polypeptide overlap in molecules of opposite orientations. This is an arrangement different from other well-established ambisense genomes such as the S ambisense segments of phlebo- and uukuviruses (see the following sections) and the S and L segments of arenaviruses (see references in Bishop, 1986, 1990b; Salvato and Shimomaye, 1989; Iapalucci *et al.*, 1989), in which the ORFs are separated by a noncoding intergenic region.

No complete sequence data of the L segment are available for phlebo-, uuku-, and nairoviruses. From gel analysis it appears that the *Nairovirus* L segment is much larger than its counterpart from other genera. Hopefully, cloning and sequencing will provide definitive answers to the question of its coding capacity.

The L protein must be involved in the RNA-dependent RNA polymerase activity, but no homology was detected at the nucleotide and amino acids levels, either between the *L* gene of the two viruses (Schmaljohn, 1990) or between any polymerase of other virus families such as Arena-, Rhabdo-, and Paramyxoviridae. The L protein of Bunyamwera virus, however, shares a low level of homology with influenza A PB1 protein, the enzyme involved in RNA polymerization (Braam *et al.*, 1983). In addition to its enzymatic activity, the L protein has been shown to contribute to virulence in mice (Rozhon *et al.*, 1981; Janssen *et al.*, 1986) and mosquitoes (Beaty *et al.*, 1981).

B. M RNA Segment

Except for nairoviruses, the complete M segment of at least one representative of each genus has been cloned and sequenced. A general characteristic of the family is that the M segment contains only one large ORF in the viral complementary-sense RNA. No ORF of significant size was detected in the virus-sense RNA. Except for phlebo- and uukuviruses, the M polypeptide does not exhibit significant homology between viruses from different genera. This reflects the fact that, in general, there is no serological cross-reactivity among different genera.

The predicted M polypeptide is the precursor to the two envelope glycoproteins and, depending on the virus, the NS_M protein (for bunya- and phleboviruses). Such a large polypeptide precursor has never been detected in infected cells. This suggests that the processing, probably due to cellular signalases, is cotranslational.

In addition to their structural role, the glycoproteins exhibit biological functions. They bear neutralizing epitopes; they possess an acid pH-dependent fusion activity and probably contain an attachment site for a putative cellular receptor; they are involved in the induction of immunity. In addition, the M segment and its products appear to be the major determinants for neurovirulence in mice (Shope *et al.*, 1981; Gonzalez-Scarano *et al.*, 1985a; Janssen *et al.*, 1986; Endres *et al.*, 1990) and for infection of mosquitoes (Beaty *et al.*, 1981, 1982; Sundin *et al.*, 1987).

Neutralizing epitopes have been identified in only G1 of bunyaviruses (Gonzalez-Scarano *et al.*, 1982; Kingsford *et al.*, 1983; Kingsford and Boucquey, 1990; Grady *et al.*, 1983) but in both G1 and G2 of Rift Valley fever virus (Keegan and Collett, 1986; Battles and Dalrymple, 1988), Punta Toro virus (Pifat and Smith, 1987; Pifat *et al.*, 1988) and Hantaan virus (Yamanishi *et al.*, 1984; Dantas *et al.*, 1986; Arikawa *et al.*, 1989). In many cases it has been shown that some (but not all) neutralizing monoclonal antibodies also inhibit hemagglutination and can protect mice passively against a challenge with infectious virus. The role of the glycoproteins in inducing immunity and protection has been documented for protein G2 (but not G1) of Rift Valley fever virus (Collett *et al.*, 1987; Dalrymple *et al.*, 1989; Schmaljohn *et al.*, 1989) and for both G1 and G2 of Hantaan virus (Pensiero *et al.*, 1988; Schmaljohn *et al.*, 1990).

La Crosse and Hantaan virions have been shown to exhibit acid-dependent fusion activity. This activity is thought to play an important role in attachment to susceptible cells and in the initial steps of infection. The same activity is probably responsible for the hemagglutination of red blood cells. For Hantaan virus the glycoprotein(s) involved in this phenomenon has not been determined (Tsai *et al.*, 1984; Arikawa *et al.*, 1985; Okuno *et al.*, 1986). In La Crosse virion, although G1 protein undergoes a conformational change at the pH of fusion (Gonzalez-Scarano, 1985; Gonzalez-Scarano *et al.*, 1985b), it seems likely that G2 is the fusion protein (Pobjecky *et al.*, 1989). In fact, Ludwig *et al.* (1989) proposed a model in which G2 would normally be hindered by G1, except after proteolytic digestion, which degrades G1 but leaves G2 intact. Thus, during vertebrate infections, when the virus is present in the bloodstream, surrounded by neutral pH and a high protein environment, G1 remains intact in the particle, playing the principal role. On the other hand, during arthropod infection, especially in the mosquito midgut, where the protease content is high, G1 would be degraded and G2 would be exposed to the environment and ready to function.

1. Bunyavirus

The M segments of two members of each of the California and Bunyamwera serogroups have been sequenced: La Crosse (Grady *et al.*, 1987), snowshoe hare (Eshita and Bishop, 1984), Bunyamwera (Lees *et al.*, 1986), and Germiston (Pardigon *et al.*, 1988). In the viral complementary sense the M segment encodes a unique ORF composed of 1433–1441 amino acids, depending on the virus.

The predicted polypeptide is the precursor to the glycoproteins G1 (108–125 kDa) and G2 (35–41 kDa) and the NS_M protein (11–16 kDa). The question of the localization of the three proteins within the large polypeptide precursor has been solved by Fakazerley *et al.* (1988), who have used Snowshoe hare virus as a working model. The authors determined the positions of the proteins G1 and G2 within the precursor by direct sequencing of the amino and carboxy termini of these proteins and placing them in the theoretical sequence. They also analyzed the reactivity of the glycoproteins with antibodies prepared against specific peptides, and found that the order is $G2–NS_M–G1$, from the amino to the carboxy terminus of the precursor (see Fig. 2). A similar organization was found in the Germiston virus M segment (Gerbaud *et al.*, 1991).

Comparison of the M polyproteins of Bunyamwera, La Crosse, and Germiston viruses with that of Snowshoe hare virus indicates that the sequence at the carboxy termini of G2 proteins can be aligned and is conserved: . . . Lys–Ser–Leu–Arg–Ala/Val–Ala–Arg (Elliott, 1990). This provides indirect evidence for the exact location of the carboxy termini of G2 in the Bunyamwera, La Crosse, and Germiston virus M precursors. These data also indicate that the cleavage generating G2 occurs after an arginine residue, but the protease has not been determined. Alignment of the M polypeptide does not allow a good prediction of the amino terminus of G1 because the level of similarity is relatively low in this region. Comparison of the four polypeptides indicates that many of the cysteine residues are conserved and that G2 and G1 proteins exhibit an overall 66% and 40% homology, respectively. Several potential N-linked glycosylation sites are present in the M

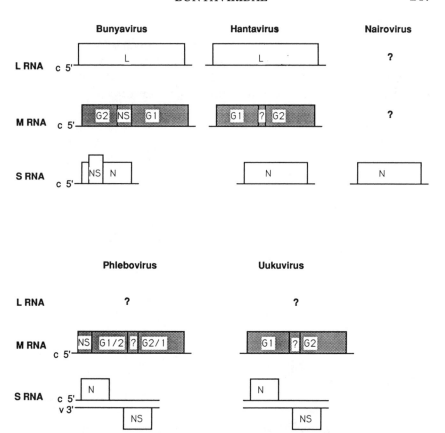

FIG. 2. Schematic representation of the genomic organization. Open boxes represent open reading frames present in the viral complementary-strand RNA (C) or the virus strand RNA (V). The 3' and 5' ends are indicated. The putative M polypeptides generating the mature envelope glycoproteins and eventually NS_M, are represented by solid boxes. Putative intergenic peptides are indicated by thin question marks. The RNA segments and the encoded ORFs are not drawn in proportion to real size.

sequences; some (if not all) of these sites are recognized by glycosylating enzymes, since synthesis of both G1 and G2 envelope proteins is sensitive to tunicamycin. Protein G2 has two glycosylation sites conserved in the four viruses, whereas G1 has only one. A second site is conserved in G1 when compared within the same serogroup.

The hydropathy profiles of the four polyproteins are similar, allowing speculation on the processing of the polyprotein. The ORF begins with a hydrophobic region, which acts as a signal peptide for the translocation of the polypeptide. The hydropathy profile of G2 clearly reveals the presence of a hydrophobic domain almost in the center of the molecule, which should serve as a transmembrane region. Protein G1 follows NS_M and is preceded by a signallike sequence. We do not know whether the cleavage at this site and at the carboxy terminus of G2 generates NS_M or whether further processing occurs. The profile of G1 reveals a hydrophobic domain at its carboxy terminus, which should anchor the protein into the membrane. To date, we do not know how G1 and G2 are positioned at the surface of the virion and what the molecular structure of the glycoprotein spike is. However, in spite of the presence of many cysteine residues in both G1 and G2, it seems that G1 and G2 of La Crosse virus are not linked by disulfide bonds (Gonzalez-Scarano and Nathanson, 1990).

2. Hantavirus

The complete M RNA sequences of Hantaan virus and other related viruses causing hemorrhagic fevers with renal syndrome have been reported (Schmaljohn et al., 1987b; Yoo and Kang, 1987; Giebel et al., 1989; Arikawa et al., 1990). A single long ORF is detected in the viral complementary-sense RNA; it has the potential to encode a polypeptide of 125–126 kDa. To localize the proteins within the polypeptide precursor of Hantaan virus, Schmaljohn et al. (1987b) determined the amino-terminal sequence of the mature glycoproteins G1 and G2 purified from virions. The amino- and carboxy-terminal halves of the polyprotein represent the G1 and G2 glycoproteins, respectively. Whereas it has been shown that the carboxy terminus of G2 extends to the carboxy terminus of the ORF, the carboxy terminus of G1 has not been precisely located. Although there is no space between G1 and G2, for a polypeptide corresponding to the NS_M protein of bunyaviruses, it has been proposed that a small intergenic region of less than 6 kDa could exist between the carboxy terminus of G1 and the amino terminus of G2. This assumption is consistent with the size estimation of proteins G1 (68 kDa) and G2 (55 kDa) and with the reactivity of G1 protein with antibodies directed against specific peptides (Schmaljohn et al., 1987b). The hydropathy profile indicates the presence of a short hydrophobic stretch following the first ATG of the ORF, which probably acts as a signal sequence. A hydrophobic region, probably with a similar function, precedes the amino terminus of G2.

Five potential glycosylation sites are present in Hantaan virus G1 protein and only two in G2, confirming the observation that glycoprotein G1 is highly glycosylated (Schmaljohn et al., 1986a). The oligosaccharides contribute approximately 7000 and 3000 to the molecular weights of Hantaan virus G1 and G2 glycoproteins, respectively. Three potential tyrosine sulfation sites were also observed (Giebel et al., 1989).

The M segments of other serologically related viruses representing various degree of severity of the disease have also been sequenced. These viruses are the Hällnäs strain B1 of Nephropathia Epidemica virus, which is closely related to Puumala virus, Sapporo rat virus, and Prospect Hill virus. Comparison of the sequences at the amino acid level reveals similarity of 53% between Hantaan virus (strain 76–118) and the Hällnäs strain and 75% between Hantaan and Sapporo viruses, indicating that Hantaan virus is antigenically more closely related to Sapporo virus than to Puumala virus (Arikawa et al., 1990). This conclusion is consistent with the results from serological tests (Schmaljohn et al., 1985). Although proteins G1 and G2 share only an overall 43% and 55% similarity, respectively, among the three viruses, the cysteine residues as well as some of the glycosylation sites are highly conserved. Two of the sulfation sites are conserved in Hantaan virus and the Hällnäs strain B1.

3. Nairovirus

The prototype of the *Nairovirus* genus is Crimean–Congo hemorrhagic fever virus. To date, no complete sequence of nairovirus M segment has been reported.

4. Phlebovirus

The M sequences of four phleboviruses have been reported–Punta Toro virus (Ihara et al., 1985b) and three isolates of Rift Valley fever virus: two wild strains, ZH501 (Collett et al., 1985) and ZH548, and an attenuated derivative, ZH.548 M12 (Takehara et al., 1989), which is a vaccine candidate.

Within the unique ORF present in the viral complementary-sense RNA, which codes for a large polypeptide precursor, the order of the proteins has been determined to be NS_M–G2–G1 for Rift Valley fever virus and NS_M–G1–G2 for Punta Toro. The difference reflects only the relative size difference between the two glycoproteins of these viruses, since the glycoproteins are numbered according to their relative migration in polyacrylamide gels, with G2 migrating faster than G1. Because the carboxy termini of G1 and G2 have not been sequenced, it is not known whether a small intergenic region is located between the two envelope proteins.

The predicted NS_M polypeptides of Punta Toro virus and Rift Valley fever virus are 30 and 17 kDa, respectively. In Punta Toro-infected cells the expected 30-kDa polypeptide can be detected with highly specific antibodies (I. Jones, personal communication). The predicted 17-kDa polypeptide has never been detected during infection with Rift Valley fever virus. Instead, two nonstructural proteins have been identified: a 78-kDa glycoprotein and a 14-kDa nonglycosylated polypeptide. Expression of the M polypeptide from cloned cDNAs via recombinant virus vectors in eukaryotic cells or in an *in vitro* system has greatly helped elucidate the biogenesis of the Rift Valley fever virus glycoproteins (Kakach *et al.*, 1988, 1989; Wasmoen *et al.*, 1988; Suzich and Collett, 1988; Collett *et al.*, 1989; Suzich *et al.*, 1990).

Between the first AUG initiation codon and the amino terminus of G2 are four in-frame AUG codons which have drawn attention (Collett *et al.*, 1985). Synthesis of the 14-kDa polypeptide starts at the second in-frame AUG and contains only the preglycoprotein region. The first AUG initiates synthesis of the 78-kDa polypeptide, which encompasses the preglycoprotein region as well as the G2 coding sequence. Thus, synthesis of these proteins appears to involve the use of at least the first two initiation codons of the ORF. Production of G2 does not result from cleavage of the 78 kDa, but is probably generated from a precursor initiated in the preglycoprotein region at an initiation codon other than the first AUG. The 14-kDa polypeptide and G2 protein are probably generated from the same precursor. Although the 14-kDa preglycoprotein region contains one potential N-linked glycosylation site, this site is not recognized in the 14-kDa protein. The same site present in the 78-kDa polypeptide is utilized there. Thus, initiation at the first AUG codon appears to predetermine utilization of the glycosylation site, which also precludes the proposed cleavage at the junction between the preglycoprotein region and G2 sequences. This suggests that the cleavage at the signal peptide site generating G2 must be masked in the 78-kDa polypeptide which, therefore, cannot serve as a precursor for G2 protein. Suzich *et al.* (1990) proposed that the use of either one of these two initiation codons might serve for controlling posttranslational modifications. Production of G1 appears to be more complex, since a fraction of this glycoprotein may be produced by an internal initiation, downstream of the preglycoprotein region, via a pathway distinct from that followed by G2 (Suzich *et al.*, 1990).

Comparison of the M sequences of Punta Toro and Rift Valley fever viruses indicates considerable similarity between the glycoproteins, but not between the NS_M proteins.

Comparison of the sequences of the wild strains and the attenuated virus should provide a clue for better understanding of the determinants of virulence. Within the M segments of the attenuated Rift Valley fever virus and its virulent patent, there are 12 nucleotide and seven amino acid changes. Interestingly, one mutation in the attenuated virus has changed a C in U, creating a new AUG codon upstream of the polyprotein ORF. To date, it is not known which of the muta-

tions contributes to attentuation. During the serial mutagenesis performed to produce the M12-attenuated variant, additional mutations
have been introduced in the L and S segments. Some of them also
contributed to attenuation. This has been shown by analyses of reassortants obtained during coinfection between M12 and the virulent
wild-type virus (Saluzzo and Smith, 1990). Reassortants containing
either the L, M, or S segment of M12 and the rest of the genome from
the wild-type parent are attenuated.

5. Uukuvirus

Uukuniemi virus is the prototype of the *Uukuvirus* genus; its M
segment has been cloned and sequenced. A single ORF comprising
1008 amino acids (113,500 kDa) was found in the viral complementary
sense (Rönnholm and Pettersson, 1987). This polypeptide corresponds
to the previously identified p110 precursor synthesized in reticulocyte
lysate, which could be processed into glycoproteins after the addition
of microsomal membranes (Ulmanen *et al.*, 1981). It is noteworthy that
it is the only M-specific mRNA isolated from cells infected with
Bunyaviridae which could be translated in an *in vitro* system.

By determining the amino-terminal sequences of purified G1 and
G2 and comparing them with the predicted protein sequence, the gene
order was shown to be G1–G2. There is no space either at the amino
terminus of the polypeptide for a polypeptide equivalent to NS_M of
phleboviruses or between G1 and G2 for a polypeptide equivalent to
NS_M of bunyaviruses. However, since the exact location of the carboxy
terminus of G1 is not known, a small intergenic peptide (of less than 6
kDa) might exist between the two glycoproteins. Despite the absence
of a polypeptide corresponding to the preglycoprotein region of *Phlebovirus*, the M products of Uukuniemi, Punta Toro, and Rift Valley
fever viruses show some similarity, which is more pronounced in the
carboxy-terminal region of the polyprotein. The positions of the cysteine residues are remarkably well conserved. This observation, confirmed later by analyses of the S segment, indicates that uuku- and
phleboviruses are evolutionary related.

C. S RNA Segment

While the M segments (and possibly the L segments) of all members
of the family share many common features as to gene organization and
expression, the S segment reveals distinct characteristics which clearly define each genus. The S segments of hanta- and nairoviruses code
only for the N protein, and those of bunya-, phlebo-, and uukuviruses
code for two proteins: N and NS_S. The bunyavirus proteins are ex-

pressed from a bicistronic mRNA, whereas those of uuku- and phleboviruses are expressed from two distinct monocistronic mRNAs of opposite polarities, transcribed from an RNA with an ambisense arrangement. Data supporting these statements are presented in this and the following section.

1. Bunyavirus

Early during the studies on Bunyaviridae, it was demonstrated that the S segment codes for two proteins: the nucleocapsid N and the nonstructural protein NS_S. Snowshoe hare and La Crosse virus S RNA segments were the first ones to be cloned and sequenced (Bishop et al., 1982; Cabradilla et al., 1983; Akashi and Bishop, 1983). At the present time four more bunyavirus S segments have been completely sequenced: Aino virus, which belongs to the Simbu serogroup (Akashi et al., 1984), and Germiston (Gerbaud et al., 1987b), Bunyamwera (Elliott, 1989a), and Maguari (Elliott and McGregor, 1989), which belong to the Bunyamwera serogroup. In every case the two proteins are expressed from two overlapping ORFs present in two different frames of the viral complementary-sense RNA. The first ORF codes for the N protein; the second, for the NS_S protein. The N proteins of these viruses are similar in size (233–235 amino acids, 26–27 kDa) and show an overall 40% similarity, or 70% when compared pairwise within the same serogroup (Elliott, 1989a). The N proteins contain many basic amino acid residues, the positions of which are conserved. In contrast, the NS_S proteins are more variable in size 91–109 amino acids, 10.5–11 kDa) and share only 25% similarity. The function of the NS_S protein is still unknown.

The N and NS_S proteins are translated from a unique mRNA. Such bicistronic mRNAs are rare among eukaryotic cellular mRNAs. However, initiation at the second AUG has been explained by the "leaky scanning model": Some ribosomes scanning the 5′-noncoding region must bypass the first AUG initiation codon and then continue scanning until they find the second AUG codon, which initiates the second ORF (Kozak, 1986a,b).

In addition to the two ORFs for which translation products are clearly identified, a third ORF of 60, 75, and 75 amino acids was detected in the S segments of La Crosse, Germiston, and Maguari viruses, respectively. The products predicted from Germiston and Maguari sequences show only 36% similarity. Most probably, these ORFs are not expressed, since the expected polypeptides could not be detected in infected cells (Elliott and McGregor, 1989; Bouloy et al., 1990).

The S cDNAs of Snowshoe hare (Urakawa et al., 1988), Maguari (Elliott and McGregor, 1989), and Germiston (M. Bouloy et al., unpublished observations) viruses have been inserted into the genome of baculoviruses. Recombinant baculoviruses furnish a very powerful means of expressing large amounts of the N protein, which can be useful for diagnostic purposes, since N is the complement-fixing antigen (Lindsey et al., 1977). When compared with the expression of N, the level of NS$_S$ expressed from the complete S cDNA is low (Urakawa et al., 1988; M. Bouloy et al., unpublished observations) or undetectable (Elliott and McGregor, 1989). In the case of Snowshoe hare virus, the level of expression of NS$_S$ can be increased by removal of the AUG and a small region of the N coding sequence (Urakawa et al., 1988).

2. Hantavirus

Hantaviruses present a simply organized S segment. It contains only one ORF in the viral complementary sense, encoding the nucleoprotein. Sequencing the complete S segment of Hantaan virus reveals that the ORF is composed of 429 amino acids corresponding to a polypeptide of 48 kDa (Schmaljohn et al., 1986b). This confirms the size estimation from SDS–polyacrylamide gels, indicating that the N protein of Hantaan virus is approximately twice the size of the N proteins of other genera (Elliott et al., 1984; Schmaljohn and Dalrymple, 1984; Schmaljohn et al., 1983).

The S segments of the following hantaviruses have been cloned and sequenced: Sapporo rat virus (Arikawa et al., 1990); Nephropathia Epidemica virus, strain Hällnäs B1 (Stohwasser et al., 1990); and Prospect Hill virus (Parrington and Kang, 1990). The organization of the S segments of the three viruses is similar to that of Hantaan virus. Comparison of the N proteins indicates an overall 58% similarity between the four viruses, 83% between Sapporo rat virus and Hantaan virus and only 62% between Hantaan and Prospect Hill viruses.

Additional small ORFs are present in the viral complementary sense of Hantaan virus, Hällnäs strain B1, and Prospect Hill virus or in the virion-sense RNA of Sapporo rat virus. They are probably insignificant, since their predicted products could not be detected.

The need for antigens for diagnostic methods, which are easy to obtain and handle, has spurred the development of proteins expressed via recombinant viruses. For this purpose the N protein of Hantaan virus has been expressed in insect cells infected by recombinant baculoviruses. Both the baculovirus-expressed N protein and the authentic one reacted similarly with specific sera (Schmaljohn et al., 1988, 1990). In addition, when inoculated into hamsters, the N protein induced an immune response which protected the animals against a challenge with infectious virus.

3. Nairovirus

Little is known about the coding and replication strategy of nairo-viruses. However, these viruses have gained interest recently since one representative, Dugbe virus, has been cloned and sequenced (Ward et al., 1990). The S segment contains only one ORF of 49.4 kDa in the viral complementary sense, which corresponds to the N protein. A small ORF of 5.9 kDa is found in the virion strand, but, most probably, is not expressed.

No significant level of homology was observed in the amino acid sequences of the N proteins of the nairovirus and hanta-, phlebo-, uuku-, or bunyaviruses. However, the S segments of hanta- and na-iroviruses appear to have a similar organization.

4. Phlebo- and Uukuviruses

Sequencing of the S segment of several members of phlebo- and uukuviruses reveals a similar organization, different from that observed with bunya-, hanta-, and nairoviruses. The sequences of the S segments of Punta Toro (Ihara et al., 1984); Sandfly fever Sicilian, the prototype of the *Phlebovirus* genus (Marriott et al., 1989); Sandfly fever Toscana, Rift Valley fever (Giorgi et al., 1991); and Uukuniemi viruses (Simons et al., 1990) have been reported. The S segment contains two ORFs: one is present in the viral complementary sense and codes for the N protein; the other is detected in the virion sense and codes for the NS_S protein. As discussed later, the N and NS_S proteins are coded by two subgenomic mRNAs of opposite polarity. When this was established for Punta Toro, the first *Phlebovirus* to be sequenced (Ihara et al., 1984), this mode of expression had already been described for several members of the Arenaviridae family and called the "ambisense strategy" (for reviews see Bishop, 1986, 1990b).

The N proteins of the five viruses are comparable in sequence (30–50% similarity when compared pairwise), but the NS_S protein does not seem to be conserved. The ORFs coding for the N and NS_S proteins are separated by an intergenic region containing either A/U-rich (Punta Toro virus and Uukuniemi virus) or C/G-rich (Sandfly fever virus Sicilian and Toscana and Rift Valley fever virus) sequences. In the case of Punta Toro virus, the intergenic region can form a large hairpin structure (Emery and Bishop, 1987).

The ambisense strategy is not restricted to animal viruses, since the S segment of the plant virus, Tomato spotted wilt virus, possesses such an arrangement (de Haan et al., 1990).

IV. Replication

Most of the studies reported to date concern the transcription processes leading to the synthesis of mRNA molecules; little is known about the synthesis of genomic or antigenomic RNA molecules. This section deals with the information acquired by analyzing viral mRNAs isolated from infected cells or synthesized in an *in vitro* system composed of purified virus as the source of template and enzyme.

A. Characteristics of mRNA Molecules

Among bunyaviruses (Bishop *et al.*, 1983; Patterson and Kolakofsky, 1984; Bouloy *et al.*, 1984, 1990; Eshita *et al.*, 1985; Cunningham and Szilagyi, 1987; Gerbaud *et al.*, 1987a; Hacker *et al.*, 1990), phleboviruses (Ihara *et al.*, 1985a; Collett, 1986; Emery and Bishop, 1987; Marriott *et al.*, 1989), and hantaviruses (Schmaljohn *et al.*, 1987a) which have been studied, it appears that all the mRNAs are subgenomic, do not contain a poly(A) tail at their 3' ends, are capped and possess 10- to 18-base non-virus-coded extensions at their 5' ends. These nonviral sequences are probably utilized as primers to initiate transcription, as in the case with influenza virus transcriptase (Bouloy *et al.*, 1978; Krug, 1981). A detailed analysis of the 5'-terminal sequences of Germiston virus S mRNAs indicates that the primers are heterogeneous in sequence, have a high content of C and G residues and a U or C at position −1 (Bouloy *et al.*, 1990). Unlike during the myxovirus cycle, transcription and replication are not affected by actinomycin D or α-amanitin (Vezza *et al.*, 1979). All steps of RNA synthesis occur in the cytoplasm (Rossier *et al.*, 1986).

The features of mRNAs transcribed from the classical negative-stranded segments also apply to mRNAs of ambisense genomes, whether the mRNA is transcribed from genomes or antigenomes. Indeed, in the case of the phlebo- and uukuvirus S ambisense segments, both genome and antigenome molecules are transcribed into two distinct mRNAs of opposite polarities (Ihara *et al.*, 1985a; Emery and Bishop, 1987; Simons *et al.*, 1990). The N protein is translated from an mRNA which represents the copy of the genome, whereas NS_S is synthesized from a virion-sense mRNA transcribed from the antigenome. In Punta Toro virus-infected cells the two species of mRNA have been identified and shown to possess heterogeneous sequences at their 5' ends (Ihara *et al.*, 1985a). Two subgenomic S mRNAs of opposite polarity have also been described in Sandfly fever virus Sicilian- and Uukuniemi virus-infected cells (Marriott *et al.*, 1989; Simons *et al.*, 1990).

A surprising observation has been reported by Hacker *et al.* (1990), indicating that La Crosse virus synthesizes RNA molecules which are transcribed from the antigenome and initiated with RNA primers (e.g., mRNAs). However, these molecules do not represent discrete size class; they are incomplete copies terminating anywhere on the templates. It was thought that this type of transcription would apply exclusively to ambisense genomes. Thus, it was not expected to occur in bunyaviruses and the role of the so-called "anti-mRNAs" is unclear. It is possible that these molecules represent a third class of unusual transcripts. Two of them have already been described: antigenomes with 5' extensions and mRNAs lacking primers but initiated with a 5'-triphosphate end (Bouloy *et al.*, 1984; Gerbaud *et al.*, 1987a; Raju and Kolakofsky, 1987).

For the nonambisense segments precise mapping of the 3' ends of the mRNAs indicated that transcription terminates some 60–120 nucleotides upstream of the 5' end of the template in a region usually rich in uridine. Up to now, the termination signal recognized by the transcriptase has not been determined. Comparison of the sequences upstream or downstream of the termination signal did not indicate any significant similarity or any common feature (Eshita *et al.*, 1985; Bouloy *et al.*, 1990).

For the ambisense S segments of phlebo- and uukuviruses, transcription of N and NS_S mRNAs terminates in the intergenic region which separates the two ORFs. The intergenic region of Punta Toro virus S segment is composed of a long inverted complementary sequence which can be folded into a clear base-paired hairpin structure. Emery and Bishop (1987) have demonstrated that transcription terminates close to the terminal loop of the stem structure. The exact locations of the 3' ends of the mRNAs of Sandfly fever virus Sicilian and Toscana, Rift Valley fever virus, and Uukuniemi virus have not yet been defined. However, the intergenic region does not reveal any potential for the formation of a stable secondary structure. Thus, presently it is unclear which elements contained in the intergenic region act as a transcription termination signal in ambisense segments.

B. Mechanisms of Transcription

For many negative-stranded RNA viruses primary transcription has been studied using two approaches: (1) analysis of the viral mRNAs synthesized in infected cells in which protein synthesis has been inhibited by specific drugs (e.g., cycloheximide, anisomycin, or puromycin) and (2) development of an *in vitro* assay containing mono- and divalent salts, ribonucleoside triphosphates, and purified virions (for a review see Banerjee, 1987). When applied as such to study Bunyaviridae transcription, these methods did not provide satisfactory answers. Indeed, in spite of the presence of an RNA-dependent RNA polymerase activity in virions (Bouloy *et al.*, 1975; Bouloy and Hannoun, 1976; Ranki and Pettersson, , 1975; Schmaljohn and Dalrymple, 1983; Patterson *et al.*, 1984), the RNA transcripts synthesized *in vitro* were not identical to authentic mRNAs. Furthermore, several laboratories found that transcription is almost completely inhibited when protein synthesis is blocked (Abraham and Pattnaik, 1983; Patterson and Kolakofsky, 1984; Gerbaud *et al.*, 1987a).

In fact, it is now clearly demonstrated that bunyaviruses (and probably other genera of the family) are unique among negative-stranded RNA viruses in that efficient transcription in many vertebrate cells requires simultaneous translation. This requirement has been studied in detail for La Crosse virus (Bellocq *et al.*, 1987; Bellocq and Kolakofsky, 1987). In *in vitro* reactions transcription which efficiently synthesizes authentic mRNAs requires the addition of rabbit reticulocyte lysate as well as conditions that permit coupling of translation and transcription. In the absence of ribosomes, or, in most cases, when translation is blocked, only short incomplete transcripts are synthesized. However, it should be noted that under specific conditions (Mn^{2+} replacing Mg^{2+}) and in the absence of lysate, Germiston virus transcriptase is able to synthesize full-length S transcripts similar to antigenome (Gerbaud *et al.*, 1987a). However, these conditions might be somehow artifactual, like those described for influenza virus, which could also utilize Mn^{2+} cations (Plotch and Krug, 1977).

In the absence of ribosomes or when they do not move on the nascent mRNA (e.g., in the presence of cycloheximide or puromycin), initiation occurs, but elongation is inhibited *in vitro* (Bellocq *et al.*, 1987; Bellocq and Kolakofsky, 1987). Working with Germiston virus in *in vitro* reactions containing reticulocyte lysate, we observed that cycloheximide or puromycin also inhibits elongation, but not initiation. However, if one of these drugs was replaced by edeine, another protein synthesis inhibitor, we observed the opposite effect: Initiation and elongation were stimulated (Vialat and Bouloy, 1991). Edeine is known to inhibit the formation of the complex between the 40 S and 60 S ribosomal subunits, but not to prevent the 40 S subunit from binding and moving along the mRNA (Kozak and Shatkin, 1978; Kozak, 1988). Thus, scanning of the nascent chain by the 40 S subunit must facilitate the motion of the transcriptase so that a larger amount of RNA is synthesized. The reason for the cotranslational requirement is unclear. Kolakofsky and colleagues (Bellocq *et al.*, 1987; Bellocq and Kolakofsky, 1987) proposed an explanation: The template and the nascent chain bind together and prevent the polymerase from moving and elongating the chains. Our results confirm this hypothesis and emphasize the role of the 40 S ribosomal subunit.

Knowledge of the exact mechanism utilized by bunyavirus transcriptase to initiate transcription is based mainly on experiments reported by Patterson *et al.* (1984), showing that La Crosse virus contains a cap-dependent endonuclease activity which cleaves Alfalfa mosaic virus RNA 4 and incorporates the resulting capped oligonucleotides at the 5' end of the transcripts, as does influenza virus (Plotch *et al.*, 1981). For a long time this was the only (indirect) evidence that mRNAs are capped, but recently the presence of a methylated cap structure in La Crosse virus mRNAs was demonstrated (Hacker *et al.*, 1990).

In many vertebrate cells treated with puromycin or cycloheximide (Abraham and Pattnaik, 1983; Raju and Kolakofsky, 1986a,b; Gerbaud *et al.*, 1987a), viral transcription is inhibited. This confirms the results obtained in *in vitro* systems lacking reticulocyte lysate, which indicates that ongoing protein synthesis is required. Such a requirement was not observed *in vivo* by Vezza *et al.* (1979) and Eshita *et al.* (1985), who showed that transcription occurs in the presence of protein synthesis inhibitors. Recently, Raju *et al.* (1989) provided a tentative explanation for these apparently conflicting results. Indeed, Raju *et al.* showed that this requirement is dependent on a cellular factor present in many vertebrate cells, but not in C6/36 mosquito cells. Although the nature of the factor has not been identified, it is possible that Vezza *et al.* (1979) and Eshita *et al.* (1985) carried out their experiments with baby hamster kidney (BHK) cells which did not contain the factor involved in requirement for translation.

C. Secondary Transcription

Unlike mRNAs, but like genomes, antigenomes contain a 5'-triphosphate end (Obijeski *et al.*, 1980; Raju and Kolakofsky, 1987; Bouloy *et al.*, 1984; Gerbaud *et al.*, 1987a), indicating that syntheses of genome/antigenome and mRNAs are initiated by distinct mechanisms. The enzyme responsible for the synthesis of antigenomes has not been identified. Because the 3' and 5' ends are conserved and complementary, it is assumed that the same enzyme synthesis antigenome and genome. Although it is tempting to assume that NS_S protein acts as a factor modulating the transcriptase, no evidence has been reported to support this hypothesis.

For ambisense segments subgenomic virion-sense mRNAs coding for NS_S are synthesized on antigenomic templates. They are synthesized at the late stage of infection, and their level of accumulation is considerably lower than that of N mRNA (Ihara *et al.*, 1985a; Simons *et al.*, 1990). As demonstrated in Punta Toro virus-infected cells, synthesis of NS_S mRNA requires prior synthesis of antigenome and probably replication, since only N mRNA could be detected when the cells were treated with protein synthesis inhibitors (Ihara *et al.*, 1985a). The temporal regulation of NS_S mRNA excludes that NS_S plays a role in the early steps of infection, unless this protein is associated with the virion particle. Although no conclusion could be drawn as to the role of NS_S, it should be noted that this protein was found to be associated with Punta Toro virions and nucleocapsids (Overton *et al.*, 1987). Although NS_S has not been detected in Uukuniemi virions, an alternative is utilized by the latter virus: The virions package full-length S RNA segments of both polarities in a ratio of about 10:1. Thus, N and NS_S mRNAs and their products could be synthesized simultaneously during primary transcription. Packaging of RNA strands of opposite polarities seems to be specific to the ambisense S segment; it is not observed for the M RNA segment (Simons *et al.*, 1990). Thus, for Punta Toro and Uukuniemi viruses we cannot exclude that NS_S has a role early during infection. The two situations have been schematically represented in Fig. 3A and B. However, more data are necessary to know whether this is true for other uuku- and phleboviruses and eventually to establish whether the difference between

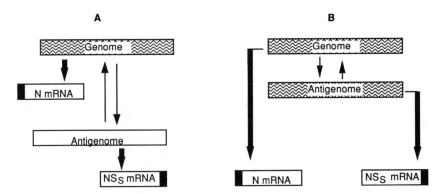

FIG. 3. Transcription replication of the S ambisense segment. Molecules packaged into virions are represented by boxes with wavy lines. Open boxes indicate molecules which are present intracellulary. The RNA primers present at the 5' ends of mRNAs are represented by solid boxes. Fat and thin arrows indicate transcription and replication, respectively. (A) Virions contain only the S genomic strand. N mRNAs and antigenomes are synthesized first; NS_S mRNAs are synthesized in a late step of transcription. (B) Virions contain the genomic and antigenomic strands as described for Uukuniemi virus (Simons *et al.*, 1990). Syntheses of N and NS_S mRNAs can occur simultaneously.

phlebo- and uukuviruses resides in the incorporation of the NS_S protein or in the packaging of the S antigenome and genome, respectively, into the virion.

At this point there is no consensus on the location of the NS_S protein. In Karimabad virus-infected cells it remains in the cytoplasm (Smith and Pifat, 1982). In Punta Toro virions it is associated with nucleocapsids (Overton et al., 1987), in Rift Valley virus-infected cells it is phosphorylated and forms intranuclear inclusions (Swanepoel and Blackburn, 1977; Struthers and Swanepoel, 1982; Struthers et al., 1984; J. F. Smith, personal communication), and in cells infected with Tomato spotted wilt virus it forms flexible filaments and paracrystalline rods (Kormelink et al., 1991).

V. MORPHOGENESIS

Maturation of Bunyaviridae occurs by budding at smooth membrane vesicles in the Golgi complex (Murphy et al., 1973; Ellis et al., 1988). Only occasionally, budding was observed at the plasma membrane. This was reported for rat hepatocytes infected with Rift Valley fever virus (Anderson and Smith, 1987) and for BHK21 cells infected with La Crosse virus (Madoff and Lenard, 1982). Electron-microscopic studies indicate that, when budding is observed, ribonucleoproteins are also found at the same site, suggesting that budding is induced by a transmembranal recognition between the viral glycoproteins and the N protein (Smith and Pifat, 1982).

The viral glycoproteins seem to play an important role in determining the site of maturation. This has been demonstrated by expressing the glycoproteins via recombinant vaccinia viruses and baculoviruses. When synthesized from cloned M cDNA, in the absence of interactions with other viral proteins, G1 and G2 proteins of Hantaan, Rift Valley fever, and Punta Toro viruses are targeted to the Golgi complex (Pensiero et al., 1988; Matsuoka et al., 1988; Wasmoen et al., 1988). For Rift Valley fever virus Wasmoen et al. (1988) demonstrated that the 14-kDa protein itself does not determine the Golgi localization, since G1 and G2 are still directed to this organelle, even when expressed from an M polyprotein in which the preglycoprotein coding region has been deleted. In addition, they showed that the 14-kDa polypeptide is distributed in the Golgi complex and the reticulum. Studies on themosensitive mutants of Uukuniemi virus defective in viral maturation also indicate that G1 and G2 are localized in the Golgi despite the absence of virion formation (Gahmberg, 1984; Gahmberg et al., 1986a).

To better understand the different processes involved in morphogenesis, it is of interest to study how glycoproteins are transported to the Golgi apparatus. For this purpose two drugs, tunicamycin and monensin, as well as endoglycosidase, have been used. Tunicamycin inhibits N glycosylation and causes a reduction in the electrophoretic mobility of the envelope glycoproteins of all the Bunyaviridae studied so far, indicating that all of them contain asparagine N-linked oligosaccharides. Monensin inhibits the intracellular transport of membrane glycoproteins by blocking the release of secretory vesicles. It does not affect the electrophoretic migration of the polypeptides, but inhibits Bunyaviridae virion formation (Kuismanen et al., 1982, 1985; Kuismanen, 1984; Schmaljohn et al., 1986a). The nature of these oligosaccharides can give an insight into the maturation events (for reviews see Klenk and Rott, 1980; Kornfeld and Kornfeld, 1985). Generally, two classes are found on mature glycoproteins: high-mannose and complex types. The high-mannose type is added in the rough endoplasmic reticulum (RER). Such oligosaccharides are sensitive to endoglycosidase H (endo H). Later, during the transport of the glycoproteins through the Golgi complex, the mannose residues are trimmed, and complex sugars are added. At this step the complex oligosaccharides become resistant to treatment with endo H, indicating that the glycoproteins have been processed in the Golgi apparatus. In Uukuniemi and Inkoo viruses three types of oligosaccharides have been detected. Protein G2 contains oligosaccharides of the high-mannose type, whereas G1 contains oligosaccharides of the complex and intermediate types (Pesonen et al., 1982). The intermediate type represents a novel class in the biogenesis of oligosaccharides. Mature proteins G1 and G2 of La Crosse virus were also shown to contain mainly oligosaccharides of the complex type as well as covalently attached fatty acids (Madoff and Lenard, 1982). Similarly, the oligosaccharides of Germiston virion glycoproteins are mostly of the complex type (Gerbaud et al., 1991). On the other hand, Hantaan virus glycoproteins contain oligosaccharides which are mostly, but not entirely, of the high-mannose type (Schmaljohn et al., 1986a).

It has been shown for several viruses that, during the transport from the RER to the Golgi complex, a large fraction of the envelope proteins G1 and G2 accumulates intracellularly in the RER, remaining endo H sensitive, whereas a small fraction is transported relatively slowly to the Golgi complex, 2- or 3-fold slower than the vesicular stomatitis virus glycoprotein G (Madoff and Lenard, 1982; Kuismanen et al., 1982; Persson and Pettersson, 1991). To explain the slow transport and the relatively low efficiency of the processing, it has been proposed that accumulation of the glycoproteins in the Golgi complex could impair the normal organization and function of the Golgi complex.

However, Pettersson and colleagues (Persson and Pettersson, 1991; Gahmberg et al., 1986b; Kuismanen et al., 1982), using Uukuniemi virus as a model, showed that this is not so. Gahmberg et al. (1986b) analyzed the effect of Uukuniemi infection in cells coinfected by Semliki Forest virus, which matures at the plasma membranes. They clearly demonstrated that the Golgi complex has retained its functional integrity to glycosylate and transport Semliki Forest virus glycoproteins to the plasma membrane. In a recent study by Persson and Pettersson (1991), it appears that maturation of G1 and G2 results in the formation of heterodimers. During their intracellular transport G1 and G2 are associated with the BiP protein and with the protein disulfide isomerase, two enzymes involved in the folding and disulfide bond formation of secretory and membrane proteins. Protein G1 seems to be transported faster than G2 to the site of budding (Kuismanen, 1984), but the apparent difference in transport kinetics is due to the fact that G1 folds much faster than G2. As a consequence G1 cannot dimerize with G2 synthesized from the same precursor, but with a G2 protein which had been synthesized 20–45 min earlier (Persson and Pettersson, 1991).

VI. Interactions with Host Cells

Most of the studies undertaken to better understand the molecular biology of these viruses were carried out in permissive vertebrate cell cultures. The members of the Bunyaviridae which are arboviruses can also infect and replicate in insect cells. Except for hantaviruses, which are not transmitted by arthropods and which establish persistent infection in susceptible mammalian cells (Schmaljohn and Dalrymple, 1984; Schmaljohn and Patterson, 1990), most infections of vertebrate cells cause a clear cytopathic effect which leads to cell death. In contrast, infection of insect cells does not cause any damage to the cells. This must be related to the inapparent infection of mosquitoes which occurs in nature. Mosquito cells infected with bunya- and phleboviruses have been shown to be able to multiply and be passaged many times, while still producing viruses (Newton et al., 1981; Nicoletti and Verani, 1985; Carvalho et al., 1986; Elliott and Wilkie, 1986; Rossier et al., 1988; Delord et al., 1989). During the numerous passages in mosquito cells, interfering particles are generated, which interfere with the development of the standard virus (Lyons and Heyduck, 1973; Elliott and Wilkie, 1986).

To gain insight into the molecular basis of the difference between vertebrate and insect cell infection, the levels of RNA and protein syntheses in both systems have been investigated (Lyons and Heyduk, 1973; Newton et al., 1981; Verani et al., 1984; Elliott and

Wilkie, 1986; Rossier *et al.*, 1988). In mammalian cells infection is rapid and synthesis of viral mRNA and genome reaches a maximum approximately at 6 hours postinfection. At this time there is a shut-off of host macromolecular synthesis which results from a general instability of all mRNAs. Later (14 hours postinfection) a cytopathic effect becomes visible (Raju and Kolakofsky, 1988). In contrast, in mosquito cells replication takes place more slowly. A maximal rate of RNA and protein syntheses is observed around 24 hours postinfection, but later, while persistent infection is being established, the virus down-regulates its replication. However, the level of intracellular S mRNA remains high. Recently, Hacker *et al.* (1989) provided an explanation. They found that a particular phenomenon occurs in mosquito cells: The N protein encapsidates its own mRNA. The N assembly site is thought to be the conserved sequences at the 5' ends of antigenomes. If this is so, the same site exists in mRNAs, but the latter is a poor substrate, presumably because of the presence of the 5'-nontemplated primer extensions. However, at some step during infection, when the rate of genome synthesis decreases, there must be a pool of unassembled N protein which increases to the point at which the N protein begins to bind to its own mRNA. As a result translation of the viral, but not cellular, mRNAs is inhibited.

VII. Conclusion

Within the past few years our understanding of the genome organization and replication of Bunyaviridae has evolved rapidly. It appears that the S segments and their translation products are subjected to variations in size, coding capacity, and expression strategy. Thus, because of the specificity of each genus and for clarity, it has been recommended to reserve the term "Bunyavirus" for viruses which belong to the *Bunyavirus* genus, whereas other members of the Bunyaviridae family should be designated by their generic names.

Sequence data have revealed unexpected relationships between phlebo- and uukuviruses, which, in addition, utilize the ambisense strategy, as does the Arenaviridae family. It has been proposed, but not yet adopted, to take into account these new data in order to create subfamilies (Bishop, 1985). Several important pieces of information which should be helpful for such a restructuring are still missing. For example, the S segment of a *Nairovirus* has been cloned and sequenced recently, revealing an organization similar to that of hantaviruses. It would be of interest to determine the organization of the rest of the genome as well as the extent of similarity with other genera.

To date, the sequences of two polymerases of Bunyamwera and

MICHÈLE BOULOY

Hantaan viruses are available. It would also be of great value to determine the gene organization of the L segments from phlebo- and uukuviruses. What is their degree of similarity with viruses of other genera and with arenaviruses? Our understanding of the replication process must be improved further. *In vitro* systems have permitted the elucidation of many aspects of transcription mechanisms, but similar systems have not been successfully developed to synthesize antigenome or genome. The role of NS_S may be crucial. It has been proposed that NS_S might be involved in replication, but, to date, there is no experimental evidence for any function. Should we expect a similar function for the NS_S protein of bunyaviruses and that of phlebo- or uukuviruses? Many other questions, such as the mechanism of encapsidation and the determination of promoter sequence in the genome, await elucidation. These provide substance for future research.

The California serogroup offers an attractive model to study the molecular determinants of virulence. It has been shown that neuroinvasiveness, which is the ability to invade the central nervous system after peripheral infection, is mainly determined by the M segment (Gonzalez-Scarano *et al.*, 1985a; Janssen *et al.*, 1986). Production of neuroattenuated variants should help to determine the molecular and genetic bases of virulence (Endres *et al.*, 1990).

From a biological point of view, Bunyaviridae is a family which comprises several severe pathogens (e.g., Crimean–Congo hemorrhagic fever, Rift Valley fever, and Hantaan virus) against which no vaccines are yet available. Several attempts have been made to express viral proteins from cloned cDNA and to study their antigenic and immunogenic properties. Schmaljohn *et al.* (1990) have expressed the three major proteins (G1, G2, and N) of Hantaan virus via recombinant baculovirus or vaccinia virus. These expressed proteins appear to be good immunogens and to protect animals against a challenge with infectious virus. Similar results are reported on the immunogenic properties of the glycoproteins of Rift Valley fever virus expressed via recombinant baculoviruses (Schmaljohn *et al.*, 1989). There is also a life-attenuated strain of Rift Valley fever virus which seems to be a good vaccine candidate for livestock and humans. This virus carries a mutation capable of independently attenuating the virus in each of the three segments (Saluzzo and Smith, 1990). Hopefully, studies such as these will be continued in the future and lead to the development of human vaccines.

I am grateful to S. Gerbaud and I. Kuchenthal for critical reading of the manuscript. I would also like to thank all my colleagues who provided me with their recent publications and allowed me to cite personal communications. The work carried out in my laboratory was supported in part by INSERM grant 861003.

REFERENCES

Abraham, G., and Pattnaik, G. (1983). Early RNA synthesis in Bunyamwera virus-infected cells. *J. Gen. Virol.* **64,** 1277–1290.

Akashi, H., and Bishop. D. H. L. (1983). Comparison of the sequences and coding of La Crosse and snowshoe hare bunyavirus S RNA species. *J. Virol.* **45,** 1155–1158.

Akashi, H., Gay, M., Ihara, T., and Bishop, D. H. L. (1984). Localized conserved regions of the S RNA gene products of bunyaviruses are revealed by sequence analyses of the Simbu serogroup Aino viruses. *Virus Res.* **1,** 51–63.

Anderson, G. W., Jr., and Smith, J. F. (1987). Immunoelectron microscopy of Rift Valley fever viral morphogenesis in primary rat hepatocytes. *Virology* **161,** 91–100.

Arikawa, J., Takashima, I., and Hashimoto, N. (1985). Cell fusion by haemorrhagic fever with renal syndrome (HFRS) viruses and its application for titration of virus infectivity and neutralizing antibody. *Arch. Virol.* **86,** 303–313.

Arikawa, J., Schmaljohn, A. L., Dalrymple, J. M., and Schmaljohn, C. S. (1989). Characterization of Hantaan virus envelope glycoprotein antigenic determinants defined by monoclonal antibodies. *J. Gen. Virol.* **70,** 615–624.

Arikawa, J., Lapenotiere, H. F., Iacono-Conners, L., Wang, M., and Schmaljohn, C. S. (1990). Coding properties of the S and M genome segments of Sapporo rat virus: Comparison to other causative agents of hemorrhagic fever with renal syndrome. *Virology* **176,** 114–125.

Banerjee, A. K. (1987). Transcription and replication of rhabdoviruses. *Microbiol. Rev.* **51,** 66–87.

Battles, J. K., and Dalrymple, J. M. (1988). Genetic variation among geographic isolates of Rift Valley fever virus. *Am. J. Trop. Med. Hyg.* **39,** 617–631.

Beaty, B. J., and Bishop, D. H. L. (1988). Bunyavirus–vector interaction. *Virus Res.* **10,** 289–302.

Beaty, B. J., Rozhon, E. J., Gensemer, P., and Bishop, D. H. L. (1981). Formation of reassortant bunyaviruses in dually infected mosquitoes. *Virology* **111,** 662–665.

Beaty, B. J., Miller, B. R., Shope, R. E., Rozhon, E. J., and Bishop, D. H. L. (1982). Molecular basis of bunyavirus *per os* infection of mosquitoes: Role of the middle-sized RNA segment. *Proc. Natl. Acad. Sci. U.S.A.* **79,** 1295–1297.

Beaty, B. J., Sundin, D. R., Chandler, L. J., and Bishop, D. H. L. (1985). Evolution of bunyaviruses by genome reassortment in dually infected mosquitoes (*Aedes triseriatus*). *Science* **230,** 548–550.

Bellocq, C., and Kolakofsky, D. (1987). Translational requirement for La Crosse virus S-mRNA synthesis: A possible mechanism. *J. Virol.* **61,** 3960–3967.

Bellocq, C., Raju, R., Patterson, J. L., and Kolakofsky, D. (1987). Translational requirement of La Crosse virus S-mRNA synthesis: In vitro studies. *J. Virol.* **61,** 87–95.

Bishop, D. H. L. (1979). Genetic potential of bunyaviruses. *Curr. Top. Microbiol. Immunol.* **86,** 1–33.

Bishop, D. H. L. (1985). The genetic basis for describing viruses as species. *Intervirology* **24,** 79–93.

Bishop, D. H. L. (1986). Ambisense RNA genome of arenavirus and phlebovirus. *Adv. Virus Res.* **31,** 1–51.

Bishop, D. H. L. (1990a). Bunyaviridae and their replication. I. Bunyaviridae. *In* "Virology" (B. N. Fields, D. M. Knipe, R. M. Chanock, J. L. Melnick, M. S. Hirsh, T. P. Monath, and B. Roizman, eds.), pp. 1155–1173. Raven, New York.

Bishop, D. H. L. (1990b). Arenaviridae and their replication. In "Virology" (B. N. Fields, D. M. Knipe, R. M. Chanock, J. L. Melnick, M. S. Hirsh, T. P. Monath, and B. Roizman, eds.), pp. 1231–1243. Raven, New York.

Bishop, D. H. L., and Shope, R. E. (1979). Bunyaviridae. Compr. Virol. 14, 1–156.

Bishop, D. H. L., Calisher, C. H., Casals, J., et al. (1980). Bunyaviridae. Intervirology 14, 125–143.

Bishop, D. H. L., Gould, K. G., Akashi, H., and Clerx-Van Haaster, C. M. (1982). The complete sequence and coding content of snowshoe hare bunyavirus small (S) viral RNA species. Nucleic Acids Res. 10, 3703–3713.

Bishop, D. H. L., Gay, M. E., and Matsuoko, Y. (1983). Nonviral heterogeneous sequences are present at the 5' ends of one species of snowshoe hare bunyavirus complementary RNA. Nucleic Acids Res. 11, 6409–641.

Bouloy, M., and Hannoun, C. (1976). Studies on Lumbo virus replication. I. RNA dependent RNA polymerase associated with virions. Virology 69, 258–268.

Bouloy, M., Krams-Ozden, S., Horodniceanu, F., and Hannoun, C. (1973/1974). Three segment RNA genome of Lumbo virus (Bunyavirus). Intervirology 2, 173–180.

Bouloy, M., Colbere, F., Krams-Ozden, S., Vialat, P., Garapin, A. C., and Hannoun, C. (1975). Mise en evidence d'une activité RNA polymérase RNA dependante dans le virus Lumbo. C. R. Hebd. Seances Acad. Sci., Ser. D 280, 213–215.

Bouloy, M., Plotch, S. J., and Krug, R. M. (1978). Globin mRNAs are primers for the transcription of influenza viral RNA in vitro. Proc. Natl. Acad. Sci. U.S.A. 75, 4886–4890.

Bouloy, M., Vialat, P., Girard, M., and Pardigon, N. (1984). A transcript from the S segment of the Germiston bunyavirus is uncapped and codes for the nucleoprotein and a nonstructural protein. J. Virol. 49, 717–723.

Bouloy, M., Pardigon, N., Vialat, P., Gerbaud, S., and Girard, M. (1990). Characterization of the 5' and 3' ends of viral messenger RNAs isolated from BHK21 cells infected with Germiston virus (Bunyavirus). Virology 175, 50–58.

Braam, J., Ulmanen, I., and Krug, R. M. (1983). Molecular model of a eukaryotic transcription complex: Functions and movements of influenza P proteins during capped RNA primed transcription. Cell 34, 609–618.

Cabradilla, C. D., Holloway, B. P., and Obijeski, J. F. (1983). Molecular cloning and sequencing of the La Crosse virus S RNA. Virology 128, 463–468.

Carvalho, M. G., Frugulhetti, I. C., and Revello, M. A. (1986). Marituba (Bunyaviridae) virus replication in cultured Aedes albopictus cells and in L-A9 cells. Arch. Virol. 90, 325–335.

Chandler, L. J., Beaty, B. J., Baldridge, G. D., Bishop, D. H. L., and Hewlett, M. J. (1990). Heterologous reassortment of bunyaviruses in Aedes triseriatus mosquitoes and transovarial and oral transmission of newly evolved genotypes. J. Gen. Virol. 71, 1045–1050.

Clerx, J. P. M., and Bishop, D. H. L. (1981). Qalyub virus, a member of the newly proposed Nairovirus genus (Bunyaviridae). Virology 108, 361–372.

Clerx, J. P. M., Casals, J., and Bishop, D. H. L. (1981). Structural characteristics of nairoviruses (genus Nairovirus, Bunyaviridae). J. Gen. Virol. 55, 165–178.

Collett, M. S. (1986). Messenger RNA of the M segment RNA of Rift Valley fever virus. Virology 151, 151–156.

Collett, M. S., Purchio, A. F., Keegan, K., et al. (1985). Complete nucleotide sequence of the M RNA segment of Rift Valley fever virus. Virology 144, 228–245.

Collett, M. S., Keegan, K., Hu, S.-L., Sridhar, P., Purchio, A. F., Ennis, W. H., and Dalrymple, J. M. (1987). Protective subunit immunogens to Rift Valley fever virus from bacteria and recombinant vaccinia virus. In "The Biology of Negative Strand Viruses" (B. W. J. Mahy and D. Kolakofsky, eds.), pp. 321–329. Elsevier, Amsterdam.

Collett, M. S., Kakach, L. T., Suzich, J. A., and Wasmoen, T. L. (1989). Gene products and expression strategy of the M segment of the phleobovirus Rift Valley fever virus. In "Genetics and Pathogenicity of Negative Strand Viruses" (D. Kolokofsky and B. Mahy, eds.). Elsevier, New York.

Cunningham, C., and Szilagyi, J. F. (1987). Viral RNAs synthesized in cells infected with Germiston bunyavirus. Virology 157, 431–439.

Dalrymple, J. M., Hasty, S. E., Kakach, L. T., and Collett, M. S. (1989). Mapping of protective determinants of Rift Valley fever virus using recombinant vaccinia viruses. In "Vaccines '89" (R. A. Lerner, H. Ginsberg, R. Chanock, and F. Brown, eds.), pp. 371–375. Cold Spring Harbor Lab., Cold Spring Harbor, New York.

Dantas, J. R., Okuno, Y., Asada, H., et al. (1986). Characterization of glycoproteins of viruses causing hemorrhagic fever with renal syndrome (HFRS) with monoclonal antibodies. Virology 151, 379–384.

de Haan, P., Wagenmakers, L., Peters, D., and Goldbach, R. (1989). Molecular cloning and terminal sequence determination of the S and M RNAs of tomato spotted wilt virus. J. Gen. Virol. 70, 3469–3473.

de Haan, P., Wagenmakers, L., Peters, D., and Goldbach, R. (1990). The S RNA segment of tomato spotted wilt virus has an ambisense character. J. Gen. Virol. 71, 1001–1007.

Delord, B., Poveda, J. D., Astier-Gein, T., Gerbaud, S., and Fleury, H. J. A. (1989). Detection of the bunyavirus Germiston in Vero and Aedes albopictus C6/36 cells by in situ hybridization using cDNA and asymmetric RNA probes. J. Virol. Methods 24, 253–264.

Donets, M. A., Chumakov, M. P., Korolev, M. B., and Rubin, S. G. (1977). Physiochemical characteristics, morphology and morphogenesis of virions of the causative agent of Crimean hemorrhagic fever. Intervirology 8, 294–308.

Elliott, R. M. (1985). Identification of nonstructural proteins encoded by viruses of the Bunyamwera serogroup (family Bunyaviridae). Virology 143, 119–126.

Elliott, R. M. (1989a). Nucleotide sequence analysis of the small (S) RNA segment of Bunyamwera virus, the prototype of the family Bunyaviridae. J. Gen. Virol. 70, 1281–1285.

Elliott, R. M. (1989b). Nucleotide sequence analysis of the large (L) genomic RNA segment of Bunyamwera virus, the prototype of the family Bunyaviridae. Virology 173, 426–436.

Elliott, R. M. (1990). Molecular biology of the Bunyaviridae. J. Gen. Virol. 71, 501–522.

Elliott, R. M., and McGregor, A. (1989). Nucleotide sequence and expression of the small (S) RNA segment of Maguari bunyavirus. Virology 171, 516–524.

Elliott, R. M., and Wilkie, M. L. (1986). Persistent infection of Aedes albopictus C6/36 cells by Bunyamwera serogroup (family Bunyaviridae), Virology 150, 21–32.

Elliott, L. H., Kiley, M. P., and McCormick, J. B. (1984). Hantaan virus: Identification of virion proteins. J. Gen. Virol. 65, 1285–1293.

Ellis, D. S., Shirodaria, P. V., Fleming, E., and Simpson, D. I. H. (1988). Morphology and development of Rift Valley fever virus in Vero cell cultures. J. Med. Virol. 24, 161–174.

Emery, V. C., and Bishop, D. H. L. (1987). Characterization of Punta Toro S mRNA species and identification of an inverted complementary sequence in the intergenic region of Punta Toro phlebovirus ambisense S RNA that is involved in mRNA transcription termination. Virology 156, 1–11.

Endres, M. J., Jacoby, D. R., Janssen, R. S., Gonzalez-Scarano, F., and Nathanson, N. (1989). The large viral RNA segment of California serogroup bunyaviruses encodes the large viral protein. J. Gen. Virol. 70, 223–228.

Endres, M. J., Valsamakis, A., Gonzalez-Scarano, F., and Nathanson, N. (1990). Neuroattenuated bunyavirus variant: Derivation, characterization and revertant clones. *J. Virol.* **64,** 1927–1933.

Eshita, Y., and Bishop, D. H. L. (1984). The complete sequence of the M RNA of snowshoe hare bunyavirus reveals the presence of internal hydrophobic domains in the viral glycoprotein. *Virology* **37,** 227–240.

Eshita, Y., Ericson, B., Romanovski, V., and Bishop, D. H. L. (1985). Analyses of the mRNA transcription processes of snowshoe hare bunyavirus S and M RNA species. *J. Virol.* **55,** 681–689.

Fakazerley, J. K., Gonzalez-Scarano, F., Strickler, J., Dietzschold, B., Karush, F., and Nathanson, N. (1988). Organization of the middle RNA segment of snowshoe hare bunyavirus. *Virology* **167,** 422–432.

Foulke, R. S., Rosato, R. R., and French, G. R. (1981). Structural polypeptides of Hazara virus. *J. Gen. Virol.* **65,** 169–172.

Gahmberg, N. (1984). Characterization of two recombination–complementation groups of Uukuniemi virus temperature-sensitive mutants. *J. Gen. Virol.* **65,** 1079–1090.

Gahmberg, N., Kuismanen, E., Keranen, S., and Pettersson, R. F. (1986a). Uukuniemi virus glycoproteins accumulate in and cause morphological changes of the Golgi complex in the absence of virus maturation. *J. Virol.* **57,** 899–906.

Gahmberg, N., Pettersson, R. F., and Kääriäinen, L. (1986b). Efficient transport of Semliki Forest virus glycoproteins through a Golgi complex morphologically altered by Uukuniemi virus glycoproteins. *EMBO J.* **5,** 3111–3118.

Gerbaud, S., Pardigon, N., Vialat, P., and Bouloy, M. (1987a). The S segment of Germiston bunyavirus genome: Coding strategy and transcription. *In* "The Biology of Negative Strand Viruses" (B. W. J. Mahy and D. Kolakofsky, eds.), pp. 191–198. Elsevier, Amsterdam.

Gerbaud, S., Vialat, P. Pardigon, N., Wychowski, C., Girard, M., and Bouloy, M. (1987b). The S segment of Germiston virus RNA genome can code for three proteins. *Virus. Res.* **8,** 1–13.

Gerbaud, S., Pardigon, N., Vialat, P., and Bouloy, M. (1991). Manuscript in preparation.

Giebel, L. B., Stohwasser, R., Zöller, L., Bautz, E. K. F., and Darai, G. (1989). Determination of the coding capacity of M genome of Nephropathia Epidemica virus, strain Hällnäs B1 by molecular cloning and nucleotide sequence analysis. *Virology* **172,** 498–505.

Giorgi, C., Acardi, L., Nicoletti, L., Gro, M. C., Takehara, K., Hildich, C., Morkawa, S., and Bishop, D. H. L. (1991). Sequences and coding strategies of the S RNAs of Toscana and Rift Valley fever viruses compared to those of Punta Toro, Sicilian sandfly fever and Uukuniemi viruses. *Virology* **180,** 733–753.

Gonzalez-Scarano, F. (1985). La Crosse virus G1 glycoprotein undergoes a conformational change at the pH of fusion. *Virology* **140,** 209–216.

Gonzalez-Scarano, F., and Nathanson, N. (1990). Bunyaviruses. *In* "Virology" (B. N. Fields, D. M. Knipe, R. M. Chanock, J. L. Melnick, M. S. Hirsh, T. P. Monath, and B. Roizman, eds.), pp. 1195–1228. Raven, New York.

Gonzalez-Scarano, F., Shope, R. E. Calisher, C. H., and Nathanson, N. (1982). Characterization of monoclonal antibodies against the G1 and N proteins of La Crosse and Tahynya, two California serogroup bunyaviruses. *Virology* **132,** 222–225.

Gonzalez-Scarano, F., Janssen, R. S. Najjar, J. A. Pobjecky, N., and Nathanson, N. (1985a). An avirulent G1 glycoprotein variant of La Crosse bunyavirus with defective fusion function. *J. Virol.* **64,** 757–763.

Gonzalez-Scarano, F., Pobjecky, N., and Nathanson, N. (1985b). La Crosse bunyavirus can mediate pH-dependent fusion from without. *Virology* **54,** 757–763.

Gonzalez-Scarano, F., Beaty, B. J., Sundin, D., Janssen, R., Endres, M. J., and Nathanson, N. (1988). Genetic determinants of the virulence and infectivity of La Crosse virus. *Microb. Pathog.* **4,** 1–7.

Grady, L. J., Srihongse, S., Grayson, M. A., and Deibel, R. (1983). Monoclonal antibodies against La Crosse virus. *J. Gen. Virol.* **64**, 1699–1704.

Grady, L. J., Sanders, M. L., and Campbell, W. P. (1987). The sequence of the M RNA of an isolate of La Crosse virus. *J. Gen. Virol.* **68**, 3057–3071.

Hacker, D., Raju, R., and Kolakofsky, D. (1989). La Crosse virus nucleocapsid protein controls its own synthesis in mosquito cells by encapsidating its mRNA. *J. Virol.* **63**, 5166–5174.

Hacker, D., Rochat, S., and Kolakofsky, D. (1990). Anti-mRNAs in La Crosse bunyavirus-infected cells. *J. Virol.* **64**, 5051–5057.

Iapalucci, S., Lopez, N., Rey, O., Zakin, M., Cohen, G., and Franze-Fernandez, M. T. (1989). The 5' region of Tacaribe virus L RNA encodes a protein with a potential metal binding domain. *Virology* **173**, 357–361.

Ihara, T., Akashi, H., and Bishop, D. H. L. (1984). Novel coding strategy (ambisensegenomic RNA) revealed by sequence analyses of Punta Toro phlebovirus S RNA. *Virology* **136**, 293–306.

Ihara, T., Matsuura, Y., and Bishop, D. H. L. (1985a). Analyses of the mRNA transcription processes of Punta Toro phlebovirus (Bunyaviridae). *Virology* **147**, 317–325.

Ihara, T., Smith, J., Dalrymple, J. M., and Bishop, D. H. L. (1985b). Complete sequences of the glycoprotein and M RNA of Punta Toro phlebovirus compared to those of Rift Valley fever virus. *Virology* **144**, 246–259.

Ishikawa, K., Omura, T., and Tsuchizaki, T. (1989). Association of double and single stranded RNAs with each of the four components of rice stripe virus. *Nippon Shokubutsu Byori Gakkaiho* **55**, 315–323.

Janssen, R. S., Nathanson, N., Endres, M. J., and Gonzalez-Scarano, F. (1986). Virulence of La Crosse virus is under polygenic control. *J. Virol.* **59**, 1–7.

Kakach, L. T., Wasmoen, T. L., and Collett, M. S. (1988). Rift Valley fever virus M segment: Use of recombinant vaccinia viruses to study phlebovirus gene expression. *J. Virol.* **62**, 826–833.

Kakach, L. T., Suzich, J. A., and Collett, M. S. (1989). Rift Valley fever virus M segment: Phleobovirus expression strategy and protein glycosylation. *Virology* **170**, 505–510.

Kakutani, T., Hayano, Y., Hayashi, T., and Minobe, Y. (1990). Ambisense segment 4 of rice stripe/possible evolutionary relationship with phleboviruses and uukuviruses (Bunyaviridae). *J. Gen. Virol.* **71**, 1427–1432.

Karabatsos, N. (1985). "International Catalogue of Arboviruses Including Certain Other Viruses of Vertebrates." Am. Soc. Trop. Med. Hyg., San Antonio, Texas.

Keegan, K., and Collett, M. S. (1986). Use of bacterial expression cloning to define the amino acid sequences of antigenic determinants on the G2 glycoprotein of Rift Valley fever virus. *J. Virol.* **58**, 263–270.

Kingsford, L., and Boucquey, K. H. (1990). Monoclonal antibodies specific for La Crosse virus that react with other California serogroup viruses. *J. Gen. Virol.* **71**, 523–530.

Kingsford, L., Ishizawa, L. D., and Hill, D. W. (1983). Biological activities of monoclonal antibodies reactive with antigenic sites mapped on the G1 glycoprotein of La Crosse virus. *Virology* **129**, 443–455.

Klenk, H. D., and Rott, R. (1980). Co-translational and post-translational processing of viral glycoproteins. *Curr. Top. Microbiol. Immunol.* **90**, 19–48.

Klimas, R. A., Thompson, W. A., Calisher, C. H., Clark, G. G., Grimstad, P. R., and Bishop, D. H. L. (1981). Genotypic varieties of La Crosse virus isolated from different geographic regions of the continental United States and the evidence for a naturally occurring intertypic recombinant La Crosse virus. *Am. J. Epidemiol.* **114**, 112–131.

Kormelink, R., Kitajima, E. W., de Haan, P., Zuidema, D., Peters, D., and Goldback, R. (1991). The nonstructural (NS) encoded by the ambisense S RNA segment of tomato spotted wilt virus (TSWV) is associated with fibrous structures in infected plant cells. *Virology* (in press).

Kornfeld, R., and Kornfeld, S. (1985). Assembly of asparagine-linked oligosaccharides. *Annu. Rev. Biochem.* **54**, 632–664.

Kozak, M. (1986a). Regulation of protein synthesis in virus infected animal cells. *Adv. Virus Res.* **31**, 229–292.

Kozak, M. (1986b). Bifunctional messenger RNAs in eukaryotes. *Cell* **47**, 481–483.

Kozak, M. (1988). A profusion of controls. *J. Cell Biol.* **107**, 1–7.

Kozak, M., and Shatkin, A. J. (1978). Migration of 40S ribosomal subunits on messenger RNA in the presence of edeine. *J. Biol. Chem.* **253**, 6568–6577.

Krug, R. M. (1981). Priming of influenza viral RNA transcription by capped heterologous RNAs. *Curr. Top. Microbiol. Immunol.* **93**, 125–150.

Kuismanen, E. (1984). Posttranslational processing of Uukuniemi virus glycoproteins G1 and G2. *J. Virol.* **51**, 806–812.

Kuismanen, E., Hedman, K., Saraste, J., and Pettersson, R. F. (1982). Uukuniemi virus maturation, accumulation of virus particles and viral antigens in the Golgi complex. *Mol. Cell. Biol.* **2**, 1444–1458.

Kuismanen, E., Saraste, J., and Pettersson, R. F. (1985). Effect of monensin on the assembly of Uukuniemi virus in the Golgi complex. *J. Virol.* **55**, 813–822.

Lee. H. W., Lee, P. W., Baek, L. J., Song, C. K., and Seong, I. W. (1981). Intraspecific transmission of Hantaan virus, etiologic agent of Korean hemorrhagic fever in the rodent *Apodemus agrarius*. *Am. J. Trop. Med. Hyg.* **30**, 1106–1112.

Lees, J. F., Pringle, C. R., and Elliott, R. M. (1986). Nucleotide sequence of the Bunyamwera virus M RNA segment: Conservation of structural features in the Bunyavirus glycoprotein gene product. *Virology* **148**, 1–14.

Lepault, J., Booy, F. P., and Dubochet, J. (1983). Electron microscopy of frozen biological suspensions. *J. Microsc. (Oxford)* **128**, 89–102.

Lindsey, H. S., Klimas, R., and Obijeski, J. F. (1977). La Crosse virus soluble cell culture antigen. *J. Clin. Microbiol.* **6**, 618–626.

Ludwig, G. V., Christensen, B. M., Yuill, T. M., and Schultz, K. T. (1989). Enzyme processing of La Crosse virus glycoprotein G1: A bunyavirus vector infection model. *Virology* **171**, 108–113.

Lyons, M. J., and Heyduk, J. (1973). Aspects of the developmental morphology of California encephalitis virus in cultured vertebrate and arthropod cells and in mouse brain. *Virology* **54**, 37–52.

Madoff, D. H., and Lenard, J. (1982). A membrane glycoprotein that accumulates intracellularly: Cellular process of the large glycoprotein of La Crosse virus. *Cell* **28**, 821–829.

Marriott, A. C., Ward, V. K., and Nuttall, P. A. (1989). The S RNA segment of sandfly fever Sicilian virus: Evidence for an ambisense genome. *Virology* **169**, 341–345.

Martin, M. L., Lindsey-Regnery, H., Sasso, D. R., McCormick, J. B., and Palmer, E. (1985). Distinction between Bunyaviridae genera by surface structure and comparison with Hantaan virus using negative stain electron microscopy. *Arch. Virol.* **86**, 17–28.

Matsuoka, Y., Ihara, T., Bishop, D. H. L., and Compans, R. C. (1988). Intracellular accumulation of Punta Toro virus glycoproteins expressed from cloned cDNA *Virology* **167**, 251–260.

McCormick, J. B., Palmer, E. L., Sasso, D. R., and Kiley, M. P. (1982). Morphological identification of the agent of Korean haemorrhagic fever (Hantaan virus) as a member of the Bunyaviridae. *Lancet* **1**, 765–767.

Milne, R. G., and Francki, R. I. B. (1984). Should tomato spotted wilt virus by considered as a possible member of the family Bunyaviridae? *Intervirology* **22**, 72–76.

Murphy, F. A., Harrison, A. K., and Whitfield, S. G. (1973). Morphologic and morphogenetic similarities of Bunyamwera serological supergroup viruses and several other arthropod-borne viruses. *Intervirology* **1**, 297–316.

Newton, S. E., Short, N. J., and Dalgarno, L. (1981). Bunyamwera virus replication in cultured *Aedes albopictus* (mosquito) cells: Establishment of a persistent infection. *J. Virol.* **38**, 1015–1024.

Nicoletti, L., and Verani, P. (1985). Growth of phlebovirus Toscana in a mosquito (*Aedes pseudoscutellaris*) cell line (AP-61): Establishment of a persistent infection. *Arch. Virol.* **85**, 35–45.

Obijeski, J. F., and Murphy, F. A. (1977). Bunyaviridae: Recent biochemical developments. *J. Gen. Virol.* **37**, 1–14.

Obijeski, J. F., Bishop, D. H. L., Murphy, F. A., and Palmer, E. L. (1976a). Structural proteins of La Crosse virus. *J. Virol.* **19**, 985–997.

Obijeski, J. F., Bishop, D. H. L., Palmer, E. L., and Murphy, F. A. (1976b). Segmented genome and nucleocapsid of La Crosse virus. *J. Virol.* **20**, 664–675.

Obijeski, J. F., McCauley, J., and Skehel, J. J. (1980). Nucleotide sequences at the termini of La Crosse virus RNAs. *Nucleic Acids Res.* **8**, 2431–2438.

Okuno, Y., Yamanishi, K., Takahashi, Y., Tanishita, O., Nagai, T., Dantas, J. R., Okamoto, Y., Tadano, M., and Takahashi, M. (1986). Haemagglutination-inhibition test for haemorrhagic fever with renal syndrome using viral antigen prepared from infected tissue culture fluid. *J. Gen. Virol.* **67**, 149–156.

Overton, H. A., Ihara, T. R., and Bishop, D. H. L. (1987). Identification of the N and NS$_S$ proteins coded by the ambisense S RNA of Punta Toro phlebovirus using monospecific antisera raised to baculovirus expressed N and NS$_S$ proteins. *Virology* **157**, 338–350.

Pardigon, N., Vialat, P., Girard, M., and Bouloy, M. (1982). Panhandles and hairpin structures at the termini of Germiston virus RNAs (Bunyavirus). *Virology* **122**, 191–197.

Pardigon, N., Vialat, P., Gerbaud, S., Girard, M., and Bouloy, M. (1988). Nucleotide sequence of the M segment of Germiston virus: Comparison of the M gene product of several bunyaviruses. *Virus Res.* **11**, 73–85.

Parrington, M. A., and Kang, C. Y. (1990). Nucleotide sequence analysis of the S genome segment of Prospect Hill virus: Comparison with the prototype Hantavirus. *Virology* **175**, 167–175.

Parsonson, I. M., and McPhee, D. A. (1985). Bunyavirus pathogenesis. *Adv. Virus Res.* **30**, 279–316.

Patterson, J. I., and Kolakofsky, D. (1984). Characterization of La Crosse virus small-genome segment transcripts. *J. Virol.* **49**, 680–685.

Patterson, H., Holloway, B., and Kolakofsky, D. (1984). La Crosse virions contain a primer-stimulated RNA polymerase and a methylated cap-dependent endonuclease, *J. Virol.* **52**, 215–222.

Patterson, J. I., Kolakofsky, D., Holloway, B. P., and Obijeski, J. F. (1983). Isolation of the ends of La Crosse virus small RNA as a double-stranded structure. *J. Virol.* **45**, 882–884.

Pensiero, M. N., Jennings, G. B., Schmaljohn, C. S., and Hay, J. (1988). Expression of the Hantaan virus M genome segment by using a vaccinia virus recombinant. *J. Virol.* **62**, 696–702.

Persson, R., and Pettersson, R. F. (1991). Formation and intracellular transport of a heterodimeric viral spike protein complex. *J. Cell Biol.* (in press).

Pesonen, M., Kuismanen, F., and Pettersson, R. I. (1982). Monosaccharide sequence of protein-bound glycans of Uukuniemi virus. *J. Virol.* **41**, 390–400.

Pettersson, R. F., and Kääriäinen, L. (1973). The ribonucleic acids of Uukuniemi virus, a non-cubical tick-borne arbovirus. *Virology* **56**, 608–619.

Pettersson, R. F., and Von Bonsdorff, C.-H. (1975). Ribonucleoproteins of Uukuniemi virus are circular. *J. Virol.* **15**, 386–392.

Pettersson, R. F., and Von Bonsdorff, C.-H. (1987). Bunyaviridae. *In* "Animal Virus Structure" (M. V. Nermut and A. C. Steven, eds.), pp. 147–157. Elsevier, Amsterdam.

Pettersson, R. F., Hewlett, M. J., Baltimore, D., and Coffin, J. M. (1977). The genome of Uukuniemi virus consists of three unique RNA segments. *Cell* 11, 51–64.

Pifat, D. Y., and Smith, J. F. (1987). Punta Toro virus infection of C57Bl–6J mice: A model for phlebovirus-induced disease. *Microb. Pathog.* 3, 409–422.

Pifat, D. Y., Osterling, M. C., and Smith, J. J. (1988). Antigenic analysis of Punta Toro virus and identification of protective determinants with monoclonal antibodies. *Virology* 167, 442–450.

Plotch, S. J., and Krug, R. M. (1977). Influenza virion transcriptase: Synthesis in vitro of large, polyadenylic-acid containing complementary RNA. *J. Virol.* 21, 24–34.

Plotch, S. J., Bouloy, M., Ulmanen, I., and Krug, R. M. (1981). A unique cap(m⁷GpppXm)-dependent influenza virion endonuclease cleaves capped RNAs to generate the primers that initiate viral transcription. *Cell* 23, 847–858.

Pobjecky, N., Nathanson, N., and Gonzalez-Scarano, F. (1989). Use of the resonance energy transfer assay to investigate the fusion function of La Crosse virus. *In* "Genetics and Pathogenicity of Negative Strand Viruses" (D. Kolakofsky and B. W. J. Mahy, eds.), pp. 24–32. Elsevier, Amsterdam.

Porterfield, J. S., Casals, J., Chumakov, M. P., *et al.* (1973/1974). Bunyaviruses and Bunyaviridae. *Intervirology* 2, 270–272.

Porterfield, J. S., Casals, J., Chumakov, M. P., *et al.* (1975–1976). Bunyaviruses and Bunyaviridae. *Intervirology* 6, 13–24.

Pringle, C., Lees, J. F., Clark, W., and Elliott, R. M. (1984). Genome sub-unit reassortment among bunyaviruses analyzed by dot hybridization using molecularly cloned complementary DNA probes. *Virology* 135, 244–256.

Raju, R., and Kolakofsky, D. (1986a). Inhibitors of protein synthesis inhibit both La Crosse virus S-mRNA and S genome synthesis in vivo. *Virus Res.* 5, 1–9.

Raju, R., and Kolakofsky, D. (1986b). Translational requirement of La Crosse virus S-mRNA synthesis: In vivo studies. *J. Virol.* 61, 96–103.

Raju, R., and Kolakofsky, D. (1987). Unusual transcripts in La Crosse virus-infected cells and the site for nucleocapsid assembly. *J. Virol.* 61, 667–672.

Raju, R., and Kolakofsky, D. (1988). La Crosse virus infection in mammalian cells induces mRNA instability. *J. Virol.* 62, 27–32.

Raju, R., and Kolakofsky, D. (1989). The ends of LaCrosse virus genome and antigenome RNAs within nucleocapsid are base paired. *J. Virol.* 63, 122–128.

Raju, R., Raju, K., and Kolakofsky, D. (1989). The translational requirement for complete La Crosse virus mRNA synthesis is cell type dependent. *J. Virol.* 63, 122–128.

Ranki, M., and Pettersson, R. F. (1975). Uukuniemi virus contains an RNA polymerase. *J. Virol.* 16, 1420–1425.

Rönnholm, R., and Pettersson, R. F. (1987). Complete nucleotide sequence of the M RNA segment of Uukuniemi virus encoding the membrane glycoproteins G1 and G2. *Virology* 160, 191–202.

Rossier, C., Patterson, J., and Kolakofsky, D. (1986). La Crosse virus small genome mRNA is made in the cytoplasm. *J. Virol.* 58, 647–650.

Rossier, C., Raju, R., and Kolakofsky, D. (1988). La Crosse virus gene expression in mammalian and mosquito cells. *Virology* 165, 539–548.

Rozhon, E. J., Gensemer, P., Shope, R. E., and Bishop, D. H. L. (1981). Attenuation of virulence of a bunyavirus involving an L RNA defect and isolation of LAC/SSH/LAC and LAC/SSH/SSH reassortants. *Virology* 111, 125–138.

Saluzzo, J. F., and Smith, J. F. (1990). Use of reassortant viruses to map attenuated and temperature sensitive mutations of the Rift Valley fever virus MP12. *Vaccine* 8, 369–375.

Salvato, M., and Shimomaye, E. (1989). The completed sequence of lymphocytic choriomeningitis virus reveals a unique RNA structure and a gene for a zinc finger protein. *Virology* 173, 1–10.

Samso, A., Bouloy, M., and Hannoun, C. (1975). Presence de ribonucleoproteines circulaires dans le virus Lumbo (Bunyavirus). *C. R. Hebd. Seances Acad. Sci., Ser. D* **280**, 779–782.

Schmaljohn, C. S. (1990). Nucleotide sequence of the L genome segment of Hantaan virus. *Nucleic Acids Res.* **18**, 6728.

Schmaljohn, C. S., and Dalrymple, J. M. (1983). Analysis of Hantaan virus RNA: Evidence for a new genus of Bunyaviridae. *Virology* **131**, 482–491.

Schmaljohn, C. S., and Dalrymple, J. M. (1984). Biochemical characterization of Hantaan virus. *In* "Segmented Negative Strand Viruses" (R. W. Compans and D. H. L. Bishop, eds.), pp. 117–124. Academic Press, Orlando, Florida.

Schmaljohn, C. S., and Patterson, J. L. (1990). Bunyaviridae and their replication. *In* Virology" (B. N. Fields, D. M. Knipe, R. M. Chanock, J. L. Melnick, M. S. Hirsh, T. P. Monath, and B. Roizman, eds.), pp. 1175–1194. Raven, New York.

Schmaljohn, C. S., Hasty, S. E., Harrison, B. A., and Dalrymple, J. M. (1983). Characterization of Hantaan virions, the prototype of hemorrhagic fever with renal syndrome. *J. Infect. Dis.* **148**, 1005–1012.

Schmaljohn, C. S., Hasty, S. E., Dalrymple, J. M., *et al.* (1985). Antigenic and genetic properties of viruses linked to hemorrhagic fever with renal syndrome. *Science* **227**, 1041–1044.

Schmaljohn, C. S., Hasty, S. E., Rasmussen, L., and Dalrymple, J. M. (1986a). Hantaan virus replication: Effects of monensin, tunicamycin and endoglycosidases on the structural glycoproteins. *J. Gen. Virol.* **67**, 707–717.

Schmaljohn, C. S., Jennings, G., Hay, J., and Dalrymple, J. M. (1986b). Coding strategy of S genome segment of Hantaan virus. *Virology* **155**, 633–643.

Schmaljohn, C. S., Jennings, G. B., and Dalrymple, J. M. (1987a). Identification of Hantaan virus messenger RNA species: *In* "Biology of Negative Strand Viruses" (B. W. J. Mahy and D. Kolakofsky, eds.), pp. 116–121.Elsevier, Amsterdam.

Schmaljohn, C. S., Schmaljohn, A. L., and Dalrymple, J. M. (1987b). Hantaan virus M RNA: Coding strategy, nucleotide sequence, and gene order. *Virology* **157**, 31–39.

Schmaljohn, C. S., Sugiyama, K., Schmaljohn, A. L., and Bishop, D. H. L. (1988). Baculovirus expression of the small genome segment of Hantaan virus and potential use of the expressed nucleocapsid protein as a diagnostic antigen. *J. Gen. Virol.* **69**, 777–786.

Schmaljohn, C. S., Parker, M. D., Ennis, W. H., Dalrymple, J. M., Collett, M. S., Suzich, J. A., and Schmaljohn, A. L. (1989). Baculovirus expression of the M genome segment of Rift Valley fever virus and examination of antigenic and immunogenic properties of the expressed proteins. *Virology* **170**, 184–192.

Schmaljohn, C. S., Chu, Y. K., Schmaljohn, A. L., and Dalrymple, J. M. (1990). Antigenic subunits of Hantaan virus expressed by baculoviruses and vaccinia virus recombinants. *J. Virol.* **64**, 3162–3170.

Shope, R. E., Rozhon, E. J., and Bishop, D. H. L. (1981). Role of the middle-sized bunyavirus RNA segment in mouse virulence. *Virology* **114**, 273–276.

Simons, J. F., Hellman, U., and Pettersson, R. F. (1990). Uukuniemi virus S segment: Ambisense coding strategy, packaging of complementary strands into virions and homology to members of the genus *Phlebovirus. J. Virol.* **64**, 247–255.

Smith, J. F., and Pifat, D. Y. (1982). Morphogenesis of sandfly fever viruses (Bunyaviridae family). *Virology* **121**, 61–81.

Stohwasser, R., Giebel, L. B., Zöller, L., Bautz, E. K. F., and Darai, G. (1990). Molecular characterization of the RNA S segment of Nephropathia Epidemica virus strain Hällnäs B1. *Virology* **174**, 79–86.

Struthers, J. K., and Swanepoel, R. (1982). Identification of a major non-structural protein in the nuclei of Rift Valley virus-infected cells. *J. Gen. Virol.* **60**, 381–384.

Struthers, J. K., Swanepoel, R., and Shepard, S. P. (1984). Protein synthesis in Rift Valley fever virus-infected cell. *Virology* **134**, 118–124.

Sundin, D. R., Beaty, B. J., Nathanson, N., and Gonzalez-Scarano, F. (1987). A G1 glycoprotein epitope of La Crosse virus: A determinant of infection of *Aedes triseriatus*. *Science* **235**, 591–593.

Suzich, J. A., and Collett, M. S. (1988). Rift Valley fever virus M segment: Cell free transcription and translation of virus-complementary RNA. *Virology* **164**, 478–486.

Suzich, J. A., Kakach, L. T., and Collett, M. S. (1990). Expression strategy of a phlebovirus: Biogenesis of proteins from the Rift Valley fever virus M segment. *J. Vriol.* **64**, 1549–1555.

Swanepoel, R., and Blackburn, N. K. (1977). Demonstration of nuclear immunofluorescence in Rift Valley fever virus-infected cells. *J. Gen. Virol.* **34**, 557–561.

Takehara, K., Min, M. K., Battles, J. K., Sugiyama, K., Emery, V. C., Dalrymple, J. F., and Bishop, D. H. L. (1989). Identification of mutations in the M RNA of a candidate vaccine strain of Rift Valley fever virus. *Virology* **169**, 452–457.

Talmon, Y., Prasad, B. V. V., Clerx, J. P. M., Wang, G. J., Chin, W., and Hewlett, M. J. (1987). Electron microscopy of vitrified–hydrated La Crosse virus. *J. Virol.* **61**, 2319–2321.

Toriyama, S., and Watanabe, Y. (1989). Characterization of single and double stranded RNAs in particles of rice stripe virus. *J. Gen. Virol.* **70**, 505–511.

Tsai, T. F., Tang, Y. W., Hu, S. L., Ye, K. L., Chen, G. L., and Xu, Z. Y. (1984). Hemagglutination-inhibiting antibody in hemorrhagic fever with renal syndrome. *J. Infect. Dis.* **150**, 895–898.

Turell, M. J., Saluzzo, J. F., Tammariello, r. F., and Smith, J. F. (1990). Generation and transmission of Rift Valley fever viral reassortants by the mosquito *Culex pipiens*. *J. Gen. Virol.* **71**, 2307–2312.

Ulmanen, I., Seppälä, P., and Pettersson, R. F. (1981). In vitro translation of Uukuniemi virus-specific RNAs: Identification of a nonstructural protein and a precursor to the membrane glycoproteins. *J. Virol.* **37**, 72–79.

Urakawa, T., Small, D. A., and Bishop, D. H. L. (1988). Expression of snowshoe hare bunyavirus S RNA coding proteins by recombinant baculoviruses. *Virus Res.* **11**, 303–317.

Ushijima, H., Clerx-Van Haaster, C. M., and Bishop, D. H. L. (1981). Analyses of Patois group bunyaviruses and existence of immune precipitable and nonprecipitable nonvirion protein induced in bunyavirus-infected cells. *Virology* **110**, 318–332.

Verani, P., Nicoletti, L., and Marchi, A. (1984). Establishment and maintenance of persistent infection by the phlebovirus Toscana in Vero cells. *J. Gen. Virol.* **65**, 367–375.

Vezza, A. C., Repik, P. M., Cash, P., and Bishop, D. H. L. (1979). In vivo transcription and protein synthesis capabilities of bunyaviruses: Wild-type snowshoe hare virus and its temperature-sensitive group I and group I/II mutants. *J. Virol.* **31**, 426–436.

Vialat, P., and Bouloy, M. (1991). Manuscript in preparation.

Von Bonsdorff, C.-H., and Pettersson, R. (1975). Surface structure of Uukuniemi virus. *J. Virol.* **16**, 1296–1307.

Ward, V. K., Marriott, A. C., El-Ghor, A., and Nuttall, P. A. (1990). Coding strategy of the S RNA segment of Dugbe virus (*Nairovirus*: Bunyaviridae). *Virology* **175**, 518–524.

Wasmoen, T. L., Kakach, L. T., and Collett, M. S. (1988). Rift Valley fever virus M segment: Cellular localization of M segment-encoded proteins. *Virology* **186**, 275–280.

Watret, G. E., and Elliott, R. M. (1985). The proteins and RNAs specified by Clo-Mor virus, a Scottish nairovirus. *J. Gen. Virol.* **60**, 2513–2516.

Watret, G. E., Pringle, C. R., and Elliot, R. M. (1985). Synthesis of bunyavirus-specific proteins in a continuous cell line (XTC-2) derived from *Xenopus laevis. J. Gen. Virol.* **66,** 473–482.

White, J. E., Shirey, F. G., French, G. R., Huggins, J. W., Brand, O. M., and Lee, H. W. (1982). Hantaan virus, aetiological agent of Korean haemorrhagic fever, has Bunyaviridae-like morphology. *Lancet* **1,** 768–771.

Yamanishi, K., Dantas, J. R., Takahashi, M., Yamanouchi, T., Domae, K., Takahashi, Y., and Tanishita, O. (1984). Antigenic differences between two viruses, isolated in Japan and Korea, that cause haemorrhagic fever with renal syndrome. *J. Virol.* **52,** 231–237.

Yoo, D., and Kang, C. Y. (1987). Nucleotide sequence of the M segment of the genomic RNA of Hantaan virus 76–118. *Nucleic Acids Res.* **15,** 6299–6300.

INDEX